Menggang Li · Qiusheng Zhang
Juliang Zhang · Yisong Li
Editors

Proceedings of 2015 2nd International Conference on Industrial Economics System and Industrial Security Engineering

Springer

Editors
Menggang Li
China Center for Industrial Security Research
Beijing Jiaotong University
Beijing
China

Qiusheng Zhang
School of Economics and Management
Beijing Jiaotong University
Beijing
China

Juliang Zhang
School of Economics and Management
Beijing Jiaotong University
Beijing
China

Yisong Li
School of Economics and Management
Beijing Jiaotong University
Beijing
China

ISBN 978-981-287-654-6 ISBN 978-981-287-655-3 (eBook)
DOI 10.1007/978-981-287-655-3

Library of Congress Control Number: 2015952033

Springer Singapore Heidelberg New York Dordrecht London

Printed on acid-free paper

Springer Science+Business Media Singapore Pte Ltd. is part of Springer Science+Business Media (www.springer.com)

Proceedings of 2015 2nd International Conference on Industrial Economics System and Industrial Security Engineering

Preface

This volume contains the proceedings of the 2015 Industrial Economics System and Industrial Security Engineering (IEIS'2015) held in Barcelona, Spain, hosted by the China Center for Industrial Security Research of Beijing Jiaotong University (CCISR), International Center for Informatics Research of Beijing Jiaotong University (ICIR), and School of Economics and Management of Beijing Jiaotong University (BEM); in cooperation with the Universitat Politècnica de Catalunya BarcelonaTech (Spain) and University of Reading (UK); and supported by the National Natural Science Foundation of China (NSFC), K.C. Wong Education Foundation (Hong Kong), Sino-EU Doctoral School for Sustainability Engineering (Program in Logistics, Information, Management and Service Science), EU FP7 (7th Framework Programme), Beijing Logistics Informatics Research Base, and Key Laboratory of Logistics Management and Technology of Beijing.

This conference provides a forum for scholars and practitioners all over the world to discuss problems in industrial economics as well as industrial security theories and practices. It aims to provide insights in solving problems in national economy, social development, and economic security arenas. The conference has two simultaneous tracks that cover different aspects including "industrial economics" and "industrial security." Papers published in each track describe state-of-art research works that are often oriented toward real-world applications and highlight the benefits of related methods and techniques for the emerging field of industries.

IEIS'2015 received 130 paper submissions from several countries and regions. Sixty-two papers were accepted and published after undergoing strict peer review. The total acceptance ratio is 47.7 %. Additionally, a number of invited talks, presented by internationally recognized experts in different areas, positively contribute to reinforcing the overall quality of the conference and to provide a deep understanding of related areas.

The program for this conference required the dedicated effort of many people. First, we thank the authors, whose research and development efforts are recorded here. Second, we thank the members of the Program Committee and the additional reviewers for valuable help with the expert review of all submitted papers. Third, we thank the invited speakers for their invaluable contribution and the time spent preparing their talks. Fourth, we thank the Special Session chairs whose collaborations with IEIS are much appreciated. Finally, many thanks are given to our colleagues at Beijing Jiaotong University and Universitat Politècnica de Catalunya BarcelonaTech for their hard work in organizing this event.

We hope you all enjoyed an exciting conference and an unforgettable stay in Barcelona, Spain, or Beijing, China. We hope to meet you again next year for the IEIS'2016 at the Henley Business School of Johannesburg, South Africa, the details of which will soon be available at: http://icir.bjtu.edu.cn/ieis2016.

Barcelona, Spain

Menggang Li
Qiusheng Zhang
Juliang Zhang
Yisong Li

Organizing Committees

Honorary Chairmen

Academician Shoubo Xu, Beijing Jiaotong University, China
Professor Bin Ning, IEEE Fellow, Beijing Jiaotong University, China

General Chairs

Qiusheng Zhang, Beijing Jiaotong University, China
Menggang Li, Beijing Jiaotong University, China
Runtong Zhang, Beijing Jiaotong University, China
Zhongliang Guan, Beijing Jiaotong University, China
Vicenc Fernandez, Universitat Politècnica de Catalunya BarcelonaTech, Spain

Program Chairs

Zhenji Zhang, Beijing Jiaotong University, China
José María Sallán Leyes, Universitat Politècnica de Catalunya BarcelonaTech, Spain
Juliang Zhang, Beijing Jiaotong University, China
Xianliang Shi, Beijing Jiaotong University, China

Financial Chair

Shifeng Liu, Beijing Jiaotong University, China

Secretariat

Yisong Li, Beijing Jiaotong University, China
Dan Chang, Beijing Jiaotong University, China
Lin Ta, Beijing Jiaotong University, China

Publication Chairs

Mincong Tang, Beijing Jiaotong University, China
Guowei Hua, Beijing Jiaotong University, China
Irene Trullas, Universitat Politècnica de Catalunya BarcelonaTech, Spain
Honglu Liu, Beijing Jiaotong University, China

Organization Chairs

Xiaochun Lu, Beijing Jiaotong University, China
Francisco Vallverdu Bayes, Universitat Politècnica de Catalunya BarcelonaTech, Spain

Special Session/Workshop Chairs

Hongjie Lan, Beijing Jiaotong University, China
Vicenç Fernandez Alarcon, Universitat Politècnica de Catalunya BarcelonaTech, Spain
Xiaomin Zhu, Beijing Jiaotong University, China

Publicity Chairs

Francisco Vallverdu Bayes, Universitat Politècnica de Catalunya BarcelonaTech, Spain
Yunan Zhang, Beijing Jiaotong University, China
Juan Li, Beijing Jiaotong University, China

International Steering Committee

Chair

Yifu Lin, The Academic Board of China Center for Industrial Security Research, Beijing Jiaotong University, China

Members

Orley Ashenfelter, University of Princeton, USA
Barry W. Boehm, University of Southern California, USA
Yushu Cao, The State Council of Western China Development Leading Group Office, China
Yin-Wong Cheung, University of California, USA
Xiqing Chen, The United Front Work Department of CPC Central Committee, China
Benjamin Faber, University of California, Berkeley, USA
Peiyong Gao, Chinese Academy of Social Sciences, China
David Gonzales-Prieto, Universitat Politècnica de Catalunya BarcelonaTech, Spain
Weida He, University of Science and Technology Beijing, China
Ran Jiao, Xinhua News Agency editorial of Economic Information, China
Bei Jin, Chinese Academy of Social Sciences, China
Marlene Kim, University of Massachusetts Boston, USA
Menggang Li, Beijing Jiaotong University, China
Pumin Li, National Development and Reform Commission, China
Wenxing Li, Beijing Jiaotong University, China
Wei Liu, Peking University, China
Yanhua Liu, Ministry of Science and Technology, China
Oriol Lordan, Universitat Politècnica de Catalunya BarcelonaTech, Spain
Jurgita Raudeliuniene, Vilnius Gediminas Technical University, Lithuanian
Xiaoqiu Wu, Renmin University of China, China
Hongjun Yu, Peking University, China
Xinli Zheng, China Central Policy Research Office, China

International Program Committee

Karim Abadir, Imperial College Business School, UK
Jason Abrevaya, The University of Texas at Austin, USA
Alessandro Acquisti, Carnegie Mellon University, USA
Rishi Amrit, University of Wisconsin-Madison, USA
Jeannie Albrecht, Williams College, USA

Roberta Lamb, University of California, USA
Gregory Lewis, Oxford University, USA
Xuemei Li, Beijing Jiaotong University, China
Liwen Li, National Taiwan University, Taipei, China
Fei-peng Li, Tianjin University, China
Jeremy Lin, PJM Interconnection, USA
Winston T. Lin, The State University of New York at Buffalo, USA
John List, University of Chicago, USA
Graham Lord, Princeton University, USA
Ming Lu, Fudan University, China
M. Rajabalinejad, Delft University of Technology, Netherlands
Rongguang Ma, RenMin University of China, China
Wentao Ma, Xi'an Jiaotong University, China
Lu Ma, Guangxi University of Science and Technology, China
Guanfeng Mao, South-Central University, China
Victor Mbarika, Southern University and A&M College, USA
Ali Miili, West Virginia University, USA
A. Nagurney, University of Massachusetts, USA
Joyojeet Pal, University of Michigan, USA
Fay Cobb Payton, North Carolina State University, USA
Parag C. Pendharkar, Pennsylvania State University, USA
Edmund S. Phelps, Columbia University, USA
Shaozhou Qi, Wuhan University, China
Oliver Rubel, Lawrence Berkeley National Laboratory, Berkeley, USA
Hossein S. Samiei, International Monetary Fund, Washington, USA
Scott Schuh, Federal Reserve Bank of Boston, USA
Malik Shahzad K. Awan, University of Warwick, Coventry, UK
Yupeng Shi, Central University of Finance and Economics, China
Donghui Shi, Anhui University of Architecture, China
Fazel Shokoofeh, Montana State University-Billings, Montana, USA
Nachiappan Subramanian, Nottingham University Business School, China
Chi-wei Su, Ocean University of China, China
Shyam Sunder, Yale School of Management, USA
Wenhao Tan, Nanchang University, China
Charles S. Tapiero, Polytechnic University of New York, USA
Carlos Teixeira, The New School University, USA
Yu-Chung Tsao, Tatung University, Taipei, China
Luis G. Vargas, University of Pittsburgh, USA
David L. Veenstra, University Washington, USA
Jeff Wagner, Rochester Institute of Technology, USA
Carl E. Walstad, University of California, Santa Cruz, USA
W.B. Walstad, Northwestern Polytechnical University, China
Junqiang Wang, Northwestern Polytechnical University, China
Fei Wang, School of Business Sias International University, China
Cheng Wang, The Institute of Industrial Economic, CASS, China

Charles M. Weber, Portland State University, USA
R.W. Wies, University of Alaska Fairbanks, USA
Christopher D. Wilen, Carleton College, USA
Bart J. Wilson, Chapman University, USA
Steve Winkelman, Center for Clean Air Policy, USA
G.C. Winston, Williams College, USA
Eve Wittenberg, Harvard University, USA
J. Woelcke, World Bank, Washington, USA
Kar-yiu Wong, The University of Washington, USA
Kesheng Wu, Lawrence Berkeley National Laboratory, USA
Lida Xu, Old Dominion University, USA
Bing Xu, Zhejiang Gongshang University, China
Lei Ynag, South China University of Technology, China
Jie Yang, University of Houston-Victoria, USA
Muhamet Yildiz, Northwestern University, USA
Dahai You, Huazhong University of Science and Technology, China
Jiashen You, University of California Los Angeles, USA
Ning Yu, Renmin University China, China
Guocai Yu, Shandong University of Technology, China
Matthew Zeidenberg, Columbia University, USA
Guofang Zhai, Nanjing University, China
Xuehua Zhang, Tianjin Polytechnic University, China
J.X. Zhang, Xian Technology University China, China
Zhongxiang Zhang, Fudan University, China
Weixi Zhang, Tianjin Polytechnic University, China
Zhuo Zhang, Nanjing University of Aeronautics and Astronautics, China
Xia Zhao, China Agricultural University, China
Jing Zhao, Beijing Forestry University, China
Shenghua Zheng, Zhejiang University of Technology, China
Yaodong Zhou, Beijing Jiaotong University, China
Zhiwen Zhu, Tianjin University, China
Shubin Zhu, Jiangxi Agricultural University, China

Contents

Part IV Special Session on Cultural Industry

Part V Special Session on National Economy

Part I
Industrial Economics

The Application Research on Tourism Enterprise Implementing Precision Marketing in the Era of High-Speed Rail

Wu Yue

Abstract Mass tourism consumption has entered the market-outbreak period, and tourism economy is the typical experience economy. As customer demands become increasingly diversified, tourism enterprise has been forced to have a full understanding of customer needs and make customer-behavior predictions on the basis of big data. Meanwhile, high-speed rail has a field-polymerization effect, so it could solve the problem of scale-precision marketing and influence tourism choices of the mainstream crowd, thus helping tourism companies achieve a high response rate and customer loyalty as well as improving enterprise efficiency. This paper first introduces the necessity of using big data in precision marketing for tourism companies; next it expounds the construction of a tourism company precision-marketing system framework relying on big data; finally, it analyzes the channels required to realize precision marketing for tourism enterprise in the era of high-speed rail.

Keywords Big data · Tourism · Precision marketing · High-speed rail · System

1 Introduction

China is entering the "high-speed rail era." Twenty-eight provinces have opened high-speed railways, which basically cover all cities with >500 000 inhabitants. In 2014, the number of passengers was >800 million, and the average daily volume of passengers was >6 million. In the next 2 years, the total mileage of high-speed rail in operation will reach 8887 km at which point China will have the longest mileage and the largest construction of high-speed rail in the world [1]. High-speed rail greatly shortens travel distance between cities, and it also leads to the blowout development of tourism and doubly increases customer excursions using high-speed rail lines.

W. Yue (✉)
Faculty of Management and Economics, Dalian University of Technology,
Dalian 116024, China
e-mail: gakko@163.com

M. Li et al. (eds.), *Proceedings of 2015 2nd International Conference on Industrial Economics System and Industrial Security Engineering*,
DOI 10.1007/978-981-287-655-3_1

"The integration of tourism resources and making tourism industry stronger" has become an "important content of prominent transformation system and accelerating transformation mode of economic development" in the Twelfth Five Year Plan [2]. Under this scenario, tourism companies, with the help of high-speed rail, could implement precision marketing to determine the exact number of people, regions, and time nodes through the analysis of big data to determine potential customer demands and satisfy every customer as much as possible through a combination of products and personalized services. Thus, tourism companies could achieve precision marketing and enchance profitability as well as promote the development of economic transition.

2 Tourism Enterprise Using Big Data to Break Through the Traditional Marketing Bottlenecks

2.1 Big Data and Precision Marketing

As the basic technical conditions and tools of precision marketing, big data can be thought of as another term for of "analysis," which seeks to extract knowledge from data and convert data into the commercial advantages of intelligent activity [3]. In most literature, big data are summarized using the following four characteristics (the four V's): huge volume of data (Volume), rapid growth velocity (Velocity), structured data (Variety), and low-value density (Value).

Precision marketing was first proposed by authoritative marketing expert Philip Kotler [4], who stated that, specifically speaking, precison marketing means that "the company needs more accurate, measurable and high return on the investment of marketing communication, which needs to pay more attention to marketing communication plans of results and action, as well as to the direct sales communication investment".

In the age of big data, enterprises obtain valuable commercial information through the collection and analysis of large amounts of data. By mining these data, tourism companies could forecast market demand and eventually transfer valuable information to business insights and intelligent decision making.

2.2 Necessity of Tourism Enterprise Using Big Data to Implement Precision Marketing

At present, the problems that must be solved in the tourism industry include fierce competition, lagging network marketing, asymmetry of tourism information, tourist placement in a passive position, having a single-product line, and difficulty meeting personalized traveler needs [5]. Although high-speed rail is changing the mainstream populations' tourism selections, along with the sensing devices and mobile terminals being connected with the network, tourism companies could use new technologies—such as cloud computing, data analysis, or mobile Internet—and establish a precise consumer communication system on the basis of high-speed rail data to precisely lock in target customers by providing specialized services to maximize customer transactions, the number of customer contacts, and customer conversion rates.

3 Construction of Tourism Enterprise Precision Marketing System Based on the Big Data Processing

3.1 Mining the Correlation Among Data

Precision marketing connects data technology with creative process. It is focused on highly oriented customers and their interactive actions and maximizes the consumption potential of customers. And it will realize cross-selling, related sales, and extended sales ultimately. First, precision marketing helps marketers understand customer needs; second, it helps marketers predict when customers may respond; finally, it helps them predict customer behavior.

3.2 Construction of Tourism Enterprise's Precision Marketing System Frame

The commercial application of big data is not just to obtain and use customer information; it should consider the product itself analyzing if tourism products can meet the needs of target customers, thus in essence maximizing competitive advantages.

To achieve precision marketing, tourism enterprise specifically must process three large nodes and six continuous steps, each of which start with business demands and, on the basis of big data, performs related planning and marketing activities, evaluates and improves the results, and then continues to self-adjustment and self-improvement to achieve accurate "precise marketing" (Fig. 1).

First, the decision-making mode changes from business-driven to data-driven, which can be used to select specific tourism products, customer groups, and marketing channels depending on the results of the information provided by big data, thus completing the first key node: "target determination." In this node, more data are collected, the use of which increases the in-depth analysis of changes in

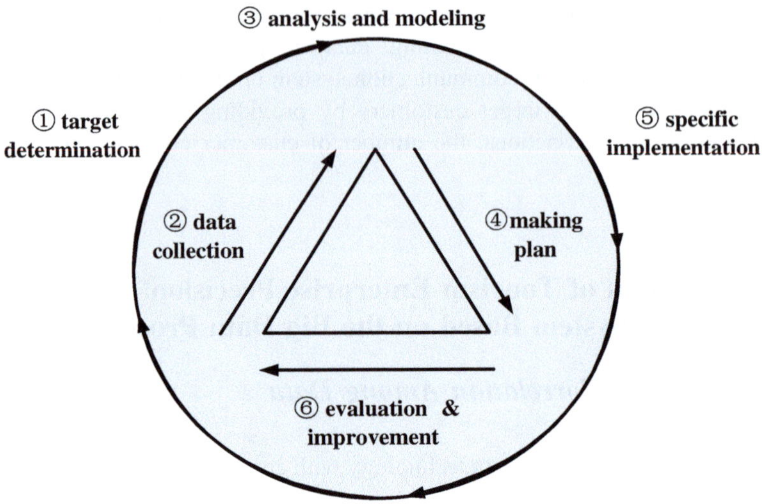

Fig. 1 Precision marketing system frame [6]

demand and updates business processes. For example, the enterprise could recommend the most appropriate tourist route via multiple channels such as SMS, micro-chat, micro-blog, etc., and record customer acceptance of recommendations to determine the next marketing goals.

The second key node is "analysis and modeling," which is used to mine the information obtained and accurately predict market demand; it can then be used to perform intelligent analysis of decision making. Each route is designed to match customer needs with the help of high-speed railway data, and the tourism enterprise can detect customer behavior and implement relative marketing strategies by providing suggestions for users at the right time and place to deliver advertisement and relevant information.

The final key node is "evaluation and improvement," which is based on a sequence of data visualization. The following important data form the basis of determining future marketing objectives:

1. Have marketing activities achieved the expected goals?
2. Which activities should be improved?
3. How should future marketing activities be guided?

4 Channels to Implement Precision Marketing for Tourism Enterprise in the Era of High-Speed Rail

4.1 Exact Market Positioning on the Basis of Big Data

More than 800 million passengers use high-speed rail, which represents a huge customer group that is bound to be the primary object of tourism enterprise. The enterprise could orient customers through high-speed rail data. For example, for customers whose characteristics include short displacement, high frequency, and short distance, a resort or vacation-village along the high-speed rail line can be recommended; for customers whose characteristics include distant displacement, long time, and low frequency density, outbound tourism and other in-depth tourism products along the rail line can be recommended. High-speed rail passengers have huge tourism consumption potential, and many tourism resources have been developed the rail line tracks, which run from the east coast to the western underdeveloped area. Meanwhile, through RFM analysis (Recently purchased time/purchase Frequency/single Monetary amount), the company could scientifically forecast regular customers' purchase amounts in the future, as well as analyse the late customer value, and then select VIP clients and design personalized products for them.

4.2 Using Big Data and Marketing Platform to Achieve Precision Marketing

While current marketing channels explosively expand and various media platforms tend to conform, high-speed rail in this media environment has unique properties. High-speed rail and train stations across the country have formed a huge "field", and in this field there are numerous high-quality media contact points. Hundreds of millions of high-level customers are attracted into this space, so the strength of contact transmission has been strengthened. Customers in this environment are more accepting of information related to tourism marketing. Tourism enterprises could classify customers to determine corresponding high-quality ones on the basis of high-speed rail data, thus making the delivery of tourism information more efficient. Meanwhile, SNS social media has a precise nature in describing the "who" of tourism, and its description of "interest and buying preferences (i.e., the "why") is also an objective of precision marketing; thus, high-speed media could use SNS media to excavate indicators—such as "comment," "contact numbers," "purchase frequency"—to design potential tourism projects and complete precision marketing plans [7].

4.3 *Personalized Product Design and Innovative Service*

Relying on high-speed rail data, tourism enterprises could design personalized products to "awaken" diverse demands, develop scenic spots along the railway, and implement route optimization and line innovation [8]. For example, Guiyang–Guangzhou high-speed railway started operation in December 2014, which greatly increases the number of tourists choosing high-speed rail for travel in the southern region. During the 2015 festival holiday period (February 18th through 24th), the number of tourists in Guangxi totalled 9.4 million tourists, a growth of 18.84 % compared with the same period last year, and total tourism income has reached 39.48 hundred million, a growth of 22.43 % compared with the same period.The tourism economy has a tendency toward blowout, This, a tourism company could design new tourism products on the basis of this new high-speed railway travel trend and customize personalized services to match different people, regions, and time nodes by means of high-speed rail data. For different audiences in different target areas and during different seasons, personalized travel routes are recommended to realize high-quality experience services, which activates the next growth point in the short-distance tourism market.

5 Summary

The key for a tourism enterprise to implement precision marketing in the era of high-speed rail lies in accurate market positioning, feasible technical support, and development of powerful product designs on the basis of big data, which is the key to reduce the trial-and-error costs and enhance marketing effects.

References

1. China is entering high-speed rail era (2015), vol 11531:A1 Economic Daily
2. China Tourism Academy (2014) China travel agency industry development report. China Tourism Press, Beijing
3. Christina H et al (2013) Big data and the creative destruction of today's business model, AT Kearney
4. Kotler P (1994) Reconceptualizing marketing: an interview with Philip Kotler. Euro Manag J 12(4):353–361
5. Chai H (2008) The application research on precision marketing in the field of tourism electronic commerce. Commer. Era 7:27–28
6. Zoratti S, Gallagher L (2013) Precision marketing maximizing revenue through relevance. Enterprise Management Press, Beijing
7. Liu Z (2007) Study on Precision Marketing Method. J. Shanghai Commun. Univ. 41:143–146
8. Cao L, Liu J (2009) Application Research on Precision Marketing in the Enterprise E-commerce. Mod. Shopp. Malls 5:149–150

Characteristics and Development of Information Industry and Its Impact on the Economy

Shan Chen

Abstract The information industry has gradually become a pillar of national economic development in various countries, but there has been disagreement about its definition and its categorical measure at home and abroad. As a constantly changing complex system, the information industry has its own law of development as well as internal characteristics. The information industry features the intellectualization of high technology and regards universality as the main goal. Another feature of the information industry is that it penetrates into and integrates with traditional industries by means of information technology and service, which makes its development have a "multiplier" effect on the economy.

Keywords Information industry · Intelligence · Universality · Infiltration and integration · Industrial measurement systems · Spillover effect

1 Introduction

The 21st century is one of communication as the information industry increases rapidly in the international context and has an increasingly high contribution to economic growth. From an international perspective, the information industry has become the largest pillar of American economics and one of the world's fastest growing industries. Since the 1990s, a strong push from the information technology revolution has led the information industry into a period of rapid development. In 2009, the scale of China's information economy topped $900 billion for the first time, $633.4 billion more than that in Britain. In 2013, the scale of China's information economy even reached $2.18 trillion, $1.58 trillion more than Japan, making it the world's no. 2 industry following the United States, whose scale of information economy was $7.49 trillion. Data show that the growth rates of the information economy in China, the United States, Japan, and Britain are significantly higher than

S. Chen (✉)
School of Economics and Management, Beijing Jiaotong University, Beijing 100044, China
e-mail: 125029529@qq.com

M. Li et al. (eds.), *Proceedings of 2015 2nd International Conference on Industrial Economics System and Industrial Security Engineering*,
DOI 10.1007/978-981-287-655-3_2

those countries' GDP growth rate, which are, respectively, 1.6, 1.5, 4.1, and 1.6 times the country's GDP growth rate. Furthermore, the information industry is playing an increasingly remarkable role in driving economic growth.

However, the information industry is an evolving complex system, and its unique internal and development characteristics differentiate it from other industries in terms of the basic law of development. How do we define the current rapid development of the information industry? How do we accurately classify and measure the size and extent of its development? Research on the characteristics of the information industry may reveal its mechanisms and rules of development, which will be conducive to clarifying the current status of the information industry as well as its advantages and disadvantages and its impact on the economy. Finally, it will be possible for us to put forward scientific suggestions and specific strategies regarding the developmental direction of the information industry.

2 What Is Information—Its Evolution and Development

The concept of information is always changing, but it evolves slowly with the development of society. Information in ancient China was known as the "news," whereas in English it was known as "information" and in Japan as "intelligence". C.E. Shannon explicitly pointed out the following: "Information is a thing used to eliminate the uncertainty of a random thing," and this definition is now widely accepted. Economic management scientists believe that "information is used to provide effective decision-making data." Our information expert, Zhong Yixin, stated that "information is the way things exist or a state of motion, and is expressed through this way or state in a direct or indirect manner." United States information management specialist F.W. Horton defined information as "processed data used to help users making decisions."

With the development of human society and the advancement of science and technology, information changes in its content, its form, and the manner in which it is conveyed. Means of transferring information evolved from oral message to letters, the telegraph, and the telephone. Today, information is transferred in huge amounts through communication networks and the Internet, which features much higher speed, larger scope, more diversified information content, and more intensified directivity. This scientific trend of information "bitization" will last until a new era, the "Internet of Everything," begins.

3 The Definition and Classification of Information Industry

Based on the theory of "knowledge industry" proposed by F. Machlup in 1962, American economist M. Porat separated out all the information and information activities included in the first, second, and third industries (based on the industry-classification method proposed by C. Clark) and labeled them as an independent industry, namely, the information industry. In his opinion, information industry has already penetrated into every sphere of the national economy; thus, information activities—in terms of production, procession, circulation, and service—cannot be simply classified into certain departments. Departments performing information activities shall be divided into explicit ones, which directly provide information products and service to the market and implicit ones, which provide information service and capital only for internal consumption but without entering the market [1]. Economic scientists at home and abroad have begun to conduct relevant research on this basis.

The American Information Industry Association believes that information is a combination of production activities that produce and provide information products and services by utilizing new technologies and innovative information processing methods.

In the opinion of our country's information expert, Wu Jiapei, the information industry is a new industry engaged in information and technology equipment manufacturing and information production, processing, storage, distribution, and service: It is composed of the information equipment manufacturing industry (hardware industry) and the information services industry (i.e., the software industry) [2].

In 1997, the United States, Canada, and Mexico jointly developed the North American Industry Classification System (NAICS) [3]. For the first time, this classification system considered the information industry as a separate industry sector and defined it in terms of statistical classification [4]. The NAICS divided the information industry into three categories: (1) production and distribution of information and cultural goods; (2) provision of methods to deliver or distribute those products, data, or communications; and (3) processing of data. In 2002, the NAICS re-divided these into four categories: (1) the publishing industry; (2) the film and recording industry; (3) the broadcasting and communications industry; and (4) the information and data-processing services industry. This standard of classification was used in a survey of these three countries [5].

Expert Si Youhe divided the information industry into three departments: (1) the IT equipment manufacturing department; (2) the information commercialization department; and (3) the quasi-information department. This classification of the information industry is regarded as reasonable and has become the three major components of China's information industry [6] as follows:

- The IT equipment manufacturing department includes microelectronic, electronic, computer, communications and network-equipment manufacturing, and information technology infrastructure construction
- The information-commercialization department includes the information-production sector, the information-communication sector, and the information-service sector.
- Quasi-information departments (information department affiliated with non-information industry).

Meiyun considered that some sectors in the implicit-information departments, which provide paid information services, should be classified into the information-service sector, whereas information products provided only on a self-service basis or provided free of charge should not be regarded as part of the information industry [7]. The definition of the information industry should also be expanded in part at different stages of historical development. Domestic information —industry development was integrated into traditional industries to create new production organization method in all aspects: production, sales, distribution, and service. The formation of a new organization and production should also be part of the information department in the information industry.

4 Characteristics of Information Industry

I. The information industry is closely related to scientific and technological development, and its intellectualization has become an increasingly prominent feature. The information industry is called a "3C" industry in foreign countries (i.e., computer, communication, and content) and features much technological content. From this perspective, it not only includes the manufacturing of information equipment and the communication, it also includes the even larger information-service industry with information resource development and utilization as its content. Information and communication technology is the product of organically combining telecommunications services, information services, and IT services and applications. It does a better job in fully and accurately reflecting the new forms by which information technology supports and promotes the "informatization" of a society.

Computers, electronics, communication equipment, and high-speed information networks all belong to the "hardware" side of the information industry, whereas the publication of news and books, posts and telecommunications, advertising, software development, and systems integration all belong to the "software" side of the information industry. If the information industry was thought of as a human being, then the hardware is like the bones supporting the body, whereas the software is like the blood circulating through the body. However, the most important part of a human being is the brain. ICT is a kind of information service industry that is based on information resources using intelligent methods and innovative approaches (i.e., the brain) to developing and utilizing various types of hardware and software resources to create new products and value.

The information industry is closely related to the development of science and technology, its development can be divided into the following three distinctive stages:

1. Television and telephone, especially the increasingly popular personal computer, is the first big step of information industry in promoting social progress. At this point in the information-transfer phase, information is mostly transmitted

directly; its form and content are very simple. During this period, the information industry is still focusing on the hardware side.

2. In the early 1980s, Internet technology developed rapidly; its rise led to the rapid development of the information industry and attained the first apex. Portals, search engines, and instant messaging appeared quickly, one after another, and it can be seen that this was the information-acquisition stage. The scope of information acquisition expanded exponentially; information content became diversified; and the information form gradually became bitization. At this point, the development of the information industry focused on the software side.
3. The recent developmental cycle of the information industry is directed at the treatment and reuse of information: big data mining, OTT applications, Cloud storage, e-commerce, and Internet finance. Information service based on ICT is no longer just the role of disseminator of information, it uses the hardware and software information resource production and creates new products and new values of the "brain."

Thus, it can be observed that the information industry is closely related to the development of science and technology, but the focus of development—whether "hardware," "software," or "brain"—has turned in an intellectual direction. The electronics and communication equipment manufacturing industry generally has a high degree of intelligence, [8] and information transfer gradually changed to intellectual development. The content and object of delivery became more selective and targeted. Information service is the best intelligent comment, either information processing or storage, and either mining or application, these cannot leave the mainline: intellectualization. The higher level of the intelligent information industry, the faster it develops and the larger it becomes the industrial standard.

The developmental element of information industry is people, and popularity (i.e., demand) is the fundamental purpose of its development. The information industry exists to develop human beings' spiritual wealth; the object of industrial activity is information; and the content is for information research, development, dissemination, and use. Information industry is high-tech and knowledge-intensive. Because information technology industry requires more sophisticated, senior, and complex labor, knowledge, technical capabilities, and the higher power requirements of scientific research have become necessary for the employee. Although the fixed-asset investment is also important for the industry, the characteristics of the information industry make the intellectual assets more critical.

Popularity is the fundamental purpose of the information industry. Whether it is information delivery or information production and service, the main purpose is to benefit people. According to CNNIC statistics, as early as 2008 the number of China's Internet users had surpassed that of the United States and become no. 1 in the world; in December 2014 the number of users reached 649 million and the Internet penetration rate was 47.9 %, which fully reflects the universality of the information industry [9].

Human society tends to be more equitable and open because of the power of information; as such this should foster more inclusive sharing and collaboration.

The popularity of information is closely related to the development of the information industry according to human needs. The desire for information transmission resulted in the emergence of the telegraph and telephone telecommunications networks. The need for long-distance information broadcasting led to the advent of television and television networks. The high-speed broadband network has reached a perfect purpose of the free and equal exchange of information. There are a variety of information services to meet the needs of individuals in all aspects of spiritual and material basic necessities. Development of the information industry should offer more opportunities for all people to enjoy more freedom and happiness. This is also in line with the goals of the World Summit on the Information Society (WSIS) for the information industry to be "people-centered, inclusive and development-oriented" [10].

The information industry has a strong capability to penetrate and integrate society. It uses information technology as a foundation and information equipment as the carrier; through the spread and digging of information resources, it gradually permeates traditional primary, secondary, and tertiary industries. Its permeability is shown by its application of rich processed information resources and service in traditional industries through advanced information technologies to promote technical innovation (often referred to as "informatization") of the industrial production or organization modes. This resembles the concept of industrialization. Most economists hold that industrialization is a process during which the proportion of the total output value and employed population of the secondary industry (i.e., industry) in the whole society's total output value and employed population is increasing continually, whereas that of the primary industry continually decreases. Our country's economist, Peigang, believes that "the national economy in a series of strategic production function process keeping change, is the process of industrialization" [11]. It is a field including two aspects of industrial and agricultural modernization and mechanization: "social productivity change", which is a series of changes caused by the significant symbol of productivity, and change in social-production tools.

Informatization is a process during which the proportion of the total output value and employed population of the information industry (i.e., the fourth industry) in the whole society's total output value and employed population continually increases. The information industry will become the leading industry of the national economy, e.g., information resources have become a new factor of production inputs to various industrial production activities as well as the process of national economic operation.

The information fuses the traditional industry, which smears fusion boundary and has new contents and forms. It is a paradigm of breakthrough in traditional industrial innovation. The development of this integration is an inevitable trend and also leads to social and economic acceleration. Many new products and services created by industrial innovation lends the traditional industry a complementary, alternative, or restructured relationship [12]. The complementary information industry, which occured during the first industry fusion, produced modern agricultural production, and the sales service system produced the food safety traceability system. The recombination of the secondary industry fusion spawned the industrial intelligent-control production line, the industrial robot, and the customized industrial manufacturing system. Fusion of the alternative third industry resulted in electronic commerce, Internet finance, and mobile payment. To generate fusion inside the information industry's own OTT service, etc. Generating fusion inside the information industry's own OTT service, etc. has fully proved that all kinds of production factors constitute the productivity of the information industry including "information equipment," "information technology," "information service," and "information resources." The productivity of information becomes a "multiplier" and promotes the development of the productive forces within other industries. From the previous analysis, it can be seen that the penetration of the information industry into traditional industries exhibits a characteristic of "mixture," whereas the fusion of products and services serve as "compound" traits.

5 The Main Method to Measure the Existing Information Industry and Its Application Analysis

The main methods to measure information industry include the following: Porat information economic measurement method, the input–output method [13], the Japanese informatization index model method [14], the Engel-Bray-Hitt multiplier method, the Hayes–Erickson logarithmic model, and the comprehensive information-industry dynamics method. [15]. Porat's method, the Engel-Bray-Hitt multiplier method and the comprehensive information industry dynamics method, etc., are a measure of the level of the information industry from the perspective of economics. Thus we can achieve the purpose of linking the information industry to the social productive forces as well as the industrial structure and labor force change. The social informatization–measurement method, such as the Japanese informatization index model method, comes from the perspective of sociology, i.e., measurement of the degree of social informatization of a country or region. The main characteristic of indexation measurement is correlation is the ability to perform longitudinal comparison of the time vector as well as compare the area vector. These two types of methods have their advantages and disadvantages, but both of them attempt to construct a comprehensive evaluation system that can completely and accurately reflect the situation of social information industry. However, they have problems as follows:

1. There are differences or overlap about the definition and scope of the information industry. Researchers have various definitions for the information industry and its range with both regional and time limitations. Today's information industry development and innovation have gradually expanded their scope, which can no longer use the old standard definitions. Original measurement system indicators may also no longer apply; thus, they must be readjusted.

2. The universality of the information industry is also an important index. The development of the information industry shows a trend to close to that of individuals, e.g. the E-mailbox number used, the number and type of APP that are used frequently, and the information services of the government and social service department. Such improvement of people's living standards and quality indicators also must be reflected in the measurement system.
3. The information industry continues to promote industrial upgrading in traditional industries, and some industries promote innovative integration such as mobile payments. That such emerging new departments continue to appear confers new problems in existed measurement method and index.

6 Development of Information Industry and the Impact on the Economy

The rapid increase of the information industry caused great changes in industrial and employment structures and has also led to a dramatic change in the quality of economic growth in each respective country. Information has become one angle of the "resource triangle" in the process of economic operation and thus is on an equal position with materials and energy. The development of information technology shifts the consumption resources by economic growth from energy and materials to information and technology. Therefore, the development of the information industry has become an important symbol of a region's, or even a country's, level of economic development and degree of social progress [16]. Because of the characteristics of information technology, its impact on the economy lies not only in its output growth, which brings direct economic growth, but also in its permeability, which may improve traditional industries' efficiency by (1) dramatically enhancing their output and remarkably reducing production costs; and (2) as well as integration, which may promote the optimization and upgrading of traditional industries and even the formation of new industries. By such reasoning, the direct economic impact and indirect economic benefits of the information industry to the national economy are diffused; however, it plays a key role in national economic development. The impact of the development of the information industry of the given economy has been the focus of academic attention, but in view of its development process, a long-term, in-depth study is required.

Neoclassical economists believed that capital, labor, and technology are the three main causes of economic growth. Specialized knowledge, technological progress, and human capital can promote economic growth, technological development, and progress and, as such, are important factors in economic growth. However, now IT and technology are the main components. Robert Solow, who considered economic growth, capital, and labor technology established a new economic-growth model. However, he did not have a clear explanation about how technology affects economic growth.

Chinese scholars Xianhua and Ji used cointegration and Granger the causality test to analyze the relationship between the information industry and economic growth. Further theoretical calculations used gray relational information industry and the national economy in the correlation between the three industries [17].

The role of information industry development in other industries has increased the productivity generated, which is known as the "spillover effect" [18]; there is no doubt that there is also overall economic growth. This spillover effect of the information industry is multifaceted; It improves the working environment and workers' productivity and it improves the quality of production factors of social production, thus contributing to economic activity in terms of more efficient, more rapid economic growth. Xinling and Li drew contributions from the two-sector model by Feder [19] to measure the information industry on economic growth [20]. Jinjin employed the GDP in 2004–2010 and the information industry output value of the panel data model to explore the impact of China's information industry on China's three major economic regions [21].

However, most scholars did not include the new information economic scale produced by the permeability and integration of information technology into traditional industries in the measurement systems. That makes the current-factors measurement of the economic impact of the information industry obviously flawed and does not conform to the development trend of the information industry. The estimation deviation of influence social and economic factors may lead to many economic-, industrial-, and policy-related issues.

7 Conclusion

Based on the characteristics of the information industry and its development, its effect on the economy should be studied from three perspectives as follows:

1. The existing information industry is at the core using current industry classification and statistic system directly to calculate the economic size and growth rate.
2. The information industry continually self-integrates into the production section of traditional industries with advanced equipment and technology. In production, sales, and service, the improvement and innovation of the method of production organization brings economic benefits and improves production efficiency. This point consists of the information industry promoting the economic growth scale while also belonging to the system. The input–output method should be used here.
3. New technology integrates with original traditional industries to bring about new products and departments in the continuous development of the information industry, which will also contribute to the development of the national economy. However, the marginal part of the new department is still fuzzy, and the definition is not clear, either, thus leading to inadequate research of the size of the economy and the growth rate as well as the impact on the economy

The sum of the three parts is the entire content of the information industry's impact on the national economy. Their development speed and scale, as well as a single part of the proportion of the GDP and the promotion of improved function, are the focus of subsequent quantitative research.

References

1. Porat MV (1965) The information economy. Government Printing Office, Washington, DC
2. Wu J (1995) Research from the practical point of view—answer Wen Ge about the conception of industry and the range of information industry. Inf Doc Work 5:8–11
3. International Standard Industrial Classification of All Economic Activities (ISICversion3) (1997) http://www.unstats.un.org/unsd/cr/registry/regcst.asp
4. North American Industry Classification System(NAICS)15 Aug 2006 (2006) http://www.census.gov/epcd/www/naics.html
5. 2002 NAICS CodesandTitles (2002) http://www.census.gov/epcd/naics02/nacicod02.txt
6. Youhe Si (2001) Information industry. Chongqing Publishing Company, Chongqing
7. Meiyun Z, Tiyun H, Yaqiong H (1997) Definitely stipulating the scope of information industry—for the need to deepen study. J China Soc Sci Tech Inf 189−192
8. Zhang G, Ye Y, Zhang Q, Ji Y, Meng Q (2002) Research on setting up Tianjin information industry. Stat Index 11(06):723–729
9. The Thirty-fifth statistics report for china Internet developing situation (2015) China Internet Network Information Center
10. Chen J, Kang Y (2004) Digital solidarity portray new picture of information society. China Telecommun Trade 1:85–87
11. Zhang P (1984) Agriculture and Industrialization. Huazhong University of Science and Technology, Wuhan
12. Zhou Z (2003) Informatization and industry convergence. Shanghai People's Publishing House, Shanghai, pp P213–P214
13. W Leontief (1966) Input-output economics. Oxford University Press, Oxford
14. J Jipeng (2002) The application of information economics. Science Press, Beijing, pp P248–P249
15. Jipeng Jing, Xin Wang (1993) The comparative study of the structure of information industry and the measurement method. Inf Sci 10(3):129–133
16. Wu J, Xie K (1999) Information technology industrialization and Economic Growth. China Industrial Economy
17. Xian-hua WU, Ji GUO (2006) Information industry and economic growth : a study based on co-integration and grey correlation theory. Stat Inf Forum 3:27–29
18. Shiliang H (2000) Shanghai information industry and economic growth. Shanghai Stat, vol 5
19. Feder G (1983) On exports and economic growth. Dev Econ 12(1–2):22–25
20. Xin-ling Z, Peng L (2005) Research on the contribution of information industry to economic growth. Stat Inf Forum, vol 06
21. Jinjin Y (2014) An empirical study on the impact and prediction of information industry upon economy and employment, Postgraduate Program, School of Mathematics and Statistics, Central China Normal University

Study on Differences of Operating-Asset Allocation of China's State-Owned and Private Enterprises

Yang Wang and XiaoYe Liu

Abstract Different enterprises own different kinds of resources, and they will carry out different business decisions depending on these different resources; thus, resource endowment is bound to affect the behavior of their resource allocation. This chapter divides China's listed companies into state-owned and private enterprises according to their government resource endowment to study the differences of operating-asset allocation. We obtained our results on the basis of data collected from 936 samples from 7 main industries of China. The empirical results suggest that there is a significant difference between state-owned and private enterprises in terms of operating-asset allocation efficiency.

Keywords Operating-asset allocation · State-owned enterprises · Private enterprises

1 Introduction

With the advances made in science and technology, our economic development is no longer governed solely by natural resources. The rational use and allocation of resources has become the inevitable choice in the bidirectional requirements of economic growth and ecological balance. China is in a critical period of industrial restructuring; thus, the rational use of resources is key to optimize the industrial structure of economic development.

Amit and Schoemaker (1993) believe that resources are elements owned or controlled by enterprises, which under certain mechanisms can be transformed into a final product or service by combining them with other assets [1]. Regarding the

Y. Wang (✉) · X. Liu
School of Management, China University of Mining and Technology, 100083 Beijing, China
e-mail: wangyang.som@cumtb.edu.cn

X. Liu
e-mail: xiaoyealan@hotmail.com

M. Li et al. (eds.), *Proceedings of 2015 2nd International Conference on Industrial Economics System and Industrial Security Engineering*,
DOI 10.1007/978-981-287-655-3_3

subject of resource use, enterprises are the most active micro-entities in China's economic development. They are not just important drivers of economic growth, they are also the main employment channels used by employees. By reasonable configuration to various resources enterprises possess, they can have great and stable development; with the optimal use of resources, China's overall economic status will be vital and healthy.

Harris (1990) believes that asset quality is the result of the business *activities* of an enterprise as well as a reflection of its *quality* [2]. Operating assets are an important part of enterprise resources, and assets that can form corporate earnings directly; their allocative efficiency cannot be ignored because the enterprise may decide that it will obtain their main business income more or less if there is sustained profitability. YuanYang put assets into three categories according to the role they have play in the performance of an enterprise: (1) assets that directly form corporate earnings; (2) assets that have no effect on corporate earnings for a certain period; and assets (fixed assets, intangible assets, etc.) that are separated from corporate earnings for a certain period [3]. Although Qiao considered an enterprise's expenditure intensity in fixed assets, intangible assets, and enterprise worker wages and benefits, which are all indicators of the allocative efficiency of an enterprise [4].

Because state-owned and private enterprises have different forms of ownership, together they will eventually provide power for the balanced long-term development of China's overall economy. State-owned enterprises focus mainly on areas of basic and strategic industries related to national interest, and their development relates to the cultivation and improvement of China's major industrial competitiveness. The rapid development of private enterprises is of great significance for the creation of employment opportunities, increase of wage income of urban and rural residents, and improvement of people's living standards. Li divided listed companies into state-owned enterprises and private enterprises by determining whether the actual controller of the enterprise is the government or private capital by measuring and classifying their political association [5]. This chapter also takes the property of control as a basis for classification. We study differences in operating-asset allocation between two types of enterprises and present the results.

2 Empirical Process

This study refers to Qiao design of indexes of operating-asset allocation efficiency to determine proxy variables [4]. Fixed assets are an important means of production and play a crucial role for enterprises. At the same time, in the knowledge-economy era, intangible assets are more advantageous than other assets in promoting enterprise development. Employee incentives helps enterprises maintain a high and sustained positive mood for employees so they will create greater value. Thus, we chose enterprise expenditure intensity in fixed assets, intangible assets, and enterprise worker wages and benefits as variables to explain enterprises' operating asset-allocation behavior. The formulas are as follows:

$$Y_1 = A_{ij}/W_{ij} \tag{1}$$

$$Y_2 = B_{ij}/W_{ij} \tag{2}$$

$$Y_3 = C_{ij}/P_{ij} \tag{3}$$

$$W_{ij} = D_{ij} + E_{ij} + F_{ij} - H_{ij} - Q_{ij} \tag{4}$$

where A_{ij} is the net value of fixed assets of each enterprise in each industry; B_{ij} is the net value of intangible assets of each enterprise in each industry; C_{ij} is the cash paid to and for employees of each enterprise in each industry; D_{ij} is the value of fixed assets of each enterprise in each industry; E_{ij} is the value of intangible assets of each enterprise in each industry; F_{ij} is the value of current assets of each enterprise in each industry; H_{ij} is the value of financial assets held for trading held by each enterprise in each industry; Q_{ij} is the net value of short-term investments of each enterprise in each industry; and P_{ij} is the value of prime operating revenue of each enterprise in each industry.

2.1 *Figures*

In this research, consecutive annual financial data from 2007 to 2012 of the listed companies in all industries are selected as samples to be studied. After classifying the entirety of industry enterprises—considering the difference in the enterprise numbers in different industries, the difference of industry status of the two types of enterprises in their industries, and the consistency and integrity of the data—we retained 936 enterprises from 7 major industries (mining, real estate, manufacturing, wholesale and retail trade, construction, agriculture and forestry, and information transmission) to continue our empirical research, which finally included 468 state-owned and 468 private enterprises. The study used SPSS21.0 software to complete the empirical analysis.

2.2 *Tables*

The statistical description of the three types of indicators of state-owned and private enterprises of the seven major industries are shown in Table 1.

Before difference analysis, we obtained normal distribution examination with Kolmogorov–Smirnov test.

Table 1 Sample descriptions

Industry	Index	State-owned					Private				
		N	MIN	MAX	MEAN	SD	N	MIN	MAX	MEAN	SD
Mining	Y_1	60	0.02	0.41	0.15	0.10	60	0.00	0.39	0.15	0.10
	Y_2	60	0.00	0.85	0.12	0.17	60	0.00	0.68	0.09	0.13
	Y_3	60	0.03	0.28	0.08	0.05	60	0.04	3.33	0.21	0.49
Real estate	Y_1	294	0.00	0.80	0.07	0.11	294	0.00	0.79	0.06	0.11
	Y_2	294	0.00	0.25	0.02	0.04	294	0.00	0.90	0.03	0.11
	Y_3	294	0.00	0.51	0.07	0.07	294	0.00	4.53	0.13	0.39
Manufacturing	Y_1	1998	0.01	0.88	0.35	0.18	1998	0.00	0.90	0..31	0.16
	Y_2	1998	0.00	0.81	0.05	0.06	1998	0.00	0.71	0.06	0.06
	Y_3	1998	0.01	23.33	0.11	0.53	1998	0.00	6.73	0.10	0.16
Wholesale and retail trade	Y_1	240	0.03	0.84	0.29	0.22	240	0.00	0.80	0.24	0.19
	Y_2	240	0.00	0.40	0.05	0.07	240	0.00	0.36	0.05	0.07
	Y_3	240	0.00	0.39	0.06	0.05	240	0.00	0.30	0.05	0.04
Construction	Y_1	60	0.02	0.41	0.15	0.10	60	0.00	0.39	0.15	0.10
	Y_2	60	0.00	0.85	0.12	0.17	60	0.00	0.68	0.09	0.13
	Y_3	60	0.03	0.28	0.08	0.05	60	0.04	3.33	0.21	0.49
Agriculture and forestry	Y_1	60	0.11	0.47	0.27	0.08	60	0.00	0.69	0.31	0.17
	Y_2	60	0.00	0.54	0.10	0.14	60	0.00	0.39	0.08	0.09
	Y_3	60	0.04	0.30	0.10	0.07	60	0.04	14.12	0.33	1.81
Information transmission	Y_1	96	0.01	0.88	0.27	0.26	96	0.02	0.51	0.15	0.10
	Y_2	96	0.00	0.08	0.03	0.02	96	0.00	0.14	0.03	0.03
	Y_3	96	0.02	0.42	0.15	0.09	96	0.04	0.53	0.20	0.13

Table 2 Normal distribution

State-owned		Y_1	Y_2	Y_3
N		2808	2808	2808
Normal parameters [a, b]	Mean	0.31	0.05	0.11
	SD	0.20	0.07	0.45
Most extreme differences	Absolute	0.06	0.22	0.41
	Positive	0.06	0.18	0.35
	Negative	−0.06	−0.22	−0.41
Kolmogorov–Smirnov Z		3.31	11.52	21.75
Asymp. sig. (two-tailed)		0.00	0.00	0.00
Private		Y_1	Y_2	Y_3
N		2808	2808	2808
Normal parameters [a, b]	Mean	0.27	0.06	0.11
	SD	0.17	0.08	0.33
Most extreme differences	Absolute	0.06	0.23	0.37
	Positive	0.05	0.18	0.32
	Negative	−0.06	−0.23	−0.37
Kolmogorov Smirnov Z		3.33	12.03	19.86
Asymp. sig. (2-tailed)		0.00	0.00	0.00

[a] means MIN, [b] means MAX

From Table 2, it can be seen that three types of indicators of all state-owned and private enterprises are not normally distributed. Thus, this study used nonparametric test to judge whether or not there are significant differences between the three indicators of the two types of enterprises. To examine the differences in the operating-asset allocation efficiency of state-owned and private enterprises, we used Mann–Whitney U test.

In organizing the data in Table 3, we constructed Table 4.

From Table 4, we can provide both a horizontal and vertical analysis.

Horizontal analysis
Although the addition of agriculture and forestry industry to all indicators was not significant, the rate of the other industries was significant at 78.9 %. This means that in most industries, state-owned and private enterprises have different operating-asset allocation behaviour.

Vertical analysis
In terms of fixed-asset expenditure intensity, there were no significant differences in the real estate, construction, and agriculture and forestry industries. Their descriptive statistics data are maintained at 6, 15, and 30 %, respectively. These three industries require many fixed assets, or they take fixed assets for the main business in their process of production and operation, so there are no significant differences between state-owned and private enterprises.

In terms of intangible-asset expenditure intensity, significant diffferences exist in all industries except for agriculture and forestry and information transmission.

Table 3 Mann–Whitney U test

Industry		Y_1	Y_2	Y_3
Mining	Mann–Whitney U	707	1371	1491
	Asymp. sig. (2-tailed)	0.00	0.02	0.10
Real estate	Mann–Whitney U	42140	37262.5	38556
	Asymp. sig. (2-tailed)	0.60	0.00	0.02
Manufacturing	Mann–Whitney U	1764095	1778754	1874615
	Asymp. sig. (2-tailed)	0.00	0.00	0.00
Wholesale and retail trade	Mann–Whitney U	25563	23711	23959
	Asymp. sig. (2-tailed)	0.03	0.00	0.00
Construction	Mann–Whitney U	1793	1336	1276
	Asymp. sig. (2-tailed)	0.97	0.01	0.01
Agriculture and forestry	Mann–Whitney U	1789	1724	1556
	Asymp. sig. (2-tailed)	0.95	0.69	0.20
Information transmission	Mann–Whitney U	3708	4238	3709
	Asymp. sig. (2-tailed)	0.02	0.34	0.02

Table 4 Summary Table

Industry	Y_1	Y_2	Y_3
Mining	Y	Y	N
Real estate	N	Y	Y
Manufacturing	Y	Y	Y
Wholesale and retail trade	Y	Y	Y
Construction	N	Y	Y
Agriculture and forestry	N	N	N
Information transmission	Y	N	Y

Y = significant; N = insignificant

In the agriculture and forestry industry, intangible assets are primarily composed of agricultural scientific and technological achievement patents. In the information-transmission industry, intangible assets are primarily composed of exclusive technology, software, trademark and licensing rights, etc. Intangible assets are an important part of an enterprises' production and operation activities no matter what their kind of enterprise.

In terms of worker wages and benefits expenditure intensity, significant differences exist in all industries except for mining, agriculture, forestry, and information transmission. This means that state-owned enterprises give more stable and average worker incentives than private enterprises.

3 Conclusion

This results of the present empirical analysis shows that there are significant differences in three kinds of indicators of state-owned and private enterprises. This means that the two kinds of enterprise have different expenditure intensities in fixed assets, intangible assets, and worker wages and benefits. With different corporate objectives and responsibilities, these enterprises will make different strategic decisions; therefore, their asset-allocation behaviour will differ. Various industries face different opportunities and constraints, but all state-owned and private enterprises in all industries operate in the same changing era. They all need innovation and to reform technological innovation, the introduction of talents, and procurement of a management system. State-owned and private enterprises should determine the right operating asset-allocation behaviour for themselves and create a beneficial cycle of "technological innovation → product → income → technological innovation." Only in this way can enterprises become more vital and have more sustainable development.

Acknowledgments This project was funded by the National Natural Science Foundation (71373262), Beijing Undergraduate Training Programs for Innovation (Z20141504), and Teaching Reform Project of China University of Mining & Technology, Beijing (j120501).

References

1. Amit R, Schoemaker PJH (1993) Strategic assets and organizational rent. Strateg Manag J 14 (1):33–46
2. Harris M, Raviv A (1990) Capital structure and the informational role of debt. J Financ 45 (2):321–349
3. Qu YY, Liu Y (2004) Discussions on the impact of the asset structure of enterprise management. 2004(3):21–21
4. Qiao, M. (2013) Research on efficiency of operating asset allocation of listed companies in mining industry. China University of Mining and Technology, Beijing, pp 11–13
5. Li, J.F. (2012) The empirical research about political connections, diversified level and corporate performance of real estate listed companies in China. Southwestern University of Finance and Economics, pp 12–13

Research on the Influence of Macroeconomic Factors on Consumer Finance Structure

Yang Wang, Linlin Du and Chenghai Zou

Abstract With the transformation of the structure of China's economic growth, the proportion of consumption in the national economy will be larger, and the effect of promoting consumer finance will be more obvious. Thus, the chapter presents empirical research on the factors affecting consumer finance to reach some conclusions about whether the structure of our nation's consumer finance is reasonable and whether macroeconomic factors affect the internal as well as the overall structure.

Keywords Consumer finance · National economy · Consumer finance structure

1 Introduction

With the transformation of China's national economic-development strategy, the role of consumer demand in promoting economic growth is becoming more important. Although there has been significant development in consumer finance, it must be determined (1) whether the consumer finance structure as well as the whole structure of the national economy is reasonable [1] and (2) how macroeconomic regulations and control the consumer finance, both of which have been neglected, should be enhanced [2].

Y. Wang (✉) · L. Du · C. Zou
School of Management, China University of Mining and Technology, Beijing 100083, China
e-mail: wangyang.som@cumtb.edu.cn

L. Du
e-mail: 1072452275@qq.com

C. Zou
e-mail: 169016861@qq.com

M. Li et al. (eds.), *Proceedings of 2015 2nd International Conference on Industrial Economics System and Industrial Security Engineering*,
DOI 10.1007/978-981-287-655-3_4

2 The Theoretical Basis of Consumer Finance Structure

The topics on which previous scholars and researchers have mainly focused can be divided into three approximate categories, namely, (1) consumer finance development and the impact of research on economic and social structures [3]; (2) the relationship between consumer credit and the macroeconomy; and (3) the relationship between credit structure and the macroeconomy [4].

Consumer finance structure can be divided into internal structure and overall structure [5] on the basis of our predecessors' research: Internal structure (the term "structure") refers to the ratio of short-term consumer loans to total consumer credit and is expressed as Y_1. Overall structure refers to the ratio of the value of total consumer credit to the value of total retail sales of social consumer goods [6] and is expressed as Y_2.

3 Empirical Research

3.1 Indicators and Data Sources

We obtained monthly data from the National Bureau of Statistics, the central bank from 2007 to 2013, as shown in Table 1.

3.2 Model Building

Presented here is the general form of the model: $X_1, X_2, \ldots, X_i$ are the explanatory variables; $a_0, a_1, \ldots, a_i$ are the unknowns (also known as the regression coefficient); Y are the explained variables; ε is the random error; and there is a multivariate regression model when $n \geqq 2$.

Table 1 Index list of explaining variables and explained variables

Explanatory variables (:year-to-year growth rate)	
X_1: economic construction expenditure X_2: people's livelihood expenses X_3: administrative expenses and defense spending	X_4: tax revenue X_5: nontax revenue X_6: M_2
Explanatory variables (:growth rate of fixed base)	
Y_1: short-term consumer credit average balance at the beginning and the end of the month	Y_2: total retail sales of social consumer goods during the month

3.3 *Descriptive Statistics and Correlation Analysis*

We performed a descriptive analysis of the index data. The results are as follows:

Tables 2 and 3 show that in Y_1, X_1, and X_5, the SD is obvious larger than X_2, X_3, X_4, and X_6, so we know that the changing range of the rate of a fixed base of Y_1, X_1, and X_5 is large, thus illustrating that during the months of economic variables, the three indicators representing the fixed base of growth changed widely.

3.4 *Regression Analysis*

To study the relationship of fiscal policy with that of short-term consumer credit, this analysis used backward step-by-step regression analysis; the results are as follows. Table 4 shows that the value of R^2 is 0.835, meaning that the explanatory variables can explain Y1 well, i.e., 83.5 %, thus showing that the regression analysis will yield good results.

The relationship between fiscal policy indexes and Y_1 is shown in Table 5; the rest of the indicators were removed because there was little correlation. The symbol of the regression coefficient, i.e., the representation of the explanatory variable, is

Table 2 Descriptive statistics

	Mean value	SD	N
Y_1	428.8420	262.12592	66
X_1	296.8403	142.16990	66
X_2	226.2359	88.71103	66
X_3	186.8914	53.70804	66
X_4	177.9094	50.72823	66
X_5	246.0300	137.07912	66

Table 3 Descriptive statistics of monetary policy

	Mean value	SD	N
Y_1	428.3000	260.94049	72
X_6	194.1835	54.91499	72

Table 4 Summary of the fiscal policy model

Model	R	R^2	Adjusted R^2	SE of the estimate
1	0.918	0.842	0.829	108.50028
2	0.917	0.840	0.830	108.16734
3	0.914	0.835	0.827	109.03320

SE: Sampling Error

Table 5 Coefficient of fiscal policy

Model		Unstandardized coefficients		Standardized coefficients	t	Sig.	Collinearity statistics	
		B	SE	Trial version			Tolerance	VIF
1	(Quantity)	−302.994	50.404		−6.011	0.000		
	X_2	1.084	0.313	0.367	3.459	0.001	0.237	4.224
	X_3	1.914	0.485	0.392	3.948	0.000	0.270	3.707
	X_5	0.524	0.122	0.274	4.303	0.000	0.656	1.523

B: Independent variable
VIF: Variance Inflation Factor

positive indicating the positive influence on the explained variable; the significance level is >0.1 meaning that X_2, X_3, and X_5 can ass the test of significance. This shows that there significant linear correlations exist among administrative management, people's livelihood spending, defense spending, tax revenue, and Y_1. We can see from Table 6 that the R^2 value is 0.957, which makes the linear relationship obvious.

Table 7 lists the results of regression analysis of Y_1 and X_6. We concluded that there is linear relationship between Y_1 and monetary X_6.

Tables 8 and 9 show that the correlation relationship between Y_1 and Y_2 is significant and that the growth of consumer credit can promote consumption.

Table 6 Monetary policy model summary

Model	R	R^2	Adjusted R^2	SE of the estimate
1	0.978[a]	0.957	0.957	54.28878

Table 7 Monetary policy factors

Model		Unstandardized coefficients		Standardized coefficients	t	p	Collinearity statistics	
		B	SE	Trial version			Tolerance	VIF
1	(Quantity)	−474.502	23.664		−20.052	0.000		
	X_6	4.649	0.117	0.978	39.627	0.000	1.000	1.000

Table 8 Y1 and Y2 relation model summary

Model	R	R^2	Adjusted 2	SE of the estimate
1	0.973[a]	0.947	0.946	11.57207

Table 9 Y1 and Y2 relation coefficient

Model		Unstandardized coefficients		Standardized coefficients	t	p	Collinearity statistics	
		B	SE	Trial version			Tolerance	VIF
1	(Quantity)	109.199	2.635		41.448	0.000		
	Y_1	0.186	0.005	0.973	35.393	0.000	1.000	1.000

3.5 Building and Testing the Structure Equation Model

This study used smartpls software to build and test the structural equation model on financial policy to obtain an analysis of the policy's reliability, validity, and significance. The numbers on the line represent the path coefficient. The equation of fiscal policy and monetary policy structure is shown in Figs. 1 and 2, respectively.

In the reliability analysis, we find that every factor of the reliability value is 0.8; thus, the reliability is good, and the internal consistency of the reliability is very good. The results are shown in Table 10.

In the analysis of validity, we see basic AVE values as shown in Table 11. In the analysis of significance, shown in Table 12, one can see that the path satisfies the two-tailed test indicating high credibility. However, the path does not meet the two-tailed test indicating low credibility. All of the path coefficients of equation model of are 0.1, so the results have higher reliability.

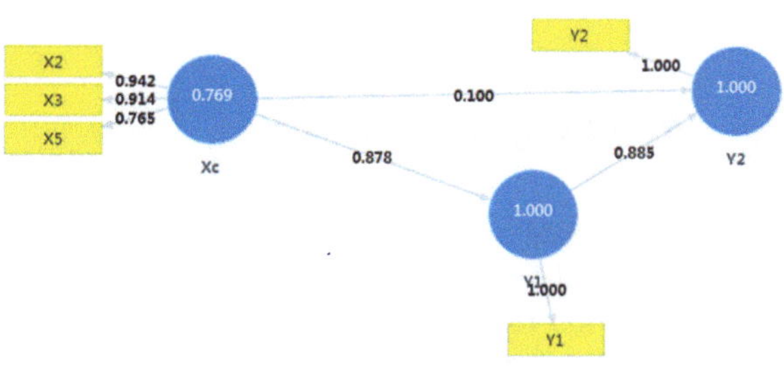

Fig. 1 Structural equation model

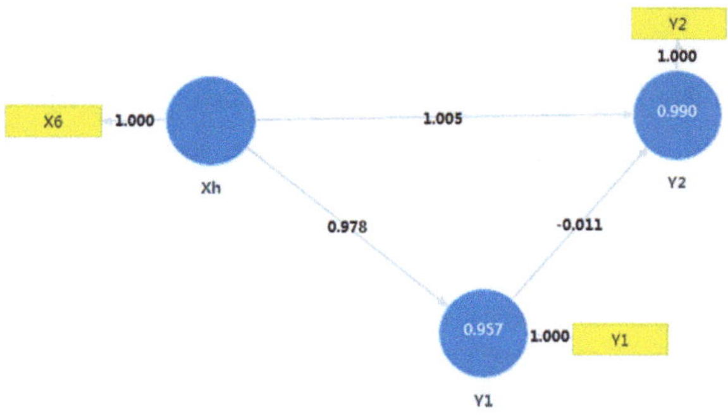

Fig. 2 Structural equation model

Table 10 Reliability of the structural equation model

	Xc	Y_1	Y_2
Composite reliability	0.908	1	1
Cronbach's alpha	0.848	1	1

Table 11 Final structure equation model AVE values

	Xc	Y_1	Y_2
AVE	0.769	1	1

AVE: Average Variance Extracted

Table 12 Coefficient paths affecting the final structure equation model

	Original sample (O)	Sample mean (M)	SE (STERR)	T	p
Xc -> Y_1	0.878	0.877	0.029	30.232	0.000
Xc -> Y_2	0.100	0.106	0.059	1.688	0.092
Y_1 -> Y_2	0.885	0.881	0.051	17.490	0.000

The results of analysis of significance are shown in Table 13. We can see that the three paths satisfy the two-tailed test indicating high credibility. However, it does not meet the two-tailed test leading to low credibility.

3.6 *Results of the Analysis*

The analysis results show that the fiscal policy affects consumer credit consumption indirectly and the monetary policy weakens the effect of short-term consumer loans on consumption However, it has a direct impact on short-term consumer credit and increased consumption, and it also confirms that the effect on the monetary policy reflects is faster and more obvious.

Table 13 Coefficient paths affecting the final structure equation model

	Original sample (O)	Sample mean (M)	SE (STERR)	T (|O/STERR|)	p
Xh -> Y_1	0.978	0.979	0.002	404.630	0.000
Xh -> Y_2	1.005	1.004	0.061	16.353	0.000
Y_1 -> Y_2	−0.011	−0.009	0.063	0.168	0.867

T: Distribution and its significance testing result

4 Conclusion and Suggestion

Through empirical testing, we found that (1) the use of macropolicy can promote the growth of consumer finance; and (2) an increase in short-term credit consumption can effectively promote the growth of consumption. Based on these conclusions, the, authors proffer the following suggestions:

- By increasing spending on people's livelihood expenses, administrative management, and defense, tax revenues can appropriately promote the increase of short-term consumer credit.
- Using the monetary policy moderately, and adjusting the supply of M2, will adjust consumption directly and quickly.

Acknowledgments This project was funded by the National Natural Science Foundation (71373262), Beijing Undergraduate Training Programs for Innovation (Z20141504), and Teaching Reform Project of China University of Mining & Technology, Beijing (j120501).

References

1. Yin S-j (2004) Consumer demand and economic growth. J Consumer Econ. 05:3–7
2. Drucker P (2007) Innovation and entrepreneurship. Machine Press, China
3. Liao L (2010) China's consumer finance study. Cambridge University Press, England
4. Xiaonan L (2006) Consumer credit's impact on consumer demand effect analysis finance and economy. Finance Trade Econ. 11:27–31
5. Liu Y-s (2006) A study of the problem of transition consumer credit in China. Ph.D. thesis, Southwest University of Finance and Economics
6. Huang H (2009) The influence of bank credit on economic growth in China research. Master's thesis, Central South University

Conclusions and Suggestions

Establish an International Market-Oriented Open Grain-Market System

Menggang Li and Shuqing Huang

Abstract Our countermeasures against a monopoly of transnational grain businesses mainly cover five aspects: (1) stimulate the development of diversified trading and enrich the international supply chain of grain; (2) intensify international agricultural cooperation and promote the connection of regional international grain markets; (3) improve the international competitiveness of China's leading grain enterprises; (4) encourage enterprises to establish foreign grain supply bases abroad; and (5) perfect the grain-reserve system and increase the proportion of long- and medium-term grain-trading contracts.

Keywords Grain market system · International grain supply chain · Foreign grain supply base

1 Stimulate the Development of Diversified Trading and Enrich the International Supply Chain of Grain

The concentration of import sources may result in excessively high import dependence on the grain markets of certain countries, as a result of which the grain supply on the domestic market of China is vulnerable to the restrictions of export policies of certain leading powers, in which case China's grain market stability would be impaired. The only way to resolve this problem is to stimulate the development of diversified trading and enrich the international supply chain of grain. The specific strategies to accomplish this are as follows: First, choose a more stable grain-import market. According to some scholars' estimation of the dependent coefficient of grain import to China in the 1990s, Argentina, Denmark, the Netherlands, Mexico, Japan, Turkey, and Spain were the most secure wheat export

M. Li (✉) · S. Huang
China Center for Industrial Security, Research,
Beijing Jiaotong University, Beijing 10044, China
e-mail: zhugezedong@163.com; 736194775@qq.com

M. Li et al. (eds.), *Proceedings of 2015 2nd International Conference on Industrial Economics System and Industrial Security Engineering*,
DOI 10.1007/978-981-287-655-3_5

markets, whereas the United States' wheat export market was the most vulnerable to supply failure. The study of the stability of the corn-import market showed that Australia, the Netherlands, Malaysia, Myanmar, South Africa, Vietnam, and New Zealand were the most stable corn export markets, whereas the United States, Argentina, Thailand, and Spain, etc., were the most fragile corn export markets where only trading factors are taken into consideration. Analysis of China's rice-import dependency showed that the Philippines, Laos, and Italy, etc., were the most stable rice export markets, whereas the United States, Japan, Korea, India, and Thailand were the most insecure rice-export markets where only trading economics were taken into account. According to the analysis, the most stable soybean export market for China was Thailand, whereas the United States' market was the most insecure. It is advisable to intensify the study of grain-import dependence to reinforce the connection with a stable market based on practical conditions. Second, enhance the import from an emerging market based on the changing trend of the international grain supply chain. The technological progress and improvement in production capacity could cause a change in the international grain-supply chain. For example, Korbsook, the Chairman of the Thai Rice Exporters Association, said in an interview with a reporter from Xinhua News Agency that Thailand, which is known as the world's largest rice exporter, was facing challenges from Vietnam, India, and Myanmar and might lose its leading position in the field of rice export. In 2012, rice export from Vietnam to China grew sharply: The export volume increased by 3 times on a year-on-year basis, and the export amount increased by 2.5 times. In the first quarter of 2012, China signed a contract with Vietnam to import approximately 1 million tons of rice. To actively promote rice export to China, the Vietnam Food Association (VFA) founded the Quality Rice Export Promotion Center in Ho Chi Minh City. The association is striving to set up and officially register a branch in Guangzhou in May. It is important to realize this changing trend in the international supply chain of grain and to increase the import volume from emerging markets to reduce dependence on traditional markets.

2 Enhance International Agricultural Cooperation and Promote the Connection of Regional International Grain Markets

Because it is large grain importer, it is of great significance for China to enhance agricultural cooperation with countries all over the world. Cooperation helps to improve the grain-production capacity of the host country, which has a significant effect on the increase in the world's grain market supply and the stable grain price of the world; this also helps to promote China's expanding international grain market. With further development of the opening up of agricultural markets, the multilateral and bilateral mechanisms for China's international agricultural cooperation experienced continuous expansion and intensification, which made a positive contribution to the development of the foreign market as well as the introduction of foreign technology for China's agriculture. By June 2011, China has established long-term stable bilateral agricultural cooperation with 91 countries and regions all over the world and signed 189 bilateral agreements on international agricultural cooperation with respect to crop farming, animal husbandry, fisheries and aquaculture, etc. China developed long-term stable agricultural cooperation with such organizations or countries as the members of Association of Southeast Asian Nations, the members of Greater Mekong Subregion (GMS), the Shanghai Cooperation Organization, the Food and Agriculture Organization of the United Nations (FAO), the World Bank (WB), the United Nations World Food Programme (WFP), and the African Union. China also signed a number of multilateral cooperation agreements on agriculture there to launch the agricultural cooperation of ASEAN with China, Japan, and Korea (10 + 3) and agricultural cooperation within Shanghai Cooperation Organization, the South South Cooperation under the framework of the Special Programme for Food Security between China and the FAO, the forum of agricultural cooperation of China with central and eastern European countries, etc. In addition, the agricultural cooperation mechanism is taking shape among BRICS. Multilateral cooperation involves crop farming, animal husbandry, aquaculture, fisheries, agricultural-products processing, agricultural machinery, etc. With respect to local international cooperation, incomplete statistics show that 11 key provinces, i.e., Heilongjiang, Liaoning, Zhejiang, Jiangxi, Shandong, Hubei, Hunan, Guangxi, Hainan, Yunnan, and Xinjiang, signed 45 agricultural cooperation agreements (or memorandums) with foreign agricultural authorities or governmental departments during the period from 1991 to 2010. The principal partners include the local state and province governments of the United States, Germany, Japan, Korea, Canada, Vietnam, and Laos, etc.; farming departments, agriculture departments; and agriculture bureaus under foreign ministries of agriculture; or the agricultural industry organization, etc. The areas of cooperation include the means of agricultural production, agricultural machinery, planting, animal husbandry and animal quarantine, etc. The cooperation is mainly launched in the form of agricultural demonstration base development, exchange visit and investigation, exchange of senior visiting students, joint technological research and technical training, etc. However, most agreements signed during cooperation involved technology, information, and joint HR development, and only a few agreements were signed regarding market access and substantive consultation and negotiation. From this point of view, cooperation with respect to the demand for and supply of grain must be intensified.

To mitigate the adverse effect of global grain market fluctuation on China, China shall reinforce its cooperation with ASEAN members Japan and Korea, promote the connection of domestic grain market with grain markets of the countries around, and obtain the complementary advantage in food import and export relations.

Because grain security is vitally important for China and the ASEAN members, the establishment of a regional international grain-market system and the connection between grain markets of different countries within the region are consistent with the common interests of both China and the ASEAN members. China substantially achieved grain self-sufficiency after its reform and opening up of agricultural trade by means of technological progress and circulation system reform etc., but China still must import grain from the international market because there is still a gap between the supply of and demand for grain as a result of the reduction of arable land, environmental pollution, consumption structure upgrade, etc. In the meantime, ASEAN members suffer from the imbalance between grain production and grain supply and demand: Thailand, Myanmar, and Vietnam represent the traditional export markets of rice, whereas Indonesia, the Philippines, Malaysia, and Singapore have extremely high external dependence on grain import. The complementarity between export and import markets creates common interests among these countries. China could enhance its cooperation with the ASEAN to develop an international cooperation mechanism for the import and export of grain. On one hand, the new hybrid rice varieties and rice-cultivation technologies enjoy great popularity in Vietnam, Cambodia, Brunei, etc., and play an important role in the steady increase in local grain production, thus improving the grain supply within the region. On the other hand, China provides Thailand and Vietnam, etc., with a stable and vast demand market, thanks to which they could maintain the stable tendency of export of their competitive products such as rice, and China's competitive agricultural products can be exported there. This complementary relationship of grain export and import enhances the common interests of China and the ASEAN members and helps assure their grain security and withstand the risk of global grain-market fluctuation. Considering such common interests, China, the ASEAN members, Japan, and Korea began to organize a conference on grain security cooperation in 2009. At the Third Roundtable Conference on Grain Security Cooperation Strategy between ASEAN and China, Japan, and Korea, which was organized by the Ministry of Agriculture of China on November 9 and 10, 2011, China put forward five suggestions on the reinforcement of the "10 + 3" regional grain security cooperation: (1) perfect the cooperation mechanism and reinforce regional grain security protection; (2) give play to comparative advantage to create a rational regional industry pattern; (3) enhance the exchange of technology and information exchange and enrich and consolidate the field and the method of cooperation; (4) promote trade facilitation and accelerate the promotion of East Asia FTA development; and (5) encourage businesses to practice international agricultural cooperation and encourage agricultural enterprises to participate in international investment and competition. All of these suggestions were unanimously endorsed by the delegates. Establishing a mechanism for connection of the regional grain market system has increasingly become the consensus of the ASEAN members—China, Japan, and Korea.

3 Give Play to the Active Role of International Grain Businesses and Improve the International Competitiveness of China's Leading Grain Businesses

In the past, when China practiced its policy of "state monopoly purchasing and selling," grain businesses were affiliated with governmental departments, as a result of which the integration of government administration with enterprise, overstaffing, and extensive management led to poor operation of the grain-circulation mechanism, and thus the national grain supply failed to meet the needs of the people. In view of these drawbacks of grain-administration and-circulation mechanisms, China separated government functions from enterprise management and carried out market-oriented reform, asset restructuring, and independent operation for grain businesses after the Reform and Opening-up. On the other hand, China introduced the competition mechanism; opened up the domestic grain market; brought in capital, technology, and people with advanced experience in management from international leading grain businesses; and gave full play to the "catfish effect" on China's leading grain businesses, thereby stimulating vital competition of the grain market in China. China made great achievements in the reform of grain-management and-circulation mechanisms and cultivated a large number of competitive leading grain enterprises with large-scale production and high market share, thereby laying a solid foundation for establishing a unified open modern grain market system with orderly competition.

At the end of November 2008, there were 11,977 grain-processing companies, including 1324 industrialized leading grain businesses, larger than the designated size in China. The 1684 industrialized leading grain businesses, with key support from the State Administration of Grain and Agricultural Development Bank of China, received a loan of 99.77 billion Yuan for the purchase of grain, development of technological transformation, and base construction, etc. In keeping with the generalization of intensive processing technology of grain and oil products, "safe grain and oil" is actively distributed to rural areas and communities. Remarkable results were achieved in grain brand building: A large number of national brand-name products were certified, and 2652 products from 1379 enterprises all over the country were called "safe grain and oil." The modern grain-circulation industry—composed of the modern grain-market system, the macroscopic grain-regulation and-control system, the grain-industrialization system, and the grain-circulation administration, enforcement supervision, and guarantee system—is finally taking shape.

However, China's grain enterprises are far behind large foreign grain corporations in terms of scale, capital, industry chain control, market share, etc., and China's state-owned and private leading grain companies are less competitive than international giant grain businesses.

With respect to industry chain control, no large Chinese grain business is considered able to contend with the four giant grain businesses. As China's largest grain source controller, Sinograin holds no grain-processing business, although it is strong enough to do so. COFCO controls no grain source and has an inadequate marketing network despite its processing business. Hualiang Group is one of the largest grain businesses in China, but its strength cannot compare with that of multinational companies.

Presented are some principal countermeasures against the monopoly of transnational grain businesses:

- First, increase access to grain processing and circulation sectors; establish a rigorous foreign business access system for wheat- and rice-, etc., processing industries; and control the rate and size of foreign investment.
- Second, foster local leading grain companies and corporations with international core competitiveness.

We should intensify the integration of the agricultural industry supply chain, accelerate the promotion of restructuring and upgrading of the grain-processing industry in China, facilitate the extension of domestic grain companies toward upstream and downstream areas of the industry chain, and encourage qualified grain companies to build competitive large leading grain-processing companies and corporations with rational management and a high level of technology through association, merger, reorganization, etc.

4 Perfect the Grain-Reserve System and Increase the Proportion of Long- and Medium-Term Grain-Trading Contracts

In view of the impact of grain price fluctuation of the international market on the Chinese market, it is advisable to perfect the existing grain buffer stock mechanism against change in the supply demand relationship in the international market. To maximize the benefits from grain import and avoid losses resulting from excessively high grain prices in the international market, as was experienced some years ago, the government shall build a considerable amount of good stocks in coastal provinces/municipalities to make more proactive use of the international market for long-term benefits.

To reduce international market risks, we should reasonably increase the proportion of long- and medium-term trading contracts. In addition, it is important to increase the transparency of trade, to avoid short-term speculation, and to improve China's international goodwill for grain import. Long- and medium-term contracts help to reduce risks. The United States fulfilled trade contracts even in the 1980s when it practiced a "grain embargo" against the USSR. Our own experience shows that temporary regulation and "fluctuation in line with market conditions" [1] are passive behaviors that cannot affect market price, but they may cause us to be at the mercy of market and to pay more when we are forced to raise the import. From this point of view, the risk of a rise in grain prices on the international market could be more satisfactorily avoided by signing long-term grain-import agreements (wheat and corn in most cases) with such leading grain exporters as Canada, Australia, and Argentina at reasonable prices based on analysis of supply demand relationship on the international grain market depending on the demand of the grain market in China.

Acknowledgments Supported by the major project of Chinese Ministry of Education Industrial Security Engineering (No. 239010522) and China's credit evaluation system research (No. 239005522).

Reference

1. Hausknecht JP, Hiller NJ, Vance RJ (2008) Work-unit absenteeism: effects of satisfaction, commitment, labor market conditions, and time. Acad Manag J, 51(6):1223–1245

Present Status of Grain Production in China and Recommendations for Regional Development of Grain Production

Menggang Li and Yu Zhao

Abstract This chapter addresses major factors that affect production in major grain-producing areas, e.g., the contradictory distribution of soil and water resources, the regional contradiction between economic development and key grain-production areas, the benefit loss of major grain-producing areas, and the inconsistency of agricultural technology input with production in major grain-producing areas. Policy recommendations for improvement are proposed on this basis.

Keywords Grain production · Regional pattern

1 Present Status of Regional Pattern of Grain Production in China

1.1 Obvious Regional Pattern Differentiation of Grain Production

This chapter investigates the grain output and acreage of major grain-producing areas, main sales areas, and production sales balance areas in China using the four sets of cross-age node data. The data in Table 1 indicate the obvious improvement in the status of major grain-producing areas, the degression of the status of main sales areas, and the substantially unchanged status of production–sales balance areas.The status of major grain-producing areas improves where grain output and

M. Li (✉) · Y. Zhao
China Center for Industrial Security Research, Beijing Jiaotong University, Beijing 10044, China
e-mail: zhugezedong@163.com

Y. Zhao
e-mail: 736194775@qq.com

M. Li et al. (eds.), *Proceedings of 2015 2nd International Conference on Industrial Economics System and Industrial Security Engineering*,
DOI 10.1007/978-981-287-655-3_6

Table 1 Grain crop yield and cultivation area in different zones

	Year	1980	1990	2000	2010
Major producing area	Grain output (×10,000 tons)	22205.0	32501.6	32607.4	41183.9
	Ratio to total grain output (%)	69.3	72.8	70.6	75.4
	Grain acreage (×1,000 ha.)	80455.7	78939.6	73142.9	78549.7
	Ratio to total acreage (%)	68.6	69.6	67.4	77.7
Production–sales balance area	Grain output (×10,000 tons)	5293.5	6897.5	9135.8	10139.2
	Ratio to total grain output (%)	16.5	15.5	19.8	18.6
	Grain acreage (×1,000 ha.)	24259.2	23256.5	26424.9	16384.8
	Ratio to total acreage (%)	20.7	20.5	24.4	16.2
Main sales area	Grain output (×10,000 tons)	4557.0	5225.2	4474.4	3324.6
	Ratio to total grain output (%)	14.2	11.7	9.7	6.0
	Grain acreage (×1,000 ha.)	12519.4	11269.8	8894.9	6191.8
	Ratio to total acreage (%)	10.7	9.9	8.2	6.1
	Total grain output (×10,000 tons)	32055.5	44624.3	46217.5	54647.7
	Total cultivation area (×1,000 ha.)	117234.3	113465.9	108462.5	101126.3

Note 1 Regarding data from 1980 and 1990, the grain output and cultivation areas of Chongqing were included in the statistics for Sichuan Province

Note 2 The data shown in table 1 are from agricultural statistics of the 60 years since the Founding of the P. R. C. and the Chinese Rural Statistical Yearbook 2011

cultivation areas increase significantly. The production status of production–sales balance areas remains unchanged: Grain output increase significantly, whereas cultivation areas decrease in a fluctuating fashion. The status of main sales areas decreases significantly, as a result of which grain output and cultivation area decreased.

1.2 Change in Key Grain-Production Areas and Supply Demand Orientation in China

China's key grain-production areas have shifted from the south to north and from the east to central part. As shown in Table 2, grain output in the southern part of China has increased by only 30 % over the past 30 years, whereas its ratio to national total grain output has decreased by 14.2 %; and grain output of the northern part increased by 130 %, whereas its ratio to national total grain output of China increased by 14.2 %. From this point of view, the key areas of China's production have shifted from the southern to the northern part of China.

As shown in Table 3, grain output of the eastern part of China slightly increased, and its ratio to the national total output significantly decreased by 11.8 %; the grain output of the central part increased by 110.6 %, and its ratio to national total output

Table 2 Changes in grain output in southern and northern parts of China

	Year	1980	1990	2000	2010
Southern part (15 provinces/municipalities)	Grain output (×10,000 tons)	19197.5	24707.9	25118.1	25006.7
	Ratio to total grain output (%)	60.0	55.4	54.3	45.8
Northern part (16 provinces/municipalities)	Grain output (×10,000 tons)	12858.0	19916.4	21099.1	29641.0
	Ratio to total grain output (%)	40.0	44.6	45.7	54.2

Table 3 Change in grain output in the eastern, central, and western parts of China

	Year	1980	1990	2000	2010
Eastern part (11 provinces/municipalities)	Grain output (×10,000 tons)	11103.5	15682.5	13969.8	14635.7
	Ratio to total output (%)	33.6	35.1	30.2	21.8
Central part (8 provinces/municipalities)	Grain output (×10,000 tons)	11669.5	17873.5	18211.5	24576.1
	Ratio to total output (%)	36.4	40.1	39.4	50.0
Western part (12 provinces/municipalities)	Grain output (×10,000 tons)	9282.5	11068.3	14036.2	15435.9
	Ratio to total output (%)	30.0	24.8	30.4	28.2

increased by 13.6 %; and the grain output of the western part increased by 66.3 %, and its ratio to the national total output changed only slightly. Along with the implementation of returning cultivated land to forests and pastures in the West Region, the western part certainly could introduce grains from the central part. It is thus clear that the shift of key areas of grain production from the eastern part to the central part is strengthening.

The supply–demand pattern of grain changes with the shift of key production areas: The pattern of grain transport from the south to the north has turned into a pattern of transportfrom the north/central part to the south/east/west part.

1.3 Concentration of Grain-Production Capacity in Advantageous Producing Areas, Core Grain-Producing Areas, and Large Grain-Producing Counties

China specified 800 large grain producing counties in 2010 including 680 in major producing areas and 120 in other areas. The 800 large grain-producing counties offer approximately 80 % of commodity grain in China; the central government assigns increasing input and award to large grain-producing counties where grain-production aggregation and grain output increase remarkably; according to statistics, the top 100 large grain-producing counties contribute 21 % of China's total grain production.

2 Principal Problems in the Current Regional Pattern of Grain Production in China

2.1 Matching Issue Between Land and Water Resources

Spatially, the southern part of China has more abundant cultivated land resources than the northern part, whereas the southern/eastern part is richer in water resources than the northern/western part. Such matching contradiction between land and water resources brings China severe challenges, which constrain the improvement of comprehensive grain-production capacity to a large extent.

2.2 Contradictions Between Economic Development and Grain Production of Major Grain-Producing Areas and Large Grain-Producing Counties

Major grain-producing areas in China are usually economically underdeveloped areas where secondary and tertiary industries suffer slow development, the flow of young labor forces toward developed areas leads to a decrease in labor input for grain production, and the increase in grain productivity is slow.

2.3 Contradiction Between Dispersed Rural Land and Mass Grain Production and Operation

First, because rural land is subject to community ownership and the family household contract responsibility system, peasants do things in their own ways and often calculate the earnings rate before planting. The government has no time to attend to, the collective is unable to, and the peasant household is not willing to participate in the development of agricultural infrastructures (especially farmland water-conservancy development). Second, the land-management scale of the peasant household is small and dispersed: In China, the per-household area of cultivated land 0.61 ha, and the cultivated land of each household must be divided into several non-neighboring parts by grade of cultivated land. Due to such small-scale household management, China's agricultural production is still in the semisubsistence and self-contained stage, and the relative efficiency of grain production is low. Third, it is impossible to employ unified planting technology standard for small-scale household management, and thus this precludes unified field management and mechanized operations; moreover, it would be impossible to assure the yield and quality of grain products. Fourth, the peasant households engaged in dispersed planting are unable to withstand market economy risks and obtrain financial support in the context of the market economy. This system and mechanism can hardly assure the constant increase in grain yield and grain farmers' income in major grain-producing areas.

2.4 The Contradiction Between Interest Lacking in Major Grain-Producing Areas and Local Grain Development

China's major grain-producing areas lag behind others respect to the development of secondary and tertiary industries; the concentration of principal agricultural areas means that manufacturers, grain-industrialization companies, and breeding businesses in major grain-producing areas are fewer than in non-major grain-producing areas. This has greatly dampened the enthusiasm of major producing areas and large producing counties in promoting local grain production.

2.5 Mismatching of Agricultural Technology Input with Grain-Production Requirements in Major Grain-Producing Areas

The so-called "line break, network failure, people dispersion, and poor public service capacity" are increasingly obvious in most major grain-producing areas, and a weak agricultural technology extension system and inadequate public service is a long-standing problem.

3 Recommendations for the Development of a Regional Grain-Production Pattern in China

3.1 Regulate Allocation of Water and Land Resources by Increasing Input

Recommendations include the following:

- Build farmland water-conservancy infrastructures and carry out farmland water-conservancy project development through large-scale input.
- Concentrate efforts to enhance small-scale farmland water-conservancy development and perfect farmland irrigation and drainage systems.
- Develop high-standard farmland that ensures stable yields despite drought or excessive rain in major grain-producing areas, core grain-producing areas, and large grain-producing counties. Enable on-demand irrigation and drainage in China and maximize the consistency of water and land resources in major grain-producing areas.

3.2 Fully Mobilize the Enthusiasm of Major Grain-Producing Areas and Large Grain-Producing Counties to Promote Intra-County Economic Development

First, it is advisable to provide preferential policies and financial support to major grain-producing areas and large grain-producing counties, enhance subsidies for large grain-producing counties, reduce the financial burden of strong agricultural counties, and roughly eliminate the phenomenon of large grain-producing counties with poor financial resources. Encourage major grain-producing areas to seek development of their food-processing industry and extend the industrial chain through policies of finance, taxation, and credit, etc., to promote economic development in major grain-producing areas. Second, encourage the masses to engage in featured large-scale agricultural production and concentrate efforts for the development of brand agriculture.

3.3 Encourage Rural-Land Contracting and Management Right Transfer and Stimulate the Development of Large-Scale Grain Production and Operation

The acceleration of rural-land contracting and management right transfer is the prior condition for developing various forms of appropriate-scale operation and promoting the transformation of grain production from a natural and semi-natural economy to socialized mass production. It is advisable to actively stimulate the development of scale business entities such as large grain-producing households, family farms, and farmers' grain cooperatives through policy guidance and to explore multi-form, multi-channel, and multi-field scale business models of grain production. Enhance the support for large grain-producing households by various means, and encourage large grain-producing households to increase their inputs of capital and technology, etc., to improve the scale and intensification level of household-run businesses. Exert great efforts to guide and support the development of farmers' special grain cooperatives and associations.

3.4 Exert Great Efforts to Increase Investment in Agricultural Technologies

First, focus on seed: Expedite development of the modern seed industry system with large businesses and bases as carriers featuring cultivation reproduction promotion integration; enhance the research of improved varieties; and promote the development, reproduction, and promotion capacity. Second, expedite the promotion of grassroots agricultural technology extension system reform and development by increasing capital investment; giving play to the promotion of public welfare of agricultural technology; and improving the public-service capacity for the promotion of agricultural technology. Third, exert great efforts for high-yield development activities and increase the per-unit area yield of grain.

3.5 *Enhance Efforts for Credit Support for Grain Production*

Inadequate input is now an important bottleneck of major grain-producing areas; the input is a prior condition for output in the case of limited self-accumulation of agricultural production. Actively explore an innovative rural financial system and reinforce the credit support for grain production through policies. First, innovate the rural financial system: Relax the rural financial access policy, and expedite the development of a rural financial system with adequate capital, complete features, perfect service, and safe operation based on the integration of commercial, cooperative, and policy-based finance. The government shall intensify the policy support for rural finance accordingly. Second, reinforce the role of financial institutions in agricultural service.

Acknowledgments Supported by the major project of Chinese Ministry of Education—Industrial Security Engineering (No. 239010522) and China's credit evaluation system research (No. 239005522).

References

1. Zhengbin Z, Ping X (2011) The development of projects for efficient use of water resources is the key to resolving the issue of southern flood and northern drought [J]. China Territ Today (2):31–32
2. Yuehui X, Jingsheng H et al. (2010) Problems in front of big grain producing counties and local governments' decision-making dilemma [J]. Acad China (11):196–203
3. Changbin H (2010) Thoroughly eliminate the phenomenon of "big grain producing counties with poor financial resources" [J]. China's Construction of Old Revolutionary Basic Area (9):4

High-Performance Composite Materials Industry Development Strategy Under Low-Carbon Economy

Changjun He, Qingxin Meng and Hui Wang

Abstract Global composites industries are mostly found in Europe and the United States; however, among developed countries in Asia, China is the main source of such industries. Composite-material production growth is one of the most significant trends in the development of industry. China has become the most active and fastest growing global composites area in Asia. In the next few years, based on the transformation of China's economic structure, its economic transformation, and the rapid development of the national economy, composite material industries will show even more development. In this chapter, the development of China's high-performance composite industry is analyzed. Problems existing in the development process, as well as the trend of development and future goals, are discussed. The development of corresponding strategies in the context of a low-carbon economy is also discussed.

Keywords Composite material · Development strategy · Low-carbon economy

C. He (✉)
Postdoctoral Programme of China Center for Industrial Security Research (CCISR), Beijing Jiaotong University, Beijing 100044, China
e-mail: ysuhcj@163.com

Q. Meng
Hebei Vocational and Technical College Building Materials, Qinhuangdao 066004, China
e-mail: kangmeihui@163.com

H. Wang
Beijing Municipal Research Institute of Environmental Protection, Beijing 100037, China
e-mail: bj-wanghui@hotmail.com

M. Li et al. (eds.), *Proceedings of 2015 2nd International Conference on Industrial Economics System and Industrial Security Engineering*,
DOI 10.1007/978-981-287-655-3_7

1 Introduction

Compared with traditional materials, composite materials are an important component of the field of new materials. They have superior performance, e.g. good design, high specific strength and modulus, good fatigue-fracture resistance, and structural and functional integration. They act as base materials for the development of modern industry, national defense, and science and technology. They are also indispensable because they provide an important material basis for the development of a revolution in new technology [1]. Composite materials have become an important dominant "player" in the field of new materials.

2 Production and Consumption

The speed of development of the composite-materials industry is much faster than the growth rate of the world economy. Global production of composite materials/composites in 2009 was approximately 9 million tons. In North America, Europe, and Asia's three largest production and application areas, the global resin matrix composite average annual growth rate was <5 %, which represents ≤7 % growth in the Asia Pacific region. China accounts for approximately 9.5 %; India for approximately 9.5 %; and Europe and North America for approximately 4 %. Personnel involved with composite materials number approximately 450,000 worldwide at present, and their output value is approximately 100 billion Euros [2, 3]. In 2009, China had an annual output of 3.23 million tons, which is greater than from Germany and Japan; ranks second in the world; and is close to that of the United States, which has the highest output in the world.

China's domestic composites have improved the level of equipment, the international production line has been brought to an advanced level. The main raw material has largely been composite material made of glass fiber: The resin quality has reached that of the international level, and the price is competitive. Nonalkali fiberglass tank furnace-drawing technology has reached an advanced international level; The world's largest annual output, 120,000 tons of nonalkali pool-kiln production, has been completed and was put into operation during the period of the "11th five-year plan;" and independent intellectual property rights of technology and equipment in our country have been widely used in all industries and exported in quick succession.

Composite material used per field distribution, both in China and internationally, is shown in Table 1.

Table 1 Composite materials used per field

Project area	Transportation (%)	Buildings and structures (%)	Industrial equipment (%)	Basic engineering (%)	Ships (%)	Other (%)
Global	24	18	26	7	6	19
China	5	32	10	39	3	11

3 Development Trends and Existing Problems

3.1 Development Trends

Global composites industries are mostly found in Europe and the United States; however, among developed countries in Asia, China is the main source of such industries. Asia's growth in the field of composites-material production is one of the most significant trends in industry development, and production is expected to double by 2015 to 10 million tons. It is believed that Asia's global composite materials market share has increased from 25 % at the beginning of this century to 51 % in 2013. With the development of advanced technology, such as wind and aerospace, the use of composite materials will increase in the field of high value added applications. China will become the most active and fastest-growing global area for composites. In the next few years, economic transformation and the rapid development of the national economy will foster the growing development of composite-materials production in China.

India also shows significant growth potential in the area of composites production. The rapid development of automobile-application fields will drive India's composites industry. In the Middle East, Malaysia, Vietnam, and Indonesia also show high growth potential in the area of composite materials.

3.2 Existing Problems

Despite its growth, China's industry-oriented integration technology, innovation and development, new products, and new technology are not strong enough. The majority of products are low-standard ones. Mainly labor-intensive, reinforced material is given priority over glass fiber. Among raw and auxiliary materials series, some high-end varieties are still relatively lacking, notably the CF and AF. The performance of continuous-strand mat material is poor. The gap with advanced equipment and automation technology abroad is larger; the capability to develop advanced equipment is weak. Industry standards and standardized work must be improved, and there is still a large gap in technical composites-material design. The capacity for technical innovation ability at the management level is not high. A lack of vigorous competition in the industry has become endemic, and the profitability of enterprises to make a profit is weak because they have difficulty finding enough money to promote technology and equipment to a higher level.

3.3 Development Goals

Powered by both independent and integrated innovation, with private enterprises as the mainstay, China will see market-oriented enterprises develop composite materials that will be used in aerospace, transportation, wind power, new energy and renewable energy, offshore oil development, and the application of high-end, smart-grid construction.

By 2015, the "twelfth five-year" period, the growth of the field of composite materials was 9 % each year; composite-materials production was 6 million tons, of which thermoset-composite production reached 3 million tons; and thermoplastic composite materials consumption was 2.3 million tons. The mechanized molding equipment manufacturing level has reached a world-class level; the product of the mechanized-forming rate is >75 %, 5 % to 8 % higher than that seen during the "eleventh five-year" period.

4 Composite-Materials Industry Development Strategy

Core competitiveness in new-materials enterprises is measured mainly in terms of resource possession, level of leading technology, market size, etc., in resources, technology, and scale, all of which can build barriers to competition. The success of new-materials industry competition lies in three key elements. Relying on the local characteristic advantage in developing a new material industrial spatial agglomeration, the eastern region mainly depends on market, talent, and technology advantages, such as Midwest China mainly relying on resource advantages and overall relying on resources and three market-oriented, technology-driven development modes such those reported in [4–6]. To speed up the new-materials industry and coupled with the low-carbon economy, new-materials industry development can be sped up in a number of ways as follows.

First, "low-carbon economy" is a new technology paradigm. A new-materials industry in terms of technological innovation offer "vertical" advantages in the context of a low-carbon economy. Development along with the technical and economic paradigm requires human beings to achieve certain economic and social development goals and select the technical and economic mode for attainment of these goals. The new-materials industry requires the multidisciplinary advantage of technological innovation, energy conservation, and emissions-reduction technology ideas from sample creation, product research, development innovation, product size determination, etc. This will form "vertical" development participation in the context of a low-carbon economy advantage.

Second, 1. A low carbon economy has the characteristics of the system of social reproduction process of low carbon. 2. New material industry in the production of low carbon and low carbonization link of circulation support low carbon economic development. 3. Low carbon economy includes production of low carbon, low circulation of carbide, distribution, low carbon and consumption of low carbon are low carbon four links such as social reproduction system. Among them, low-carbon production is the starting point; low consumption of carbide is the "finish line"; and low circulation and distribution of carbide is the mediation. The four links bear different low-carbon system economic functions, and so the process continues.

Third, the low-carbon economy must build a low-carbon industry and interactive platform such as effective policy support, new material industry in low-carbon industry support the development of low-carbon economy.

5 Conclusion

At present, the development of new materials is mainly based on advancement of the new-materials industry. Acceleration of the development of the new-materials industry should be from a national perspective, should fully consider China's new-materials industry in terms of international strategic positioning, should be combined with the future low-carbon new-materials industry development in important key areas, should integrate related resources in the field of new materials, should establish a dynamic alliance of new-materials production, and should develop an innovation system.

First, there should be various education modes to train various kinds of professional talents and strive to build a good team of production, joint research, and development. Second, we must actively build on the research and development of production, the main unit and risk-sharing, benefit sharing between enterprises, and long-term stability of the coordinated development of an "industry university institute" cooperation mechanism. Third, through the alliance-organization pattern, China should further deepen and improve the links of production, innovation, and joint cooperation. Finally, China should highlight the four production, key projects, and key techniques: (1) conserve energy and reduce emissions; (2) develop new energy and other industries, (3) increase attention paid to the union of the industry, and (4) implement the demonstration project.

References

1. Nie KB, Deng KK, Xu FJ, Wang XJ, Wu K (2015) Development of microstructure in submicron particles reinforced magnesium matrix composite processed by room temperature deformation. Mater Chem Phys 149–150:21–26

2. Mehboob H, Chang S (2014) Evaluation of the development of tissue phenol-types: bone fracture healing using functionally graded material composite bone plates. Compos Struct 117:105–113
3. Padmaraj NH, Kini MV, Pai BR, Shenoy BS (2014) Development of short areca fiber reinforced biodegradable composite materia. Procedia Eng 64:966–972
4. Meng T, Chen YM, Jin XY, Zhan SL, Qian XQ (2012) Development and design of composite cementitious material at long-time circular elevated temperature in metallurgy environment. Constr Build Mater 35:368–375
5. West J, Schandl H (2013) Material use and material efficiency in Latin America and the Caribbean. Ecol Econ 94:19–27
6. Cheng A, Chen C, Chen C (2008) A fuzzy multiple criteria comparison of technology forecasting methods for predicting the new materials development. Technol Forecast Soc Chang 75(1):131–141

An Innovation Trajectory in an Emerging Industry Under a Strategic Niche-Management Perspective

Huajun Li

Abstract In the cultivation of emerging industry, industries are faced with two bottlenecks—(1) the "Death Valley" from R&D to the market and (2) "path dependence and lock-in"—under the existing socio-technical regime, and industries must overcome these bottleneck to achieve further development. From the view of strategic-niche management (SNM), the innovation trajectory of emerging industries is through niche construction and optimizing a plan to obtain their ecological inheritance and a multilevel-perspective (MLP) interaction within and between niche innovation, the socio-technical regime, and the socio-technical landscape. This much be done to promote the first transition, i.e., from technology niche to market niche to cross "Death Valley," as well as the second transition, i.e., the one from market niche to paradigm niche to break through path dependence and lock-in.

Keywords Emerging industry · Innovation trajectory · Strategic niche management · Multilevel perspective

1 Introduction

At present, the development of emerging industry is facing many problems that lead industries into "Death Valley" [1, 2] and "path dependence and lock-in" [3–5]. The emerging industrial growth process, as a coevolution of technology, market, system, and society, has become the common focus of economic and social problems. Countries around the world strengthen science and technological innovation and emerging industrial development to revitalize economy and strengthen international competition. It is particularly important and urgent to master the innovative and evolutionary law of emerging industry and then to design policy around this understanding.

H. Li (✉)
Economics and Management College, ZhaoQing University, Zhaoqing 526061, China
e-mail: lhj33118@163.com

M. Li et al. (eds.), *Proceedings of 2015 2nd International Conference on Industrial Economics System and Industrial Security Engineering*,
DOI 10.1007/978-981-287-655-3_8

Strategic-niche management (SNM) theory has been practiced abroad for nearly two decades before the rise in China. There has been much major research on technological innovation and how to achieve and lead the industry revolution until there is a reform of the socio-technical system. STM is current gradually being applied to the study of emerging industries to solve the problems posed by many major innovations. This theory provides a new analytical perspective and research methods for the study of the evolution of emerging industries.

2 Literature Review

The growth of emerging industry is the essence of the innovative evolution process with new technology as the axis. However, new technologies often face a mismatch with the established socio-institutional framework [6]; they also usually emerge as "hopeful monstrosities" for their promising future and crude performance [7]. They often have relatively low technical performance, are cumbersome and expensive, require complementary technologies that are perhaps not available or expensive to use, and require the development of infrastructure and maintenance [8]. Moreover, the existing government policy and regulatory framework may actually form a barrier to the development of new technologies as well as cultural and psychological factors.

Building on the research of technological transition and the socio-technical system, scholars introduced the Strategic Niche Management (SNM) perspective to address this problem [9, 10]. They developed the theory through a major research project of "Strategic Niche Management as a Tool for Transition to a Sustainable Transportation System" for the EC DG XII [8, 11]. In their views, SNM is defined as the "creation, development and controlled break-down of test-beds or niche (experiments, demonstration projects) for promising new technologies and concepts with the aim of learning about the desirability (for example, in terms of sustainability) and enhancing the diffusion of the new technology" [9]. These test-beds, or technological niche, where novelties emerge offer some protection because the selection criteria are very different from the existing regime, and they also provide space to build the social networks that support innovations, e.g., supply chains, user–producer relationships, etc.

In SNM work, the success of niche was defined in terms of transformation of a technological niche into a market niche and eventually a regime shift [8, 12], as well as five steps were reasonable taken (the adoption of promising new technologies, designing of experiment to test technologies, implementing of experiment, scaling of experiment, and break-down of protection) [8, 11].

Although SNM research provides evidence that there is a correlation between the design of experiments and outcomes in terms of technological and market niche development, it is also clear that internal niche development is not the only important factor; external factors also play a crucial role. Niche innovations are rarely able to bring about regime transformation without the help of broader forces and processes. Thus, some scholars have paid attention to searches that link niche internal and external processes, providing more elaborate analysis of the interaction between niches and their broader environments, under the term "multi-level perspective" (MLP), which developed in parallel with much of the SNM work discussed previously [13].

The important point of MLP is that the further success of a new technology is not only governed by processes within the niche but also by developments at the level of the existing regime and the socio-technical landscape [14]. Under the MLP model, technology transitions take place through interactions within and between landscape developments at the macro level, regimes at the meso level, and niches at the micro level [14].

The MLP model helped to further develop SNM and later became its core content. There is multiple-level of interaction and symbiosis evolution in technical innovation, which reflects the innovative evolution path and paradigm. Over the course of two decades, the studies and practices of SNM show that it is not only a policy analysis tool but also an innovative management method that contributes to study of the innovative trajectory of emerging industries.

3 The Innovative Mechanism of Emerging Industry

"Niche construction" refers to the capacity of an organism to modify its local important environments, consequently modifying both its biotic and abiotic sources of natural selection in their environment. Under the interaction of natural-selection pressure and changed selection pressure, the offspring organism has a "genetic inheritance" in traditional evolution and an "ecological inheritance" under niche construction [15, 16]. Thus, niche-construction theory may also be used to analyze emerging industries' innovative mechanisms under SNM.

In the introduction of new industries, protected space (i.e., a technology niche) is built to provide a variety of resources, policies, and measures needed in the technological R&D stage to change and improve the innovative ecology with integrative resources, collaborative innovation, knowledge exchange, multilevel learning, user testing, and other mechanisms. That new technology obtains a "technical genetic inheritance" under traditional ecology (i.e., technical characteristics obtained by its own laws and mechanisms of technological innovation and performance) and a "technical ecological inheritance" under niche construction (i.e., through the change and improvement of innovative ecology to compensate for deficiencies under traditional ecology. Finally technical features and performance are obtained such as the phenomenon of new technology being considered an unwelcome "monster" causing the user's participation, experience, testing, and learning to quickly change.

In industrial growth, new technology realizes commercialization and becomes accepted by part of society for its advantages. However, the emerging industry does not completely penetrate into the mainstream market and thus competes with old dominant industry under the existing socio-technical regime. Thus the market niche of an emerging industry must be optimized through strengthening or weakening user needs by nurturing users by providing government policy guidance, further funds, an "actors network", and other protective measures. Thus, the new technology obtains a "market genetic inheritance" under traditional ecology (i.e., market characteristics obtained by its own laws and mechanisms of technological innovation and performance) and a "market ecological inheritance" under niche optimization (i.e., through optimization and improvement of evolutionary ecology to compensate for deficiencies under traditional ecology. Finally the product's features and performance are obtained such as infrastructure, subsidiary industry, consumer culture, customer awareness, etc.

4 The Innovation Trajectory of Emerging Industry

According to niche construction and the MPL model under SNM, there are four stages in emerging industries' innovative trajectory: (1) breakthrough of the innovation niche; (2) construction and transition of a technology niche; (3) optimization and transition of market niche; and (4) development of a paradigm niche (Fig. 1).

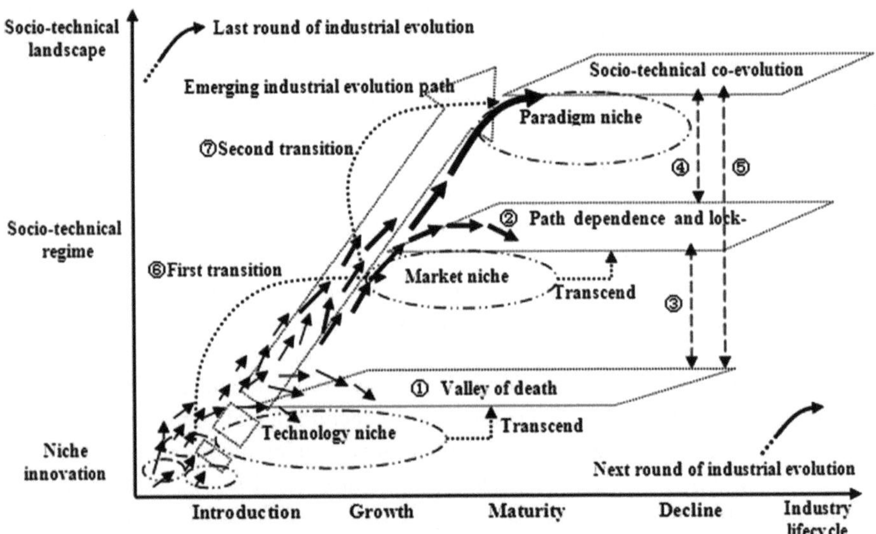

Notes:
(1) ① means industrial technical innovation failed into "valley of death";
(2) ② means industrial cultivation not enough to "path dependence and loch-in";
(3) ③, ④ and ⑤ mean interactions within and between multi-level players (niche innovation, socio-technical regime and socio-technical landscape;
(4) ⑥ and ⑦ mean twice transitions from technology niche to market niche and then to paradigm niche.

Fig. 1 Innovative trajectory in emerging industry under an SNM perspective

4.1 Innovation Niche Breakthrough

After the last round of industrial innovation evolution ended, the existing socio-technical system formed around the original technology that had been previously developed and also lacked sufficient strength to promote socio-technical landscape development. There were also needs regarding new anchors for industrial profit-driven enterprise and sustainable development. Under such circumstances, under endogenous innovation-driven and external landscape-driven processes, industry and the research community begin to research and select a new idea or technology (i.e., interactions take place in three layers, represented by ④ and ⑤ in Fig. 1).

At this stage, due to the diversity difference of innovation, all activities are in a constant state of exploration and trial-and-error testing showing the direction of technological innovation to be in disorder and scattering (see arrows [→] of different types and directions as indicated in Fig. 1). However, the innovative theme, as well as the field and content with local convergence, leads the industrial innovation niche to "brewing" and breakthrough (see small circles [○] for niche brewing in major innovative-field niches in Fig. 1). At this point, industry is in the introduction stage.

4.2 Technology Niche Construction and Transition

Due to the trajectory of technology innovation, which has a general gradual convergence, the evolutionary direction becomes clearer so that new technology can be developed from concept to prototype. However, new technology in the development process faces technical characteristic, market demand, consumer psychology, social system, and other problems. These problems, if not solved, will cause the industrial evolution to fall into "Death Valley" (represented by ① in Fig. 1).

Therefore, in this process, the technology niche as a protected space must be constructed and be matched statically and dynamically. A "static match" means that more innovation resources should be input (such as capital, human resource, facilities, and equipment, etc.). A "dynamic match" means that the innovation social network needs be constructed more widely (e.g., with users, suppliers, vendors, government, social groups, etc.), and different dimensions of learning (e.g., first- and second-order learning) must be promoted. In these matching processes, the technology niche is cultivated, which makes the new industry obtain a "genetic inheritance" and "ecological inheritance" of the technical characteristics needed to transcend "Death Valley." Finally, industrial innovation realizes the first transition from "technology niche" to "market niche" (represented by ⑥ in Fig. 1). In this stage, new technology occupies only a small part of the market, and the industry enters into the early stages of growth.

4.3 Market Niche Optimization and Transition

When industrial innovation realize the first transition from "technology niche" to "market niche," elements around the new technology innovation become aliened, and stable internal momentum increases. However, due to the immaturity of industrial technology itself as well as user preferences, the technological aspect is not highly industrialized enough to compete effectively and fully with older establish technology. Simultaneously, under the existing socio-technical regime, with increasing returns and a self-strengthening mechanism, the old technology dominated industry continues to occupy an unshakable position (represented by ③ in Fig. 1). That means that the new technology of emerging industry suffers the second bottleneck in evolution, namely path dependence and lock-in (represented by ② in Fig. 1).

Therefore, in the early stage of this process, the market niche must be optimized and cultivated for statically and dynamically matching. "Static match" means the more innovative resources and policies should be controlled, inputted, and strengthened (such as market subsidy, government procurement, industrial infrastructure, subsidiary industry, etc.). "Dynamic match" means that the social network of innovation must be optimized deeply and a connatural adjustment (such as key elements of industrial chain and innovation chain) made, and more users must participate in the testing, inspection, and experience of a new product. In these matching processes, users' cognitive preferences change, and more support is obtained from regime leaders. On these fundamentals, market niche has been cultivated and optimized to make new industry have "genetic inheritance" and "ecological inheritance" in product features to transcend the second bottleneck of path dependence and lock-in. Finally, the industrial evolution realizes the second transition from a market niche to a "regime niche" (represented by ⑦ in Fig. 1).

In this stage, a new technical system, formulated in the mainstream market, gradually replaces the old one while the existing socio-technical regime becomes unstable with a series of incremental changes, and the industry enters into the early stages of maturity as well.

4.4 Paradigm Niche Formulation and Development

With the development of a paradigm niche (this concept is not mentioned directly in the SNM research but rather is referred to as a "technological paradigm" or "technological regime"; this term is used in this chapter to describe and reflect the state of socio-technical coevolution), the industrial new technical system replaces the old one and coevolves with society, and eventually a new socio-technical regime is formulated to maintain the coevolution of technology and society as well as to promote the socio-technical landscape over a period of time (represented by ③ and ④ in Fig. 1).

In this stage, the industry enters into maturity stage for a long time, and then it faces adjustment or decline. Thus far industrial innovation has completed the entire path, and the next round is brewing.

5 Conclusions

We have discussed emerging industries' innovative trajectory based on SNM, especially the MPL model. This discussion showed that internal analysis of niche processes (technology niche, market niche, paradigm niche) is increasingly complemented by attention paid to the external environment (socio-technical landscape, socio-technical regime, niche innovation) in the innovation of emerging industries. Niches and experiments are considered crucial for bringing about industrial innovation, but that are not the sole forces in doing so. Linkages with ongoing processes at broader regime and landscape levels are important to transcend the double bottlenecks of "Death Valley" and path dependence and lock-in in the process of evolution. In summary, tools for policy analysis and innovation management must be included in the fusion process of sci-tech innovation and industrial transformation. It seems fair to conclude that SNM has already become such a tool, but further work and implementation is needed to enhance the symbiotic evolution within and between multilevel layers.

Acknowledgement We had the great fortune to discuss this article with a wide range of scholars, many heavily involved themselves in SNM and MLP research and therefore cited in the text. We thank them for their contributions. We gratefully acknowledge financial support from the National Natural Science Foundation of China (71173051), PhD Start-up in Natural Science Foundation of Guangdong Province of China (2014A030310247) and Young Talents Fund in Educational Commission of Guangdong Province of China (2014WQNCX159).

References

1. Ehlers V (1998) Unlocking our future: toward a new national science policy [R/OL]. A report to congress by the House Committee on Science
2. Moore GA (2002) Crossing the chasm: Marketing and selling high-tech products to mainstream customizers. Harper Paperbacks
3. David PA (1985) Clio and the economics of QWERTY. Am Econ Rev 75:332–337
4. Arthur WB (1989) Competing technologies, increasing returns, and lock-in by historical events. Econ J 99:116–131
5. North DC (1990) Institutions, Institutional Change and Economic Performance. Cambridge University Press, Cambridge
6. Freeman C, Perez C (1988) Structural crisis of adjustment,business cycles and investment behaviour. In: Dosi G, Freeman C, Nelson R, Silverberg G, Socte L (eds) Technical change and economic theory. Pinter, London, pp 38–66
7. Mokyr J (1990) The lever of riches: technological creativity and economic progress. Oxford University Press, USA

8. Kemp R, Schot J, Hoogma R (1998) Regime shifts to sustainability through processes of niche formation: the approach of strategic niche management. Technol Anal Strat Manag 10(2): 175–196
9. Schot J, Hoogma RJF, Elzen B (1994) Strategies for shifting technological systems. The case of the automobile system. Futures 26:1060–1076
10. Schot J, Rip A (1996) The past and future of constructive technology assessment. Technol Forecast Soc Chang 54:251–268
11. Kemp R, Rip A, Schot J (2001) Constructing transition paths through the management of niches. In: Garud R, Karnoe P (eds) Path dependence and creation. Erlbaum, Mahwah, pp 269–299
12. Hoogma R, Kemp R, Schot J, Truffer B (2002) Experimenting for sustainable transport: the approach of strategic niche management. Spon Press, London
13. Schot J, Geels FW (2008) Strategic niche management and sustainable innovation journeys: theory, findings, research agenda, and policy. Technol Anal Strateg Manag 20(5):537–554
14. Geels FW (2002) Technological transitions as evolutionary reconfiguration processes: a multi-level perspective and a case-study. Res Policy 31(8–9):1257–1274
15. Lewontin RC (1983) The organism as the subject and object of evolution. Scientia 188:65–82
16. Odling-smee FJ (1988) Niche-constructing phenotypes. In: Plotkin HC (ed) The role of behavior in evolution. MIT Press, Cambridge

Chinese Railway Transportation's Spillover Effects on Industry

Yaodong Zhou, Xinyan Ou and Jianing Liu

Abstract This chapter applies a simultaneous equation to measure the railway spillover effects on industries by three spillover channels: labor, capital, and product. Based on samples of the Chinese railway industry and 35 industries ranging from 1995 to 2012, this chapter analyzes the railway spillover effects on extent, routine, and distribution and compares changes of the spillover effect before and after 2008. Findings show that the following: (1) Railway spillover effects depend on the technology linkage and the railway's real transport output, produced by labor, capital, and the product channel, and it has positive external effect; (2) the spillover impact on industry's productivity varies from industries and spillover channels; and (3) spillover effects changed after 2008 because of the operation of high-speed railway (HSR).

Keywords Railway · Spillover effects on manufacture · Simultaneous equation

1 Introduction

As a transportation service, railway meets the users' demand of displacement mainly by providing transport activities. It plays an indispensable and fundamental role for the individual, enterprise, market, industry, and even the whole country. OECD [5] concludes that benefits from traffic infrastructure can be classified into internal and external benefits, of which the external benefit is regarded as infrastructure spillover.

Y. Zhou (✉) · X. Ou · J. Liu
School of Economics and Management, Beijing JiaoTong Univesity,
Beijing 100044, China
e-mail: ydzhou@bjtu.edu.cn

X. Ou
e-mail: 12120557@bjtu.edu.cn

J. Liu
e-mail: 14125234@bjtu.edu.cn

M. Li et al. (eds.), *Proceedings of 2015 2nd International Conference on Industrial Economics System and Industrial Security Engineering*,
DOI 10.1007/978-981-287-655-3_9

Rothengatter [6] classifies the positive externality into two kinds: providers and users. The former is the spillover benefit of a traffic service provider brought by expanding infrastructure such as network effects and economies of density. The latter refers to time saving, new modes of production and consumption, and formation of logistics form brought by users due to accessibility improvement.

From the existing research, the spillover effects of railway can be divided into three aspects including the macro effect, the spatial effect and the industrial effect.

- Macroscopic effect: This term refers to the whole economic growth contribution brought by the railway as an infrastructure investment. Before the 1980s, this effect attracted much attention and it has been applied to measure the railway's effect on the community economy. The output elasticity of investment on railway or traffic infrastructure on economic growth has become one of the important issues of measurement [1, 2, 8].
- Spatial effect: a relevant hypothesis claims that transportation infrastructure investment improved the geographic location and accessibility and finally promoted the change of industrial and regional agglomeration form. Zhang [9] use empirical result to show that the investment spatial spillover elasticity is between 0.016 and 0.106, which presents weak increasing return effect.
- Industrial effect: Analyses have also been performed about logistics, industrial inventory, and employment, etc. [3, 7].

This chapter attempts to investigate the function of railway spillover in its actual transportation operation and provide empirical analysis of Chinese railway's influence path and effect on the industry spillover effect. We group industries into four industrial clusters and find that railway spillover effect in the cluster of mining industry is far greater than in other industrial clusters. After railway technological progress in 2008, there was an increasing trend of railway spillover effect on industry. The remainder of the chapter is organized as follows: Sect. 2 describes the hypotheses and models; Sect. 3 shows the sample and method; Sect. 4 report the results; and Sect. 5 provides concluding remarks.

2 Path and Modeling of Railway Spillover

2.1 Path of Railway Transportation Spillover

According to the railway passenger and freight transportation scale and structure, we assume that there are three spillover channels: labor, capital, and product.

Labor: The labor channel embodies that expansion of space frequency, and the range of labor and personnel flow can save travel time, promote enterprises to choose labor elements in a wider range, and thus enterprises can share extra benefits of expansion of labor market scale and scope.

Capital: The capital channel refers to capital goods, i.e., fixed assets including equipment and workshop, etc. Enterprises can share transport time saving benefits through capital goods transport, and railway transport is advantageous for enterprise to reduce production preparation time;

Product: Because of the different industrial objectives, the railway transport product may include materials (such as steel, coal, etc.) and finished products. Railway transportation has fast, low-cost, and large-volume characteristics that can provide mutual flow support for different elements of business investment, which effectively reduces the transport time and cost and promotes enterprises to allocate resources effectively in the larger market space and scale.

2.2 Model Specification

Assume that a certain industry has production function form with $Y_{it} = F_{it}(X_{it})$, in which Y_{it} and X_{it}, respectively, represent output and inputs of i industry in the year t. $F_{it}(\bullet)$ stands for the production function of the industry. The total output of one industry is divided into private output and spillover. Private output is determined by the industry's own input elements, whereas spillover output comes from the external economy of railway transportation and other departments. Assume that spillover formed by other factors is constant (not considering the influence of other spillover factors), and the total output of a department function can be given by:

$$Y_{it} = A(R_{it})F(K_{it}, L_{it}) \tag{1}$$

where $A(R_{it})$ is the spillover effect, and R_{it} represents the spillover index of railway transportation. This chapter argues that railway industry spillover has stable, long-term, and technological association, so it is related to railway transport capacity and the correlation coefficient between industries. Based on the method of defining the technology of the spillover effect, railway spillover can be given by:

$$R_{it} = \omega_{it} \cdot T_t \tag{2}$$

where ω_{it} refers to the total input coefficients of the railway and other industries in t years, and T_i is the railway industry indicator, which stands for annual transportation capacity. Now supposing that production function is consistent with the property of the C-D function and taking the natural logarithm on both sides, Eq. (2) is given by

$$\ln Y_{it} = \ln A_{it} + \ln F_{it} = \alpha \ln L_{it} + \beta \ln K_{it} + \gamma \ln R_{it} \tag{3}$$

In Eq. (3), Y_i is the outputs of i industry; the spillover index is defined as R_i; and K_i and L_i refer to the direct input of each industry in terms of capital and labor. Given our assumption of existing labor, capital, and product, and taking into account the influence of time delay (assuming 1 period lag), the spillover effect can be given by Eqs. (4)–(6):

$$\ln L_{it} = c_i + \gamma_1 \cdot \ln R_{it} + \gamma_2 \cdot \ln L_{it}(-1) + \varepsilon_{it} \tag{4}$$

$$\ln K_{it} = \tau_i + \lambda_1 \cdot \ln R_{it} + \lambda_2 \cdot \ln K_{it}(-1) + \xi_{it} \tag{5}$$

$$\ln Y_{it} = \alpha_i + \beta_1 \cdot \ln K_{it} + \beta_2 \cdot \ln L_{it} + \beta_3 \cdot \ln R_{it} + \mu_{it} \tag{6}$$

Equations (4)–(5) respectively denote the labor (L) and capital (K) channels; we define γ_1, λ_1 respectively to be the value's own labor and capital elasticity coefficient of the first-order lag of 1 period. Equation (6) is product channel; β_3 represents the direct spillover coefficient of railways; and β_1 and β_2, respectively, represent the output elasticity of capital and labor for the total output. c_i, τ_i, and α_i, respectively, represent a specific section effect associated with each industry; and ε_{it}, ξ_{it}, and μ_{it}, respectively, measure the equation stochastic error. Substituting Eqs. (4) and (5) into Eq. (6), Eq. (6) can be converted to a simplified equation:

$$\ln Y_{it} = \alpha_i + (\beta_1 \cdot \lambda_1 + \beta_2 \cdot \gamma_1 + \beta_3) \cdot \ln R_{it} + \beta_1 \cdot \lambda_2 \cdot \ln K_{it}(-1) + \beta_2 \cdot \gamma_2 \cdot \ln L_{it}(-1) \tag{7}$$

From coefficient of ln R_{it} in Eq. (7), the railway aggregate spillover effect is expressed in Eq. (8), $\rho_1 = \beta_2 \cdot \gamma_1$, $\rho_2 = \beta_1 \cdot \lambda_1$, $\rho_3 = \beta_3$, which measure the degree of eventual spillover to output through the labor, capital, and product paths, respectively. ρ_4 stands for the degree of aggregate spillover to output.

$$\rho_4 = \rho_1 + \rho_2 + \rho_3 = \beta_2 \cdot \gamma_1 + \beta_1 \cdot \gamma_1 + \rho_3 \tag{8}$$

3 Data

To analyze the railway spillover effects to other industries, we need an empirical design that is able to cater industry differences to our theoretical interests. To satisfy this requirement, we construct our data primarily by using panel data for 35 industries from *China Statistical Yearbook* during the period from 1995 to 2012 including the mining industry and the manufacturing industry (Table 1).

Table 1 Conclusion of variables

Variables	Description of variables	Data source
Y	Output value of the industry	Chinese Statistical Yearbook
T	Railway transport capacity indicators	Chinese Statistical Yearbook
R	Railway technology spillover indicators	Calculated by Eq. (2)
ω	Total input coefficients	Chinese Statistical Yearbook
K	fixed assets of each industry	Chinese Statistical Yearbook
L	Number of employees	Chinese Statistical Yearbook

The indicator (Y) database reflects industry output between 1995 and 2012 and is generated by the method of Medianu [4]. The variable (K) is capital input, which is estimated in terms of perpetual inventory according to the method of [10]. The net value series of fixed assets is generated by subtraction of the original value of fixed assets between two consecutive years and then deflated by the price indices of fixed assets investment, whereas the index of labor input (L) is the average annual number of employed persons. Until now, the national statistical center has published Chinese input output matrices from 1997, 2002, and 2007. According to the China Statistical Yearbook and the theory of total input coefficients this chapter uses the total input coefficients of 1997, 2002, and 2007, respectively, as the index of three stages of 1995–1999, 2000–2004, and 2005–2012. At each stage, we assume that total input coefficients is the constant, namely, the technology association between the railway, and that other industries in each stage are stable. The index of railway transport capacity (T) is expressed by physical variables (total volume of conversion) and compared with length of railway in operation, total density of passenger and freight, it can represent the real transportation scale and level.

4 Empirical Results

Tables 2 and 3 present the regression results based on the methods of 3SLS and GMM.

The L equation presents positive spillover elasticity of railway to labor element. Table 2 shows that 1 % increases in the railway input labor workforce

Table 2 Regression result of railway spillover in 35 industries

Variable	L equation		K equation		Y equation	
Method	3SLS	GMM	3SLS	GMM	3SLS	GMM
lnR	0.0175** (2.0052)	0.0157** (2.0129)	0.0303*** (3.5523)	0.0331*** (3.0894)	−0.0166 (−0.6086)	−0.0150 (−0.5174)
lnL					0.5113*** (14.7708)	0.5149*** (11.5344)
lnL(-1)	0.9598*** (81.2134)	0.9628*** (84.3019)				
lnK					0.3651*** (9.9450)	0.3618*** (9.1236)
lnK(-1)			0.9792*** (101.1364)	0.9772*** (87.5592)		
C	0.4851*** (3.4061)	0.4531*** (3.2603)	0.0675 (1.5246)	0.0709 (1.6369)	−2.6699*** (−7.3427)	−2.7111*** (−5.5528)
R^2	0.9465	0.9465	0.9783	0.9782	0.6996	0.6996
Sargan test(p)		1.0000		1.0000		1.0000

***, **, and * denote significance at the 0.01, 0.05, and 0.10 level, respectively

increases by 0.0157–0.0175 %, then the spillover to output increases by 0.008–0.0089 %. The spillover to labor element is significantly positive, which is consistent with our expectation.

The K equation presents the positive spillover elasticity of railway to the capital element. Table 2 shows that 1 % increases in the railway input capital goods value increases by 0.0303 % to approximately 0.0331 %, and then the spillover to output increases by 0.0111 % to approximately 0.0119 %. Compared with the influence degree of the labor force, the railway has a higher spillover influence to capital goods. The results also show that in the existing stage of development, the industrial association effect between the railway and other industries is stronger. On one hand, industry has stronger dependence on capital than labor, and the contribution of capital to the output exceeds the labor for that capital goods flow velocity, and frequency, and the transportation cost reduction can promote more output formation. In contrast, by providing capital goods transport, the railway has more stability, direction, and long-term growth, and the optimization of conditions of capital goods transportation is conducive to the increase of industrial output.

Table 3 shows the eventual spillover effects of three spillover channels and indicates the differences in the degree of spillover. Because the spillover effect of a product channel is not significant, it is only provided as a reference value. The degree of eventual spillover to output through labor and capital channels is 0.0080 to approximately 0.0089 and 0.0111 to approximately 0.0119, respectively, as a 1 % increase of railway technology spillover s indicator. Without considering the spillover effect of a product channel, the degree of aggregated spillover to output is 0.0200, and when taking into account the product channel effect, the degree of aggregated spillover to output is 0.0034 to approximately 0.0049. As a result, the aggregated spillover effect is smaller when considering product channel effect.

We divide 35 industries into 4 clusters and estimate the respective degree of spillover effect between industries in the same cluster using the method of 3SLS and GMM. Table 4 shows the degree of eventual spillover effects to output.

From Table 4, we find that for cluster I, which is mainly related to mining, the product channel (ρ_3) is the main spillover channel, and the labor channel is lower than this. Cluster II comprises mainly agricultural byproducts and chemical-based industries, and railway spillover effects are present in the labor (ρ_1) and capital (ρ_3)

Table 3 Overall degree of spillover of three channels for industries

	3SLS	GMM	3SLS		GMM	
			95-08	09-12	95-08	09-12
ρ_1	0.0089**	0.0080**	0.0111**	0.0045**	0.0097**	0.0043**
ρ_2	0.0111***	0.0119***	0.0173***	0.0086	0.0187***	0.0100**
ρ_3	−0.0166	−0.0150	−0.0646**	0.0488	−0.0635**	0.0288
ρ^*	0.0200	0.0199	−0.0362	0.0045	−0.0351	0.0143
ρ_4	0.0034	0.0049	−0.0362	0.0619	−0.0351	0.0431

ρ^* refers to a statistically significant degree of spillover to output

Table 4 Overall degree of spillover degree among four industry clusters

	Cluster I		Cluster II		Cluster III		Cluster IX	
	3SLS	GMM	3SLS	GMM	3SLS	GMM	3SLS	GMM
ρ_1	0.0113	0.0113**	0.0033	0.0042**	0.0040	0.0121	0.0210***	0.0211***
ρ_2	0.0007	0.0009	0.0251***	0.0210***	0.0341***	0.0411***	0.0030	0.0026
ρ_3	0.4756***	0.4745***	0.0025	0.0016	0.2142**	0.2498**	0.1121***	0.1140***
ρ^*	0.4756	0.4858	0.0251	0.0252	0.2483	0.2909	0.1331	0.1351
ρ_4	0.4877	0.4867	0.0309	0.0268	0.2522	0.3030	0.1361	0.1376

channels. Industries in cluster III are mainly related to light industries, where the railway spillover effects are reflected in capital (ρ_2) and product (ρ_3) channels, whereas the spillover effects of the product channel are greater than those of the capital channel. In contrast, for the cluster IX, which includes mainly metal-based manufacturing industries, spillover effects are present in the labor (ρ_1) and product (ρ_3) channels. As a result, from the estimated indicator (ρ_4), the railway spillover effect to output of cluster I is the largest among the four clusters. In contrast, the spillover effects of railway have relatively small spillover effects on clusters II, III, and IX. In fact, in the Chinese railway freight transport, infrastructure and basic materials are the main transport objectives. Those cargos—including coal, coke, oil, steel, nonferrous metal, metal ores, nonmetal ores, mineral building materials, and cement—account for 78 % of all railway freight turnover in 2013. This partly explains why the mining industry becomes the main industrial sector of the railway spillover effect.

From spillover channels, different industry clusters show clearly different degrees of spillover. In the most obvious labor channel, spillover is reflected in the agricultural and machinery industries, which can be interpreted by the demand for labor in those industries being far greater than that in capital-intensive heavy industries. Similarly to the mining industry, the spillover effect is more embodied in the direct product path. For those industries whose products can be used in the consumption, information, and communication industries, spillover effects are more reflected in the capital channel.

Railway technical progress has a positive effect on railway spillover effects. In 2009, with the speed-up of railway, railway transport efficiency has obviously changed. Dividing the period 1995 to 2012 into two periods with year 2008 as the time node, we try to explore whether the railway spillover effect will change over time. Table 3 shows that estimated parameters ρ indicates that spillover effects to output of the labor channel is decreasing, whereas spillover effects of the capital channel are increasing, and those of the product channel are not significant. On the whole, the aggregated spillover effects to output (ρ^*) have changed from negative spillover effects (-0.351 to approximately -0.0362) into positive spillover effects (0.0045 to approximately 0.0143), which shows that railway spillover effect to other industries is growing. The indicator of aggregated spillover effect to output (including significant and insignificant values) using ρ_4 matrix indicates a more obvious growth trend.

Table 5 shows the result of comparison of four industrial clusters. The empirical results further validate that there is a growing tendency of railway industry spillover. The estimated parameter using the ρ^* matrix indicates that the aggregated spillover effect to output presents an increasing trend in the four industrial clusters after considering railway technology innovation (railway speed-up, separation of passenger and freight, and high-speed railway). From the perspective of the four industrial clusters, industries in cluster I have a growing and stable product spillover, which is the largest railway spillover embodiment; spillover channels of industries in cluster II have changed from labor and capital channels to labor and product channels, and product spillover has accelerated faster, which shows that the product channel spillover of agricultural byproducts and chemicals expands after technological shifts occur. The railway spillover channel of industries in cluster III are present in the capital channel where the degree of spillover increases slightly. The railway spillover channel of the industries in cluster IX (mainly including metal-based manufacturing industries) is embodied in the labor channel, and spillover effect drops slightly. Generally the evidence indicates that the railway has independent technology associations with other various industrial departments. Spillover channel of clusters II and III has changed since technical adjustments have been made to the railway.

From the respective spillover channels for the four industrial clusters, the labor spillover channel is mainly embodied in industrial clusters I, II, and IX and is stable and significant but certainly descendant. Capital spillover channel are reflected in cluster II, while spillover degree is still in a downward trend. The product spillover channel is embodied mainly in clusters I and II and show an obvious growth trend. As a result, among the three spillover paths, labor is the most stable, and capital and products have some volatility in different industries.

5 Conclusions

This chapter constructs a simultaneous equation to measure railway spillover effects on industries based on three spillover channels: labor, capital, and product. Using samples of the Chinese railway industry and 35 industries ranging from 1995 to 2012, we analyze the degree of railway spillover, channels, and sector distribution comparing changes in spillover effect before and after 2008. Note that our result is consistent with a previous spillover mechanism assumption. We find that the following:

- As one input factor, the railway affects other industry departments' output through the channels of labor, capital, and product, which reflects the external economy.
- Railway spillover effect depends on the technology association between the railway and other industries as well as the railway's practical transport level.
- Chinese railway spillover comes mainly from the labor and capital channels, whereas the degree of spillover in the product channel lacks significance. The

Table 5 Overall degree of spillover among four industry clusters during two time periods

	Cluster I		Cluster II		Cluster III		Cluster IX	
	3SLS	GMM	3SLS	GMM	3SLS	GMM	3SLS	GMM
1995–2008								
ρ_1	0.0129	0.0124*	0.0058*	0.0075**	0.0094	0.0161	0.0262**	0.0266***
ρ_2	0.0046	0.0049	0.0413***	0.0376***	0.0561***	0.0691***	0.0076	0.0083
ρ_3	0.4557***	0.4552***	−0.0424	−0.0403	−0.3765***	−0.4353***	−0.0195	−0.0141
ρ^*	0.4557	0.4676	0.0471	0.0451	−0.3204	−0.3662	0.0262	0.0266
ρ_4	0.4731	0.4725	0.0047	0.0048	−0.3111	−0.3500	0.0143	0.0208
2009–2012								
ρ_1	0.0024	0.0060**	0.0061*	0.0068**	−0.0151	−0.0117	0.0050	0.0081***
ρ_2	0.0005	−0.0010	−0.0003	−0.0014	0.0523	0.0369**	−0.0064	−0.0075
ρ_3	0.4583***	0.5761***	0.1027**	0.1388***	−0.1164	0.0605	0.0546	0.0306
ρ^*	0.4583	0.5821	0.1088	0.1456	–	0.0369	–	0.0081
ρ_4	0.4612	0.5811	0.1085	0.1442	−0.0792	0.0856	0.0532	0.0312

spillover elasticity of the capital channel is higher than that of the labor channel. The degree of aggregated railway spillover on industry's output is 0.02 through the labor and capital channels, which showed an increasing trend after 2008.

- Differences exist in the spillover routine and extent among four the clustered industries. As the representative of the cluster I industry sector, the mining industry shows the strongest and the most stable spillover effect. The labor channel is the most stable and widely radiating channel, which also has a positive spillover effect on the other sectors; and the capital and product channels show significant spillover differences among different industries.

Acknowledgments I appreciate the support of Joint Project of scientific research and graduate student training by the Beijing Board of Education-City traffic environment externality in Beijing: empirical study and governance mechanism (NO. B13H100010).

References

1. Canning D, Pedroni P (1999) Infrastructure and long run economic growth. Center for analytical economics of Cornell University, Working paper
2. Carlino GA, Voith R (1992) Accounting for differences in aggregate state productivity. Reg Sci Urban Econ 22(4):597–617
3. Duffy-Deno KT, Dalenberg DRD (1993) The municipal wage and employment effects of public infrastructure. Urban Stud 11:1577–1589
4. Medianu D, Whalley J (2012)water availability as a constraint on china's future growth. NBER working papers 18124, National Bureau of Economic Research, Inc. pp 1–56
5. Quinet E (1994) Internalising the Social Costs of Transport. OECD, Paris
6. Rothengatter W (1993) Externalities of transport. In: Polak J, Heertje A (eds) European transport economics, vol 91. Blackwell Publishers, Oxford
7. Shirley C, Winston C (2004) Firm Inventory Behavior and the Returns from Highway Infrastructure Investments. J Urban Econ 11(2):398–415
8. Tatom JA (1995) Public Capital and Private Sector Performance. Can Bus Econ 1(6):3–15
9. Zhang X (2012) Has transport infrastructure promoted regional economic growth? With an analysis of the spatial spillover effects of transport infrastructure. Soc Sci China 3:60–77
10. Zhang J, Wu G, Zhang J (2004) The estimation of China's provincial capital stock: 1952–2000. Econ Res J (China) 10:35–44

Part II
Industrial Security

Impact of Resources Constraint on Industrial Security: A Case Study of Japan

Dong Tong and Juan Li

Abstract Data on the constraint of resources over time were used to estimate the impacts of resources and nonnatural resources on industrial security and industrial development in Japan. The results showed that the factors—such as stock and flow of natural resources, the self-sufficiency of natural resources, the price of natural resources, the consumption elasticity of natural resources and labor cost—had threatened the industrial security of Japan. However, factors such as the consumption intensity of natural resources, the stock and flow direction of labor, the quality of work force and the increasing bring-up ratio has not threatened the industrial security of Japan.

Keywords Industrial security · Industrial development · Resources constraint

1 Introduction

It is well known that the three main factors of resources, including natural resources, social economic resources, and technological resources, play important roles in the production process. As a resource-poor country, Japan has limited storage of oil, coal, and natural gas. Therefore, >95 % of its energy resources, such as iron ore, oil, and rubber, are imported.

D. Tong (✉)
Institute for Culture Industrial Security Research, Beijing Institute of Graphic Communication, Beijing 102600, China
e-mail: td1202013@sina.cn

J. Li
School of Economy and Management,
Beijing Jiaotong University, Beijing 100044, China
e-mail: limily009@sohu.com

M. Li et al. (eds.), *Proceedings of 2015 2nd International Conference on Industrial Economics System and Industrial Security Engineering*,
DOI 10.1007/978-981-287-655-3_10

The importance of economic security, especially industrial security, has attracted the attention of several researchers. In the field of the theoretical research, Li [1] first defined industrial security and described the influencing elements of industrial security systematically. The industrial security analytical methods and industrial security model are built in the framework of industrial economics theory.

Industrial security is an important component of economic security. "Industrial security" refers to a status in which the survival and the development of a given industry have not been threatened by foreign investment, international trade, the constraint of resources, social demand, and industrial transfer. In resource-poor countries, the constraint of resources could not only influence industry security but also hamper national economic security and industrial upgrading. Therefore, research on the influence of resources constraint on industrial security are of great importance to resource-poor countries. Japan is a typical resource-poor country whose industrial security and development are threatened by the constraint of resources. In this chapter, data on natural and nonnatural resources constraint over a period of 16 years were used to estimate industrial security in Japan. The results of this research may benefit the economic plan that designers formulated and promulgated, the economic development strategy in Japan, and other resource-poor countries.

2 Empirical Analysis

Resource supply is the most important factor influencing the industrial security of resource-poor countries in five main factors. There are two aspects of resources constraint of industrial security: constraint of natural resources and constraint of nonnatural resources (Table 1).

Table 1 The effect of resources constraint on industrial security

Natural resources
Stock and flow of natural resources
Self-sufficiency of natural resources
Price of natural resources
Consumption elasticity of natural resources
Consumption intensity of natural resources
Nonnatural resources
Stock and flow direction of labor
Quality of work force
Increasing bring-up ratio
Labor cost

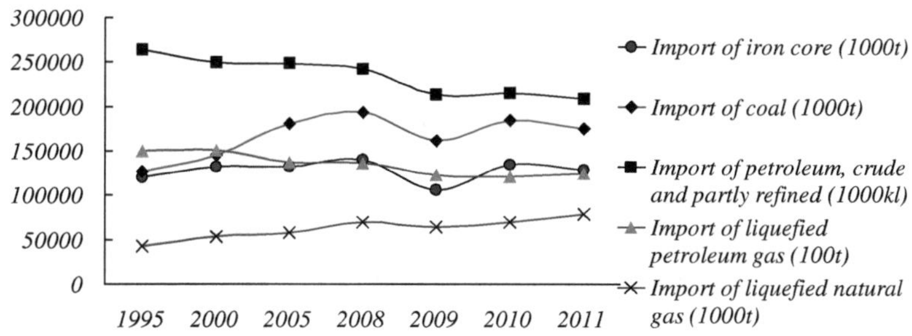

Fig. 1 The main resources imported to Japan during 1995 to 2011

The data of natural and nonnatural resources used in this chapter come from the Japan Statistical Yearbook (2007–2013) [2], and macroeconomic data were obtained from several sources such as Japan's Ministry of Economy, Trade and Industry web page and Japan's Finance Ministry web page.

Stock and flow of natural resources. As a resource-poor country, there is a wide gap between the national supply and demand for natural and nonnatural resources. Even for the available resources, the quality exceeds the demand. Therefore, importing resources is the only choice. Taking iron ore as an example, although it has the largest storage in Japan, the imported quantity reached 99.4 % of the national demand in the last several years. The imported quantity of the main natural resources of Japan during 1995 to 2011 is shown in Fig. 1. According to industrial security theory, no large change of importing main resources in one particular country indicates that the country's industrial security of this country has not been threatened. From this perspective, the stock and flow of natural resources has not threatened Japan's industrial security.

Self-sufficiency of natural resources. According to industrial security theory, the impact of the resource supply on industrial security will increase with a decrease of the self-sufficiency of the resources. During 2004 to 2010, the self-sufficiency of coal was the lowest in Japan, and all of the demand was supported by importation. The self-sufficiency of petroleum and gas are both <5 %. The self-sufficiency of wood was better, but the self-sufficiency ratio was still <30 % (Table 2.). From this perspective, self-sufficiency of natural resources has threatened the industrial security of Japan.

Table 2 Self-sufficiency of natural resources during 2004 to 2010

	2004 (%)	2008 (%)	2010 (%)
Wood	20.47	24.43	26.32
Coal	0.00	0.00	0.00
Petroleum	0.34	0.38	0.38
Gas	3.80	4.28	3.72

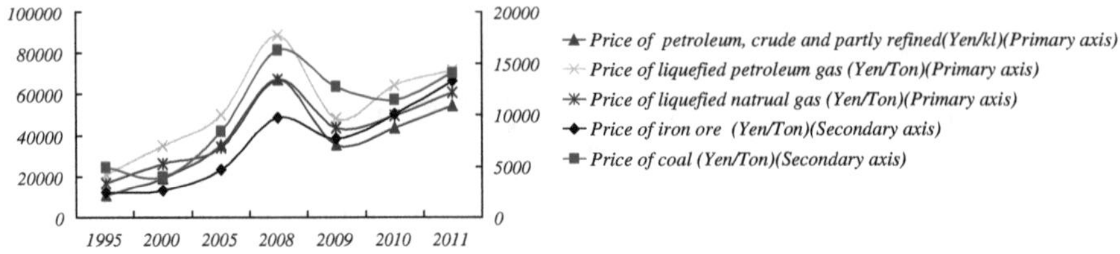

Fig. 2 Price of main resources in Japan during 1995 to 2011

Price of natural resources. Because it is a natural resource poor country, the demand for resources in Japan must be met by importation. As an independent economy, there is a close relationship between industrial security and resource price change. The prices of five resources in Japan during 1995 to 2011 are shown in Fig. 2. Except for price fluctuation in the year of 2008, which can be attributed to the financial crisis worldwide, prices of the five resources increased during this time period. The quantity of imported oil and liquefied petroleum gas decreased and their prices increased during 2005 to 2011.

Consumption elasticity of natural resources. The consumption elasticity of natural resources is shown in Fig. 3. Decrease of the consumption elasticity of wood and petroleum suggested that the use of wood and petroleum is increasing, which is helpful for industrial security [3]. The consumption elasticity curve of tungsten, pig iron, electrolytic copper, silver, coal, tin, antimony, and cobalt first decreased and then increased, with the turning point being at the year of 2009 after the worldwide financial crisis owing to the GDP declining in 2008. The consumption elasticity of

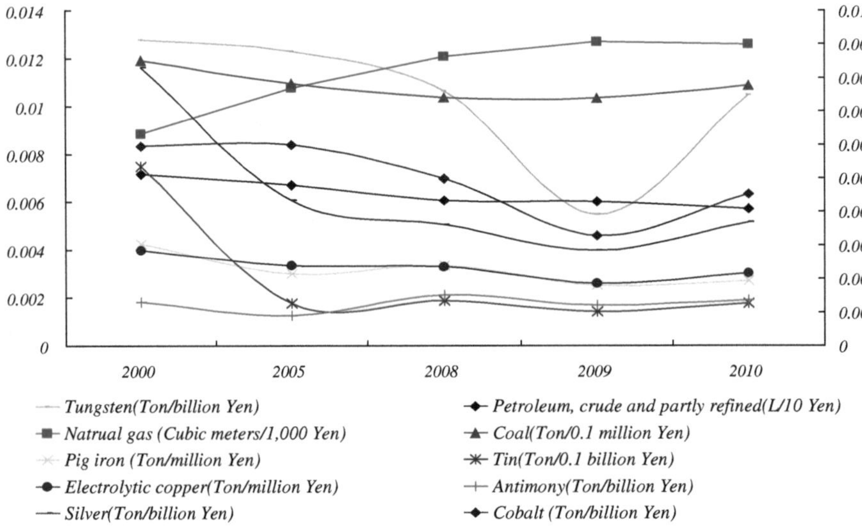

Fig. 3 Consumption elasticity of main resources in Japan during 1995 to 2011

the seven resources mentioned above generally decreased, so resource-use efficiency increased, and industrial security was well maintained during past years. However, the consumption elasticity of natural resources increased, which may threaten industrial security in Japan.

According to industrial security theory, the lower the consumption, the higher the industrial security. Therefore, industrial security in Japan has not been threatened by the consumption elasticity of natural resources in general, but the negative impact of natural gas on industrial security is concerning.

Consumption intensity of natural resources. The consumption intensity of the main natural resources, except petroleum and natural gas, is >1. For tungsten, the ratio of consumption growth rate to the economy is 15.92. That is, the enhancement of each unit of the economy will increase the consumption of tungsten by 15.92 units in Japan (Fig. 4). In view of this, the industrial security of Japan has been threatened by the consumption intensity of natural resources.

Stock and flow direction of labor. Concurrent with labor transferring from lower levels to higher ones, the industrial structure is also being upgraded. According to this theory, if the percentage of the higher industry (i.e., secondary industry and/or tertiary industry) increases, and the percentage of the lower industry (i.e., primary industry and/or secondary industry) decreases at the same time, industrial security will improve. As shown in Fig. 5, the percentage of the tertiary industry increased, whereas the percentages of primary and secondary industries decreased during 2003 to 2011. Thus, the stock and flow direction of labor benefited the industrial security of Japan [4].

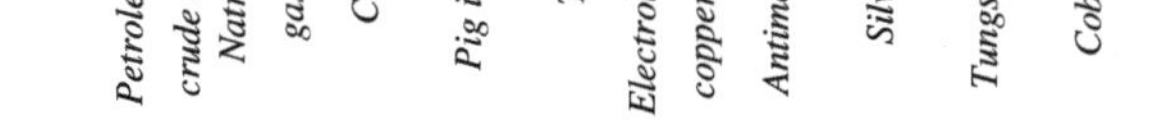

Fig. 4 Consumption intensity of the main resources in Japan during 1995 to 2011

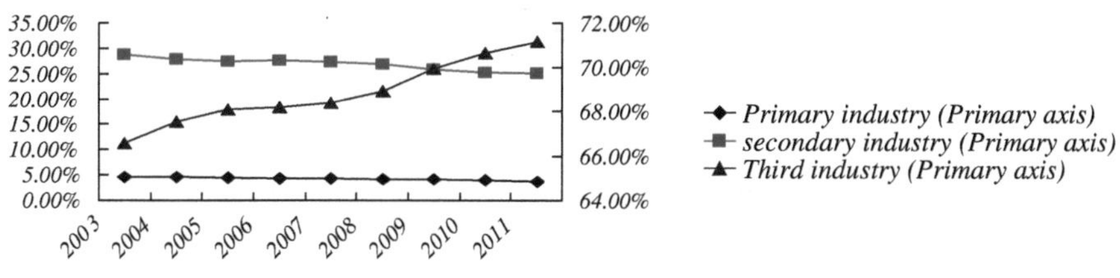

Fig. 5 Ratio of the total employment population in primary, secondary, and tertiary industries

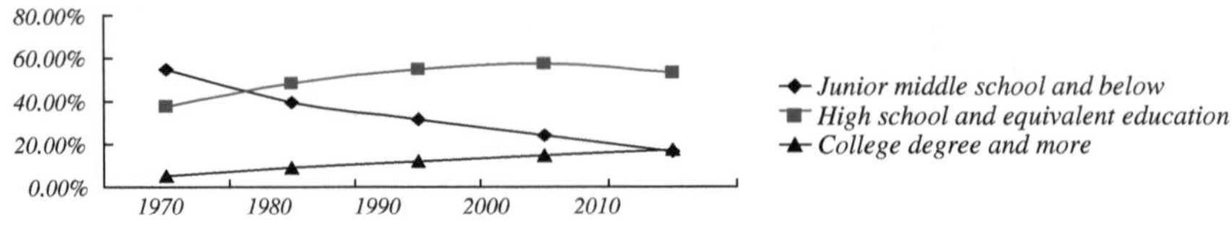

Fig. 6 Bring-up ratio of the employment-population older than 15 years during 1970 to 2010

Quality of the workforce. Economic development is closely related to the economic security of a country. During the years 1970 to 2010, for the proportion of people who are older than 15 years and who have an education level of junior middle school, educational experience is decreasing, whereas for the proportion of people in the same age group but with an educational level of high school and above,educational experience is increasing instead (Fig. 6.). As workforces are better educated, more high-tech talents engage in national development. This will help to promote industrial upgrading and maintain the industrial security of Japan.

Increasing bring-up ratio. The social burden is measured by the increasing bring-up ratio. The bring-up ratio has increased since 1990 because of the aging of individuals in Japan (Fig. 7). However, the industrial security has not been threatened by the increase of the increasing bring-up ratio. The reason is that the most appropriate way of industrial development in Japan is by transferring labor-intensive industries to developing countries and promoting an upgrade of the industrial structure. At the same time, more workforces are set free from simple work. This phenomenon is appropriate to the increase of the increasing bring-up ratio.

Labor cost. Industrial security is influenced by labor cost through the consumption structure [5]. Alteration of the consumption structure will lead to the change of commodities production and resource supply to meet demand. If the demand of higher industry produced commodities increases, the industrial structure will be upgraded. This means that with an increase of labor cost increasing, the industrial security is improved. The curve of labor cost during 1990 to 2011 is shown in Fig. 8 indicating that decreasing labor cost benefited the industrial security of Japan.

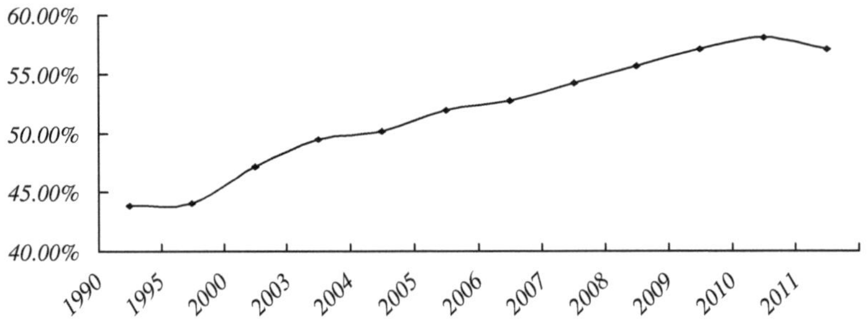

Fig. 7 The increasing bring-up ratio of Japan during 1990 to 2011

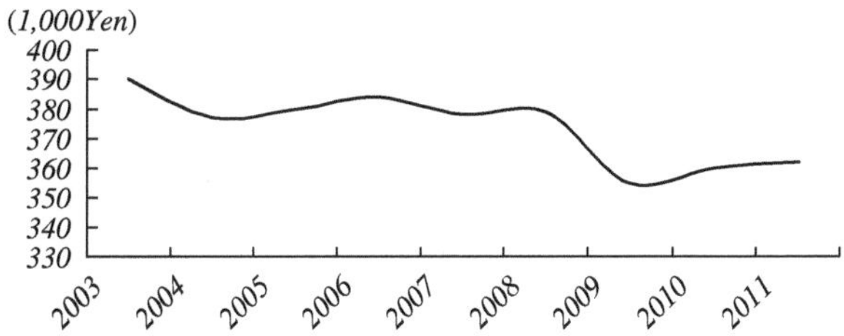

Fig. 8 Labor cost of Japan during 1990 to 2011

3 Conclusion

Our results show that different factors of natural resources and nonnatural resources could influence the industrial security of Japan. According to the theory of the "cask principle," a shortage of resources can impede the development and the security of an industry. With abundant capital, technology, and high-quality laborers, developed countries could create knowledge-intensive industries. However, policymakers of resource-poor countries such as Japan should consider the constraints on resources during the development of plans because a shortage of the resources will constrain industrial development.

Acknowledgements This research was supported by initial funding for the Doctoral Program of BIGC.

References

1. Menggang L (2006) Research on industrial security theory. Economic Science Press, Beijing, pp 189–192
2. Statistical research and training institute (2013). Japan Statistical Yearbook. Statistics Bureau, pp 23–56
3. Dan W (1999) Structural changing is the main factors impact of energy consumption in China. China Ind Econ 11:38–43
4. De BS, Vanden BJ, Opschoor JB (1998) Economic growth and emissions: reconsidering the empirical basis of environmental Kuznets curves. Ecol Econ 25:161–175
5. Stern DI (2002) Explaining changes in global sulfur emissions: an econometric decomposition approach. Ecol Econ 42:201–220

Integration of Construction Worker Safety in Design Through the Use of BIM

Jia Qi, Raja R.A. Issa and Shengyue Hao

Abstract The construction industry has incurred the most fatalities of any United States industry in the private sector in recent years. Although many factors may contribute to this statistic, one likely cause is due to designers who often lack knowledge of design for construction safety, which results in many safety hazards being built into project models. To improve the situation, this research was undertaken to identify the possible influence of Building Information Modeling technology on construction safety. After identifying the extent of the positive impact of BIM technology on safety, the research entailed the development of a design for a construction worker safety tool that efficiently makes suggestions regarding designing for safety available to designers and constructors. Particular emphasis was placed on falling accidents because falls are the most frequently occurring cause of fatalities on construction sites.

Keywords Construction safety · Prevention through design · Model-checking software

J. Qi (✉)
Postdoctoral Programme of China Center for Industrial Security Research (CCISR), Beijing Jiaotong University, Beijing 100044, China
e-mail: 283009276@qq.com

R.R.A. Issa
M.E. Rinker, Sr. School of Building Construction, University of Florida, Gainesville, FL 32611-5703, USA
e-mail: raymond-issa@ufl.edu

S. Hao
Management, Beijing Jiaotong University, Beijing 100044, China
e-mail: haoshyue@bjtu.edu.cn

M. Li et al. (eds.), *Proceedings of 2015 2nd International Conference on Industrial Economics System and Industrial Security Engineering*,
DOI 10.1007/978-981-287-655-3_11

1 Introduction

It is an alarming statistic that every workday nearly four workers are killed in the United States construction industry. In 2010, construction industry fatalities represented 16.5 % of all work-related fatalities in the United States [1]. Researchers have proposed many methods to reduce job site hazards. One technique is to incorporate safety considerations into the project during the design phase or to involve designers in considering construction worker safety during the design process. This approach is known as Prevention through Design (PtD).

In current practice, building information modeling (BIM) does not automatically address construction worker safety. Instead, construction safety is addressed by the inclusion of written notes in the design, which is a cumbersome and inefficient procedure. The aim of this chapter was to explore how BIM technology can be used to enhance construction worker safety and to develop a BIM-based PtD tool for construction safety that can automatically address formatted construction documents in industry foundation classes (IFC) for construction safety.

2 Literature Review

2.1 Background

Several approaches and tools have been developed to help designers identify design decisions that can significantly improve construction safety without compromising architectural form or function [2–4]. The first and most fundamental effort in promoting PtD is to improve communication and coordination between the different stakeholders. Although improvement of safety performance is possible based on the effort of an individual discipline, communication and cooperation among the project participants regarding the safety aspects of a project are important and necessary. It is believed that the major opportunity for improving the design and construction of facilities lies at the interface between disciplines [5]. This is most apparent especially in the case when the traditional design bid build delivery method is used, which acts as a barrier to safety-in-design process.

Another approach to promoting PtD is to form a systematic design process to eliminate or reduce design errors and increase the opportunity of incorporating PtD knowledge into construction documents. If designers are not sufficiently experienced to continuously consider safety, outside design reviews should be performed to improve the quality of the drawings.

2.2 *Automated Checking of Building Designs*

The vision of automatically checking building plans against codes emerged as early as the 1970s. However, it was not until recent years that the new generation of software technology and standards made automated checking much easier. The Health and Safety Executive (HSE) of the United Kingdom developed a software prototype to provide designers with easy access to health and safety information by establishing a means of structuring HSE's information as a knowledge-based system. In the United States, the International Code Council (ICC) has cooperated with corporations such as AEC3 and Solibri to develop a platform for its automated code checking [5]. It should be noted that checking for building-code compliance is different from applying PtD knowledge because the principle of PtD is to protect construction workers during the process of construction. Another difference between them is that currently no administrative processes require designers and constructors to fulfill design for construction safety responsibilities or to ensure that projects are in compliance with United States construction safety regulations.

2.3 *PtD Tool Development*

Ideally, a PtD software tool should perform three major functions to help designers provide high-quality building information models in a mature design process. The first function is to minimize safety noncompliance during the main design phase by informing and guiding designers about alternative design options and the consequence-reducing practices. The second function is to conduct safety noncompliance detection after the main design work has been completed by carrying out overall safety checks. The third function is to correct the safety noncompliance situation. By performing these three functions, a PtD tool would be able to help improve designs by providing appropriate knowledge at the appropriate time during design development or by checking the building information model representing the completed final design.

3 Architecture and Functionality of the PtD Software Tool

3.1 *Architecture of the Software*

The architecture of the software tool was first developed to define the scope of the tool. Several issues were considered to be important. Based on documented experiences in past research efforts, three features of the PtD software tool were closely considered when deciding when and how to intervene during the design process.

The first feature concerns the conditions under which the software tool would activate the intervention function. Systems with an active critiquing strategy would continuously monitor designs as they evolve and offer feedback, while systems with a passive critiquing strategy would only give feedback when designers specifically ask for it.

The second feature concerns the type of feedback. Applications developed for code checking usually point out specific errors in building models. The PtD software tool is different in that it suggests the opportunities for integrating safety into building models even though the documents may be errorless. The software tool should be developed to not only offer negative evaluations but also provide positive knowledge.

The third feature concerns the form in which the feedback will be reported to users. This PtD software tool includes two types of feedback—text messages and graphical markers—to indicate the locations where improvements are possible. Text messages give detailed explanations to designers to help them to better understand the design alternatives. By marking the problematic components in the proposed 3D building models, the PtD software tool explicitly shows designers the locations of the building components that need to be addressed.

Figure 1 shows the desired architecture of the PtD software tool. The horizontal axis represents the project-development process from the beginning of the design to delivery of the documents to the constructors. This process begins on the left with the design-development period when the designers create the initial building models by using BIM authoring software. It then evolves into the design review and the phase of checking for code compliance. Compliance checking of a building model is performed in these two phases. This culminates in the construction phase.

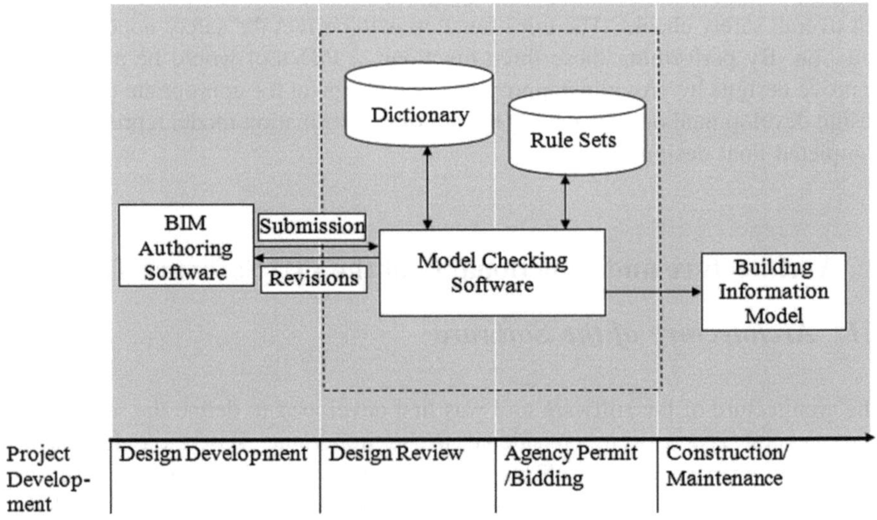

Fig. 1 Architecture of PtD tool

The design process is an iterative one. Users submit their building information models to the PtD tool and check them for worker safety noncompliance. After the report identifies the problematic building components, the users can revise their building models by returning to their BIM-authoring software.

3.2 Functionalities of the Software

After the architecture of the PtD software tool had been determined, the next step was to define the functionalities of the PtD software tool. The software tool was developed to have three main functions. The first function is to check building information models against the PtD-computable rules. The second function of is to identify the safety hazards discovered for certain building system components. The third function is to provide the designer with information to help correct the safety hazard. This is based on both the characteristics of the PtD knowledge and the reasoning process of the software tool. Many of these suggestions were difficult to encode into rules that can be used to check the properties of building system components and to check for and restrict noncompliance with worker safety rules.

4 Model-Checking Software

4.1 Selection of Model-Checking Software

Based on the previously discussed PtD tool requirements and their ability to import and export IFC files, BIMserver was selected for use as the software platform in this research. Computable rules were developed for this software tools. BIMserver is an IFC model server that centrally stores the building model and manages all access to it. The URL, http://bimserver.org, is linked to an open-source IFC model server and plays a role as an information hub that allows users to merge, filter, query, or conduct clash detection. In this research, the noncompliance of a building model is determined through the Advanced Queries function of BIMserver.

4.2 Creation of Rule Sets

The IFC schema provides most of the information necessary to derive construction features. However, it also has the most complex schema. Properties are often not directly attached to building components but are related through long indirect identification references (IDRef). The core step in compiling any PtD suggestions into a computable rule is to analyze how the objects are linked with different attributes and relationships. A solid understanding of the IFC schema and its hierarchy is necessary to perform this work and to successfully extract information from building models. This task is often extremely difficult and time-consuming because of the large size and complexity of the IFC schema.

5 Conclusions

BIMserver is proposed as building-model checking software for compliance with safety requirements. This application tool can be used to automatically check for hazards in building information models and to provide design alternatives to users. It can be used by architects during the design process or by constructors before commencing construction work. The PtD tool allows users to check for compliance with specific safety requirements, thus allowing for further improvement of the design, which results in enhanced construction site safety.

In BIMserver, the IFC specification is expressed in Java language. To write a computable rule, one must be familiar with both Java language and IFC specification. Although Java language is a popular computer language, only a few researchers and engineers are acquainted with the IFC hierarchy prescribed under the BIMserver environment.

Acknowledgments The author appreciates the support of BIMserver for their generosity in contributing the use of their software, which was invaluable to completing this research.

References

1. Bureau of Labor Statistics (2011) Number and rate of fatal occupational injuries, by industry sector. http://www.bls.gov/iif/oshcfoi1.htm
2. Toole TM (2005) Increasing engineers' role in construction safety: opportunities and barriers. J Prof Issues Eng Educ Pract 131(3):199–207
3. Gambatese JA, Hinze JW, Behm M (2005) Investigation of viability of designing for safety. The Center to Protect Workers' Rights (CPWR), Silver Spring
4. Zhou W, Whyte J, Sacks R (2012) Construction safety and digital design: a review. Autom Constr 22:102–111
5. Fischer M, Kunz J (2004) The scope and role of information technology in construction. Center for Integrated Facility Engineering (CIFE), Stanford Univ., Stanford, Calif
6. Eastman C, Lee J, Jeong Y, Lee J (2009) Automatic rule-based checking of building designs. Autom Constr 18:1011–1033

Security Impacts and Key Issues of Integration of REGs on Distribution System

Rui Li, Wei Wang and Lijie Xu

Abstract With the permeability of the renewable energy generations (REGs) becoming greater, the distribution network (DN) has become a complex interconnected network with distributed generations (DGs) and loads from a traditional radiate one. REGs bring the great opportunity to low-carbon and sustainable development of the distribution system (DS), but it also brings severe impacts to the security and stability of the operation. In this chapter, six key impacts regarding secure operation are analyzed including power transmission, power quality, frequency stability, islanding effects, protection, and power-system planning. Based on the analyses, corresponding suggestions are introduced. Particularly, the promotion of REGs' stability and controllability, as well as the coordinated planning method for DN considering the integration of REGs, are discussed in detail.

Keywords Renewable energy generation · Distribution system · Secure operation · Distribution network planning

R. Li (✉) · W. Wang · L. Xu
National Active Distribution Network Technology Research Center, Beijing Jiaotong University, Beijing 100044, China
e-mail: 13117378@bjtu.edu.cn

W. Wang
e-mail: wwang2@bjtu.edu.cn

L. Xu
e-mail: ljxu1@bjtu.edu.cn

R. Li · W. Wang · L. Xu
Collaborative Innovation Center of Electric Vehicles in Beijing, Beijing 100081, China

M. Li et al. (eds.), *Proceedings of 2015 2nd International Conference on Industrial Economics System and Industrial Security Engineering*,
DOI 10.1007/978-981-287-655-3_12

1 Introduction

In recent years, energy security and environmental concerns have caused many countries to develop REG technology. With economic stimulation and strong support of policies, REGs have become a significant complement to the generation of centralized power.

High permeable integration of REGs in the DS is of significance to the low-carbon and sustainable development of the DS, but it also changes the radiate structure of the network to an interconnected complex one with DGs and loads. As a consequence, the DS is becoming increasing larger in scale and more complicated in its operating characteristics. If there is no guarantee of technology and economy, the impacts of REGs will be threats against the security and stability of DS operation [1, 2].

2 Analyses on Security Impacts of REGs' Integration

2.1 Power Transmission

Power transmission and dispatching in a traditional power system are designed toward single-direction dynamic power flow. If the output power of DGs exceeds load demand, the extra power will reverse to the high voltage level DN, and the bidirectional power flow will traverse the distribution equipment [3].

Because the capacity of existing equipment is selected in accordance with the quality of secure operation, the bidirectional power flow caused by DGs will lead to the off-limit thermal stability of the equipment and increase potential risk, which include severe impacts on the secure operation of the DS.

2.2 Power Quality

The injection of active and reactive power of DGs to the DS will lead to a change of voltage profile in the feeder line, which will cause voltage deviation out of the safe-operation code [4]. The degree of influence is related to the capacity and location of the DG. In addition, the superimposition effect produced by fluctuation of both the loads and DGs may aggravate the degree of voltage deviation.

The DGs, such as photovoltaic (PV) generation and microturbines (MT), are integrated into the DS through the use of inverters. These electronic power devices will inject plenty of harmonic current [5], which will reduce the service life of equipment and disrupt operation of the electrical control equipment and power switches. Meanwhile, the fluctuant power output of REGs will cause the voltage fluctuation and flicker, especially the wind generations (WG) and PVs.

The majority of PVs are connected into a low-voltage DS by single-phase converters, which may cause the problem of three-phase voltage unbalance when they are connected in and uncoordinated fashion [6]. This situation has the risk to set off a chain of negative problems, such as increased network loss, decreased transformer output, and accelerated aging of equipment. In addition, this situation may cause the malfunction and/or the rejection of protection relays.

2.3 *Stability of Frequency*

In the normal state, the active power balance between supply and demand ensures frequency stability. However, the fluctuation and randomness of REG output have negative effects on the frequency stability of the system.

When the frequency exceeds the statutory limits because of surplus power output, WGs and PVs must be disconnected immediately to maintain stability. Afterward, these DGs will be connected again. However, the of DGs taking off centrally have a potential risk that the frequency adjustment of the system may not be able to meet the loss of power output. Meanwhile, the whole process is difficult to monitor and control, which may cause frequency instability.

2.4 *Island Effects*

With the increasing use of DGs, the emergency unintentional isolated mode will cause plenty of negative problems as follows:

1. Risk of personal injury: When "islanding" occurs, utility workers may be injured due to neglect of the DGs' continuing power supply.
2. Damage to customers' devices: When islanding occurs, customers' devices may be damaged by fluctuations in the power output of DGs or the emergence of the overload condition, etc.
3. Clean up of failures: After failure clean-up, the integration of DGs may cause a current surge to the system due to unsynchronized phase position.

2.5 *Protection and Control*

Because the capacities and locations of DGs vary, their impacts of DGs n the direction and levels of fault currents will differ [7, 8]. These situations may cause the malfunction and/or rejection of protection relays and even the extension of the fault and part of the DS blackout emergency.

Therefore, the impacts on protection are varied, which leads to difficulties of setting protection strategies and tuning parameters. Furthermore, when instantaneous failure occurs, the sustained current provided by DGs may lead to auto-reclosure failure and electrical security accidents.

2.6 Planning and Design

REG-based DGs have the characteristics of being random, intermittent, and unpredictable, the impacts of which on network planning and design render difficult the planning of generation expansion, DN topology, etc.

First, the randomness and volatility characters of REG power output bring great difficulty to the accurate prediction of power generation. Meanwhile, different integration locations of REGs will impact the secure and economic level of operation, which lends difficulty to the optimal planning of DG location and capacity.

DN topology planning is a combinatorial-optimization problem that involves many variables and constraints. Due to uncertainty factors inherent in DGs, the mathematical complexity of the planning model and the difficulty of the solution's algorithm both increase.

3 Suggestions Regarding Security Impacts of REG Integration

3.1 Formulation of a Comprehensive Guide or Code

Due to the series of adverse impacts of DGs, a uniform and comprehensive guide or code related to security and stability for integration of DGs should be formulated. From the point of DGs, in addition to codes about general requirements, power quality, and grid-connection test, the requirements of power control, ancillary service and the fault ride through (FRT) must be considered. Meanwhile, from the point of the power system, considering the integration of DGs, the guides for the security and stability of system operation, such as stability computation, planning and design, and operation control, should be revised and improved.

3.2 Improvement of Stability, Regulation and Control of REGs

As previously mentioned, the characteristics of REGs power output are the key problem to secure operation. Therefore, improvement of stability, regulation, and control is the prerequisite for DS and REGs [9].

1. Improve the stability of REG output: In an attempt to dampen the fluctuation of REG output, using the complementary of multiple energy features or energy storage system (ESS), the integrated generations—such as hydro-wind generation and integrated generation of WG, PV, and ESS—should be established.
2. Improve the regulation and control of REGs output: REGs should be accepted into the Supervisor Control And Data Acquisition (SCADA) and the Energy Manage System (EMS). By this means, the technology functions of REGs, such as the generation of forecasting, state estimation, on-line monitoring and diagnosis, and optimal dispatch, could be implemented.

3.3 Optimization of the Distribution Network Structure

From the point of DS, optimization of the structure is an effective way to improve the DN's capability of accommodating REGs and ensure secure operation. In view of the advantages of uninterruptible power supply capacity and power quality, the closed-loop distribution network should be adapted gradually, and the network structure, operation and protection schemes for closed-loop DN with DGs should be studied. Meanwhile, to improve the interconnected levels and construct a clear network structure, the interconnection switches should be increased in number and be optimized in their locations.

3.4 Formulation of a Coordinating Protection Strategy

To solve the coordinating operation problem of DS and islanding, a coordinating and comprehensive protection strategy should be formulated. From the viewpoint of islanding, the following three key issues should be resolved:

- How should reasonable active splitting, according to the capacities of DGs and loads, be realized?
- How should secure operation and control of unintentional islanding be realized?
- How should the secure reconnection of islanding to the DS be realized after failure clean-up? [10].

From the viewpoint of the distribution system, new protection strategy and parameters should be formulated based on intentional islanding as well as unintentional islanding to ensure consistency between islanding and the system.

3.5 Research of Source-Grid-Load Coordinating Planning

Power Forecasting and Generation Expansion Planning
The output model of DGs should be proposed considering the natural conditions, geographical conditions, and device-tripping possibility, etc. Then the credible capacity of the DGs should be evaluated and the results used to configure the capacity. Meanwhile, considering the integrated generations of multiple energy generations, the supply and storage character of integrated generations should be used as a whole.

Demand Response (DR) and Load Forecasting
To begin, the total maximum load should be forecasted according to the historical data, natural climate conditions, population and economy data, etc. Then, according the characteristics of the different loads, the FLs should be distinguished. Subsequently, the math model of the FLs is created based on the characteristics of FLs to guidance mechanisms. Finally, the effect degrees of DR on FLs are evaluated, and the result of the load forecasting based on the DR can be obtained.

Distribution Network Topology Planning
A variety of uncertain factors are introduced in the process of planning with REG-based DGs, FLs, and ESS. These must be resolved using the method of probability statistics or fuzzy mathematics [11].

In an attempt to consider operating cost of the distribution system comprehensively, the operation costs of DGs, FLs, and power purchased from main grid should be analyzed separately. Based on the analyses, reasonable capacity configurations of the three items are formulated. In this process, to promote the efficient use of REGs, a proportion of the REGs should be improved to the greatest extent possible under the constraint conditions of secure operation of the system.

In addition to the common constraint conditions—such as power balance, bus voltage, and network connectivity—the constraint conditions introduced or changed by the integration of DGs, such as the DG output, state of charge (SOC) of the ESS, and reliability of the system, should be taken into account.

In general, due to the impacts of REG integration and the interaction demand of FLs, distribution-systems planning should abandon passive management to the REGs and FLs, and the source-grid-load coordinating planning method should be considered.

4 Conclusion

The structure and operational scheme of the DS are changed by the integration of REG-based DGs. The impacts of REG-based DGs to the secure operation of the distribution system are concentrated on six key issues: power transmission, power quality, frequency stability, islanding effects, protection, and power system planning.

Based on the analyses of the six key issues, corresponding suggestions are proposed. These suggestions provide an approach to solving these issues to ensure the integration security of DGs and the stability of operation of the distribution system.

Acknowledgments This work is supported by the Fundamental Research Funds for the Central Universities (2015YJS160).

References

1. Djapic P, Ramsay C, Pudjianto D (2007) Taking an active approach. IEEE Power Energ Mag 5(4):68–77
2. Wang DT-C, Ochoa L, Harrison G (2010) DG impact on investment deferral: Network planning and security of supply. IEEE Trans Power Syst 25(2):1134–1141
3. Fan M, Zhang Z, A Su et al (2013) Enabling technologies for active distribution systems. Proc CSEE 33(22):12–18
4. Pei W, Sheng K, Kong L et al (2008) Impact and improvement of distributed generation on distribution network voltage quality. Proc CSEE 28(13):152–157
5. Ladjavardi M, Masoum MAS, Islam S (2008) Impact of a SG nonlinear model on the harmonic distortion of a distribution generation system. In: 2008 Australasian universities power engineering conference (AUPEC08), Sydney
6. Liu K, He K, Sheng W (2014) A Cooperative particle swarm optimization based state Eestimation of unbalanced three-phase distribution network. Power Syst Technol 4:1026–1031
7. Zhou W, Zhang Y, Xia C-J et al (2010) Effect of distributed generation on relay protection of distributed system. Power Syst Prot Control. 38(3):1–5
8. Sukumar B, Adly D (2006) Overview of the effect of distributed generation on protection in power distribution systems. In: IASTED 6th international conference on European power and energy systems, EuroPES 2006, Rhodes
9. Xue C, Li C, Cao Y et al (2011) An overview and prospects of grid friendly technology in smart grid. Autom Electr Power Syst 35(15):102–107
10. Ding L, Pan Z, SU Y et al (2007) Splitting and islanding of networked dispersed generators. Electr Power Autom Equip 27(7):25–29
11. Jianhua Z, Bo Z, Yuying Z et al. (2014) Key issues and research prospects of active distribution network planning. Trans China Electro Tech Soc 29(2):13–23

Distinguishing Chinese Grain Safety from the Angle of Economy Ecologization

Menggang Li and Jicheng Zhang

Abstract Through analyzing and evaluating the safety level of the grain industry in terms of system equilibrium, stability, adaptability, and fluency, problems of the grain safety were identified from the angle of economy ecologization and feasible suggestions were put forward to strengthen national resources security, promote coordination in dynamic mechanism, and improve comprehensive decision-making in grain development.

Keywords Economy ecologization · Grain safety · System · Resource

1 Identification of Food Security Issues from the Perspective of Economy Ecologization

1.1 Identification of Food Security Issues in China from the Perspective of System Adaptability

The suitability of an agro-ecological economic complex system is principally dependent on whether or not development of the economic system suits sustenance of the ecosystem and whether development of economic system endangers or undermines the self-healing and self-sustaining capacities of the ecosystem. The basic requirement of the suitability of an agro-ecological economic complex system is to closely restrict economic development within the "threshold" of the agricultural resource-carrying and environmental capacities.

M. Li (✉) · J. Zhang
China Center for Industrial Security, Research, Beijing Jiaotong University, Beijing 10044, China
e-mail: 736194775@qq.com; zhugezedong@163.com

J. Zhang
e-mail: zhengkangning@163.com

M. Li et al. (eds.), *Proceedings of 2015 2nd International Conference on Industrial Economics System and Industrial Security Engineering*,
DOI 10.1007/978-981-287-655-3_13

(1) Limited agricultural resource-carrying capacity.
China suffers from a relative shortage of agricultural resources: its cultivated land area decreased by 123 million Mu during the period from 1997 to 2009; the current per-capita area of cultivated farmland is only 1.4 Mu; the area of low-yielding land accounts for approximately one third of total cultivated land area; the per-capita water resource in China is restricted to one quarter of the world's average and features uneven spatio-temporal distribution; and the water deficiency for agriculture is more than 30 billion cubic meters.
(2) Agricultural non point source pollution impairs grain production capability.
Fertilizer input to farmland in China has increased over the years to improve the per-unit-area yield of grain. In 2009, the fertilizer input reached 54.044 million tons, and the average fertilizer input reached 444 kg/ha, at which point China became the world's biggest largest manufacturer and consumer of fertilizer. Excessive fertilizer application is likely to result in a decrease of marginal investment return. The grain output increased by 100 million tons while the fertilizer input increased by 96 % in 1984 compared with 1978; while the grain output increased by 100 million tons in 1998 compared with the year 1984, the fertilizer input increased by 266 %. In addition, excessive application of fertilizers leads to water pollution and soil degradation. The fertilizer input in China is far in excess of 225 kg/ha, i.e., the security limit recognized by the world for preventing water pollution.
(3) Improper use of agricultural wastes hampers grain development potential.
Along with the ever-increasing input of exogenous substances such as fertilizer and pesticide, more than 4 billion tons of agricultural wastes, including more then 700 million tons of crop straws and more then 3 billion tons of animal manure, fail to be properly used every year in China. The COD from animal manure is 5 times that from industrial and domestic sewage throughout China; the contribution of non point source pollution of agriculture to the pollution-induced water shortage in China is an extremely troubling issue. The failure to return waste resources fundamentally derived from farmland ecosystem to farmland ecosystem in a reasonable manner is bound to bring about a decrease in organic matter content and soil degradation in cultivated land, cause artificial interruption of the substance and energy circulation chain in the farmland ecosystem, and impair the maximum production capacity of the land, thereby impairing the foundation of China's grain security. In addition, urban and rural household garbage and industrial waste lead to pollution of the agroecological system to a greater or lesser degree. At present, China's agricultural resource-carrying capacity and environmental capacity cannot afford to withstand the high-intensity resource consumption and environmental pollution in traditional economic form.

In a word, the principal issue of food security facing China in terms of system suitability is how to save natural resources, improve resource-use efficiency, minimize input of exogenous substances, make rational use of agricultural wastes, reduce urban and rural domestic pollution as well as industrial pollution, and use limited agricultural resources and environmental capacity to meet the requirements for the security of food quantity and quality.

1.2 Identification of Food Security Issues in China from the Perspective of System Stability

The stability of the agro-ecological economic complex system finds expression in whether or not the system can operate without failure and avoid or alleviate possible fluctuation especially in the case of remarkable natural changes of the system: This is an important guarantee for a proper cycle of the agricultural ecoeconomic system. According to ecological principles, the maturity and complexity of the ecological chain, which is composed of producers, consumers, and decomposers, is directly proportional to the level of stability of the agroecological system. From the perspective of economic ecologicalization, different sectors must act as producers, consumers, and decomposers to assure the sustainable operation of the agricultural economic system. The level of their coevolution capacity is directly proportional to the level of stability of the agricultural economic system.

As the unity of natural and economic reproduction, food production is inevitably confronted with natural and market risks that affect the stability of the agricultural Ecoeconomic system.

(1) Inadequate ability of agriculture to withstand natural risks impairs the foundation of food security.
The natural risk of food production is principally rooted in two aspects: First, extreme climate change brings about natural disasters on food production. Experts predicted that China would suffer from more frequent extreme climate change in the 21st century, which would bring about increasingly serious natural risks to food production as well as huge concerns about food security. The effective measure against natural disaster is to reinforce agricultural-infrastructure development and turn cultivated land into standard farmland with "well-developed irrigation system, drainage system, ditch system and wood network." China currently has an irrigated area of only 928 million Mu, including an effective irrigated area of 868 million Mu, which accounts for less than 50 % of China's cultivated land area; there exist more than 900 million Mu of cultivated land that lack irrigation. Another natural risk to China's food production is the vulnerable farmland ecosystem: The long-term excessive application of such exogenous substances as fertilizer, pesticides, and mulching film, as well as failure to return organic matter such as straws and manure to fields, results in low content of soil organic matter, decrease of beneficial microflora, soil hardening, aggravation of plant diseases and insect pests, and other farmland ecological crises.

(2) Grain market risk amplifies the erratic factor for food security.
The market risk to food production principally roots in three aspects. First, the low comparative revenue from food production leads to low enthusiasm of grain farmers to grow grain. The principal cause is that the absolute price of grain is low, and the margin of increased grain price behind that of production means. The price of production means in China has been rising at a double-digit rate since 2007, and the net income per Mu of wheat decreased from 471 Yuan in 2010 to 465 Yuan in 2011. Moreover, the low grain-production efficiency directly results in continuous deterioration in the capacity of grain cultivation to attract labor: Only 150 million people (the proportion of old people is up to 60 %) have been engaged in the first industry in China since 2009, which has lead to increased employment costs of food production and more difficult promotion of agricultural technologies.

Second, food processing and circulation haves not been fully developed in China, and the industrial scale and level are on the low side. Compared with food production, grain circulation and processing is the most important link of the industrial chain of grain for improving the comprehensive income from grain; for example, compared with corn production, the benefit from corn processed could increase by, e.g., 1.5 times in case of starch production, 3 times in case of sorbitol production, 5 times in case of vitamin production, and more than 30 times in case of lysine production. However, because China has no sophisticated grain-circulation system and lacks the potential to profoundly tap its grain-processing capacities, the system cannot effectively motivate and indicate the planting and processing segments.

Third, there is no long-term benefit-linking mechanism between subjects of the grain industrial chain. With respect to the made-to-order agriculture generally adopted in China, the mutual suspicion among subjects such as grain farmer, processing company, and intermediary organization leads to a high occurrence rate of breach of contract. For example, the peasants collectively violated the contract at the pilot of the made-to-order agriculture in Pingdong Grain Depot, Siping, Jilin Province, in 2006. Despite this, the stable strategic alliance between the subjects in the industrial chain of grain may after all be accepted as an effective measure for improving the economic benefit for subjects and their enthusiasm for the grain industry, responding to the speculation at the grain market by domestic and foreign capitals, and assuring the long-term stability and sustainable development of China's grain security.

To sum up, the principal issue to be addressed for China's grain security in terms of system stability is how to build and improve balanced, coordinated, and sustainable agricultural ecological and economic ecological chains to improve the ability of the grain sector to carry labor and to enhance the ability of the grain sector to deal with natural as well as market risks.

2 The Feasible Solution for Improvement in China's Grain Security in Context of the Principle of Economic Ecologicalization

2.1 *Enhance the National-Resources Security Concept and Consolidate the Foundation for Grain Security*

From the perspective of the main structure of China's demand for grain, the principal pressure facing China with respect to grain supply is to meet the growing demand for feed grain; matter-of-factly, many grains are used to feed herbivores from which the demand for meat, eggs, and milk, etc., meet residents' demand for food. Provided that artificial pasture is developed on some farmland for planting feed grain, it is possible to produce more biomass and much more feed protein compared with grain as well as to conserve water resources and improve soil to a certain extent. Hence, artificial pasture could replace feed grain planting in the regions suitable for artificial pasture development to assure resource, grain, and food security in China.

In some sense, to assure China's resource security means to assure the agricultural ecological environmental security because most agricultural ecological environmental issues are the result of irrational use of resources. The fundamental way in which to improve China's resource security is to minimize resource input, improve resource use, and assure the sustainable use of resources. The principal way to achieve this is to make active use of agricultural wastes and return wastes from the agricultural ecological system back to such in a harmless manner. In addition, it is important to enhance farmland-infrastructure development, especially water infrastructure, to improve China's resource use and its ability to withstand natural disasters.

2.2 *Stimulate Intersystem Dynamic Mechanism Collaboration and Improve the Sustainability of Grain Security*

It is possible to improve the sustainability of grain security only when the grain industry meets the requirements for equilibrium, suitability, stability, and smoothness of the proper cycle of the ecological economic system. The key to a proper cycle is to assure the balanced, coordinated, and sustainable operation of the ecological chain and the economic chain of the grain sector through coevolution of dynamic mechanism between the ecological and economic systems.

First, it is important to extend and improve the agro-ecological economic chain. Within the agriculture economical system, it is advisable to return organic matter to the farmland in a harmless manner for the purpose of soil conservation and improvement, promote resource-saving agricultural technologies and modes such as ecological and circular agriculture, improve the resource-carrying capacity and environmental quality of the agricultural ecological system, assure the smooth flow of substances and energy in the farmland ecosystem, reinforce the stability of the soil ecosystem, and maximize the productivity of cultivated land. Within the agricultural economic system, it is advisable to stimulate labor-intensive agricultural industry, deepen the supply of staple food products and other high value added grain products, and comprehensively improve the ability to meet the demand of grain industry market and the labor-carrying capacity based on local conditions to improve the labor-carrying capacity and meet consumers' diverse demands for food.

Second, it is important to give play to the key role of information flow in directing the material, energy, and value flows as well as that of capitals and technologies based on the variation trend of market demand through informatization development. On one hand, it is advisable to convert information on the resource environment to price signals acceptable by market and change consumers' consumption behavior; in contrast, it is essential to transfer the information on price and quality of agricultural products to different agriculture-related subjects, change and adjust producer behavior through a benefit mechanism of the economic system, improve the motivation and capacity of economic-environment optimization, and assure benign operation of the dynamic mechanism of the ecological system through conducting effects to achieve two-way optimization of the environment and the economy. Informatization development should focus on information collection, analysis, and conduction and give full play to specific advantages of government, business, futures market, universities and colleges, and scientific research institutes with respect to informatization and perform rational division of labor so that each component performs its own functions and seeks positive interaction through collaboration.

2.3 Perfect the Integrated Decision-Making System for Grain Development and Improve the Level of Grain-Security Management

Because grain security involves such stakeholders as government, business, consumer, and farmer, it is important to make institutional arrangements that fully mobilize the enthusiasm of all stakeholders. From a macroscopic point of view, it is advisable to establish a special institution called the National Commission on Food Security for unified coordination and guidance for grain and other primary foods obtained from animals and plants throughout the entire industrial chain and the whole process; from a mesoscopic point of view, it is advisable to continuously cultivate and develop, such intermediary organizations as farmers' specialized cooperative economic organization, circulation economic organizations, and associations of processing businesses, and give play to their role as the bridge and link between government, business, and farmer. From the microcosmic point of view, it is advisable to give full advantage to different roles of farmer, circulation business, processing business, trading market, and scientific research institutions, etc., in grain and food security, integrate the efforts of all parties, give play to synergy, and promote industrial self-regulation and-management to create a good governance-based governing structure in the macro, meso, and micro respects and to comprehensively improve the competitiveness and sustainability of China's grain and food security.

References

1. Menggang L (2009) Establish a new concept of food security and protect China's grain security [J]. China Natl Cond Strength 11:6–10
2. Jikun H (2010) A study on frontier issues of agricultural economy and policies from the perspective of supply of and demand for agricultural products [J]. Econ Surv 3:1–7
3. Gaoming J (2012) On the future of agriculture in China [J]. Eng Res—Eng Interdiscip Perspect 1:7–9
4. Linfeng L, Menggang L (2012) Analysis of principal factors affecting grain production in China [J]. China Bus Mark 4:109–115
5. Xiangshu Z (2007) Basic assumptions of sustainable development economics and the characteristics of sustainable economic system [J]. J China Univ Geosci 2:27–30
6. Xiaozhen L (2010) Principal channel of grain conversion in China– staple food [J]. Agric Prod Process 5:7–12

A Study of a Dynamic Early Warning Model to Determine Grain Security in China

Menggang Li and Kangning Zheng

Abstract This chapter studies the key indicators of the impact on food security, using DS theory to build China's food-security dynamic early warning model, by calculating the weight of key indicators fusion method. This chapter verifies the validity of the model about China's data from the period 1996 to 2011. The results show that by selecting objective, scientific indicators, the fusion method used can calculate the different years in which the key indicators of weight changed and then obtain the main factors affecting food security in a particular year as well as the previous year to obtain dynamic evaluation data on the food security in the future, which is consistent with the results of the evaluation of historical data over the years, thus proving the feasibility of constructing a dynamic early warning model.

Keywords Food security · DS theory · Indicators fusion · Dynamic evaluation

1 Dynamic Early Warning Model Development

1.1 Evaluation Index Setup

In view of grain security, this chapter interprets factors affecting grain security from the macro, micro, and natural perspectives and indentifies eight indicators—such as grain-production fluctuation rate, grain reserve, and grain price fluctuation rate (Table 1)—that are likely to affect China's grain security.

M. Li (✉) · K. Zheng
Center for Industrial Security Research, Beijing Jiaotong University, Beijing 10044, China
e-mail: 736194775@qq.com; zhugezedong@163.com

M. Li et al. (eds.), *Proceedings of 2015 2nd International Conference on Industrial Economics System and Industrial Security Engineering*,
DOI 10.1007/978-981-287-655-3_14

Table 1 Grain security index system of China

Level of index	Indicators
Macro level	R1: Trend output fluctuation rate
	R2: Grain reserve
	R3: Grain-price fluctuation rate
	R4: Grain self-sufficiency
Micro level	R5: Per-capita grain cultivation area
	R6: Per-capita share of grain
	R7: Engel coefficient
Natural factors	R8: Disaster area

1.2 Dynamic Evaluation Model

(1) Evidence Theory and Information Fusion
Dempster–Shafer (DS) evidence theory is a reasoning method for processing uncertain information. By reason that the previous data necessary for evidence theory is more intuitive and easier to obtain than that for probabilistic reasoning theory, evidence theory could determine obscurity and uncertainty in a more favorable fashion. Furthermore, because the information fusion equation of DS evidence theory combines data from different experts or sources, DS evidence theory is considered a natural and powerful technique for the expression and synthesis of uncertain information and has been extensively applied to information fusion and decision-making analysis.

(2) DS Evidence Theory
DS evidence theory is established in a nonempty set Θ known as a "discernment frame" composed of a limited number of incompatible elements. .

Here, $\forall A \subseteq \Theta$; if the set function m:2Θ[0, 1] meets the following equation:

$$\begin{cases} m(\Theta) = 0 \\ \sum_{A \subseteq \Theta}^{n} m(A) = 1 \end{cases}$$

m could be called the basic probability assignment (BPA) of discernment frame Θ; the BPA reflects the level of support of evidence, i.e., m(A) for A in the discernment frame. Provided that $\forall A \subseteq \Theta$ and $m(A) > 0$, A could be called the focal element.

In practice, the basic strategy of evidence theory is to use two or more pieces of evidence to evaluate the discernment frame, respectively, and then use the Dempster combination rule to combine the basic probability assignment functions that correspond to the evidence. This technique combines the information in each piece of evidence and outputs a new basic probability assignment (the combination rule is expressed by $\oplus$). The equation of Dempster combination rule:

$$m_1(A) \oplus m_2(A) = 1/K \sum_{B \cap C = A}^{n} m_1(B) \oplus m_2(C)$$

where k is a normalization constant, the computing equation of which is given below:

$$K = \sum_{B \cap C = \emptyset}^{n} m_1(B).m_2(C) = 1 - \sum_{B \cap C = \emptyset}^{n} m_1(B).m_2(C)$$

(3) Improvement in Information Fusion Method
Where the Dempster combination rule is used to process highly conflicting information, the result would be contrary to common sense from time to time. Here is an example that shows the Dempster combination method and its drawbacks.

Under the discernment frame Θ = {X, Y, Z}, the mass function is as follows:

$$m1(X) = 0.98\, m1(Y) = 0.02\, m1(Z) = 0$$
$$m2(X) = 0\, m2(Y) = 0.02\, m2(Z) = 0.98$$

The normalization constant

$$K = \sum\nolimits_{B \cap C \neq \emptyset}^{n} m_1(B) \cdot m_2(C) = m1(X) \cdot m2(X) + m1(Y) \cdot m2(Y) + m1(Z) \cdot m2(Z)$$
$$= 0.98 \cdot 0 + 0.02 \cdot 0.02 + 0 \cdot 0.98$$
$$= 0.0004$$

Combination mass function about X

$$m_1(X) \oplus m_2(X) = 1/K \sum_{B \cap C = X}^{n} m_1(B) \oplus m_2(C) = 1/0.0004 \cdot 0.98 \cdot 0 = 0$$

Combination mass function about Y

$$m_1(Y) \oplus m_2(Y) = 1/K \sum\nolimits_{B \cap C = Y}^{n} m_1(B) \oplus m_2(C) = 1/0.0004 \cdot 0.02 \cdot 0.02 = 1$$

Combination mass function about Z

$$m_1(Z) \oplus m_2(Z) = 1/K \sum\nolimits_{B \cap C = Z}^{n} m_1(B) \oplus m_2(C) = 1/0.0004 \cdot 0 \cdot 0.98 = 0$$

Evidence 1 highly supports X, whereas evidence 2 highly supports Z; however, the conclusion is two Ys with an extremely low level of evidence support. It is observed from this example that the conclusion drawn from the Dempster combination rule is contrary to people's cognition in the case of highly conflicting pieces evidence.

Here is a simple and effective improved technique, i.e., Murphy's Dempster combination method, where the basic probability assignment function of evidence is averaged before the information fusion with Dempster combination rule is applied. Where the system comprises n pieces of evidence, the weighted average evidences are combined $n - 1$ times.

The above-noted method could be used for integrated processing of grain-security early warning indicators during two consecutive years in China to obtain a grain security level with dynamic characteristics.

2 Empirical Analysis of Grain Security Early Warning and Evaluation

2.1 Warning Limit Definition

The indicators data are be summarized in Table 2 based on relevant data in Tables 2, 3, 4, 5, 6, 7, 8 and 9 as follows:

Among the indicators, the trend output growth rate R1, the grain reserve R2, the grain self-sufficiency R4, the per-capita share of grain R5, and the per-capita grain cultivation area R6 are positive indicators that are directly proportional to the level of grain security; the Engel coefficient R7 and the disaster area R8 are negative indicators that are inversely proportional to the level of grain security; the grain price fluctuation rate R3 is an immediate indicator, the stability of which facilitates grain security.

This chapter defines the warning limit in an average principle and divides the warning level into levels of security, mild warning, medium warning, serious warning, and insecurity. In the meantime, the maximum and minimum values of each indicator are removed to eliminate the effect of extreme values of indicators that occur in some years from the definition of warning limit. Tables 3 and 4 show the indicator data-processing result and warning-limit definition criteria.

Table 5 shows the warning-limit definition results for indicators obtained based on indicator data in Table 2 and the warning-limit definition criteria in Table 4.

The indicator data of each year could be assigned to the corresponding intervals based on the warning-limit definition results, thereby analyzing the status of grain security.

Table 2 Summary of indicators data of grain security in China during the period 1996 to 2011

Year	R_1: trend output growth rate (%)	R_2: grain reserve (×10000 t)	R_3: grain price fluctuation rate (%)	R_4: grain self-sufficiency (%)	R_5: per-capita share of grain (kg)	R_6: per-capita grain cultivation area (Mu)	R_7: engel coefficient	R_8: disaster area (×1,000 ha)
1996	6.66	14155.1	−0.74	97.96	412.3	1.379	56.3	21234
1997	3.14	18985.2	−10.41	100.30	399.7	1.370	55.1	30307
1998	5.66	21239.4	−2.32	100.39	410.6	1.368	53.4	25181
1999	3.71	24752.3	−2.23	99.97	404.2	1.349	52.6	26734
2000	−6.68	27196.7	−10.26	100.10	364.7	1.284	49.1	34374
2001	−9.48	24859.9	0.79	98.19	354.7	1.247	47.7	31793
2002	−9.41	22726.9	−0.60	100.21	355.8	1.213	46.2	27160
2003	−15.34	18579.2	0.79	99.70	333.3	1.154	45.6	32516
2004	−8.43	12728.9	21.75	94.83	361.2	1.172	47.2	16297
2005	−6.28	12211.3	−0.39	95.49	370.2	1.196	45.5	19966
2006	−4.33	13043.7	0.99	95.04	378.4	1.204	43.0	24632
2007	−4.16	15186.7	1.62	95.14	379.6	1.199	43.1	25064
2008	0.40	17026.7	1.04	93.37	398.0	1.206	43.7	22283
2009	0.28	22640.0	6.45	91.56	397.8	1.225	41.0	21234
2010	2.70	28711.3	3.58	89.49	407.5	1.229	41.1	18538
2011	6.83	38996.2	6.45	90.35	424.0	1.231	40.4	12441

Table 3 Indicator data processing

Indicator	R_1 (%)	R_2	$\|R_3\|$	R_4 (%)	R_5	R_6	R_7	R_8
Maximum	6.66	28711	10.41	100.30	412.3	1.370	55.1	32516
Minimum	−9.48	12729	0.60	90.35	354.7	1.172	41.0	16297
min + p	−5.45	16725	3.06	92.84	369.1	1.222	44.5	20352
min + 2p	−1.42	20720	5.51	95.32	383.5	1.271	48.0	24407
min + 3p	2.61	24716	7.96	97.81	397.9	1.321	51.6	28461

Note (1) p = (maximum–minimum)/4
(2) Indicator R3 could be taken as a negative indicator after its absolute value is obtained

Table 4 Warning-limit definition criteria

Positive indicator		Negative indicator	
Minimum value and smaller values	Insecurity	(Minimum value and smaller values min + p)	Security
(Minimum value min + p)	Serious warning	[min + p min + 2p)	Mild warning
(min + p min + 2p]	Medium warning	[min + 2p min + 3p)	Medium warning
(min + 2p min + 3p]	Mild warning	[min + 3p maximum value)	Serious warning
(min + 3p Maximum value and larger values)	Security	Maximum value and larger values	Insecurity

Table 5 Indicator warning-limit definition results

Warning limit	Security	Mild warning	Medium warning	Serious warning	Insecurity
R_1	>2.61 %	(−1.42 % 2.61 %]	(−5.45 % −1.42 %]	(−9.48 % −5.45 %]	≤−9.48 %
R_2	>24716	(20720 24716]	(16725 20720]	(12729 16725]	≤12729
R_3	<3.06 %	[3.06 % 5.51 %)	[5.51 % 7.69 %)	[7.69 % 10.41 %)	≥10.41 %
R_4	>97.81 %	(95.32 % 97.81 %]	(92.84 % 5.32 %]	(90.35 % 92.84 %]	≤90.35 %
R_5	>397.9	(383.5 397.9]	(369.1 383.5]	(354.7 369.1]	≤354.7
R_6	>1.321	(1.271 1.321]	(1.222 1.271]	(1.172 1.222]	≤1.172
R_7	<44.5	[44.5 48.0)	[48.0 51.6)	[51.6 55.1)	≥55.1
R_8	<20352	[20352 24407)	[24407 28461)	[28461 32516)	≥32516

Table 6 Indicator normalization results during 1996 to 2011

Indicator	R_1	R_2	R_3	R_4	R_5	R_6	R_7	R_8
1996	0.9922	0.0726	0.9837	0.7769	0.8705	1.0000	0.0000	0.5986
1997	0.8336	0.2529	0.5310	0.9918	0.7321	0.9595	0.0755	0.1844
1998	0.9474	0.3371	0.9097	1.0000	0.8525	0.9514	0.1824	0.4184
1999	0.8592	0.4682	0.9138	0.9619	0.7814	0.8687	0.2327	0.3475
2000	0.3906	0.5595	0.5380	0.9731	0.3457	0.5762	0.4528	0.0000
2001	0.2645	0.4722	0.9811	0.7980	0.2355	0.4123	0.5409	0.1166
2002	0.2677	0.3926	0.9899	0.9839	0.2483	0.2630	0.6352	0.3281
2003	0.0000	0.2377	0.9812	0.9368	−0.0001	0.0000	0.6730	0.0836
2004	0.3116	0.0193	0.0000	0.4899	0.3072	0.0822	0.5723	0.8240
2005	0.4087	0.0000	1.0000	0.5505	0.4064	0.1878	0.6792	0.6565
2006	0.4968	0.0311	0.9721	0.5090	0.4978	0.2212	0.8365	0.4435
2007	0.5043	0.1111	0.9423	0.5186	0.5100	0.2012	0.8302	0.4238
2008	0.7100	0.1798	0.9696	0.3561	0.7129	0.2321	0.7943	0.5507
2009	0.7047	0.3894	0.7165	0.1897	0.7108	0.3156	0.9643	0.5986
2010	0.8139	0.6160	0.8506	0.0000	0.8186	0.3339	0.9563	0.7217
2011	1.0000	1.0000	0.7162	0.0787	0.9995	0.3422	1.0000	1.0000

2.2 *Determination of Indicator Weight*

(1) Indicator normalization

Where indicator weight is determined by the entropy weight method, the indicator data should be normalized to eliminate data dimension and consolidate the variation range of indicators. The specific method is as follows:

Equation 1 is used for a positive indicator:

$$r_{ij}^{i} = \frac{r_{ij} - \min(r_{ij})}{\max(r_{ij}) - \min(r_{ij})} \tag{1}$$

Equation 2 is used for a negative indicator:

$$r_{ij}^{i} = \frac{\max(r_{ij}) - r_{ij}}{\max(r_{ij}) - \min(r_{ij})} \tag{2}$$

where are, respectively, the maximum and minimum values of indicator, while $\max(r_{ij}), \min(r_{ij})$ is the result of indicator normalization.

Table 6 shows the indicator normalization results.

Table 7 Indicator weight

Computational item	R_1	R_2	R_3	R_4	R_5	R_6	R_7	R_8
$\Sigma P_{ij} \times \ln(P_{ij})$	−2.6192	−2.4250	−2.6901	−2.5879	−2.6212	−2.4933	−2.5869	−2.5638
e_j	0.9447	0.8746	0.9702	0.9334	0.9454	0.8993	0.9330	0.9247
1- e_j	0.0553	0.1254	0.0298	0.0666	0.0546	0.1007	0.0670	0.0753
Entropy weight	0.096	0.218	0.052	0.116	0.095	0.175	0.117	0.131

(2) Indicator Weight Computation
The indicator weight can be determines based on the data in Table 5 using the weight-computing method presented. The computation procedure and results are shown in Table 7.

2.3 *Indicator Fusion*

(1) Basic Probability-Assignment Function.
The generation of the basic probability-assignment function is a prior condition for the use of the DS information-fusion method. The basic probability-assignment function could be generated based on the warning-limit definition and the data and weights of indicators in the following steps:

1. First, based on the definition of the five warning limits (i.e., security, mild warning, medium warning, serious warning, and insecurity), set up elements X, Y, and Z as the focal elements of the basic probability-assignment function, and correspond them, respectively, to the three warning limits, i.e., security, medium warning, and serious warning. In the meantime, correspond the "mild warning" between the "security" and "medium warning" limits to XY; correspond the "serious warning" limit between the "medium warning" and "insecurity" limits to YZ; and define:

$$\begin{aligned} &m(X) \cap m(XY) = m(X) \cdot m(Y) \cap m(XY) = m(Y) \\ &m(Y) \cap m(YZ) = m(Y) \cdot m(Z) \cap m(YZ) = m(Z) \\ &m(XY) \cap m(YZ) = m(Y) \end{aligned}$$

The purpose is to reinforce the connection between indicators during information fusion, and

$$m(X) \cap m(Y) = \emptyset \, m(XY) \cap m(Z) = \emptyset$$

The rest may be deduced by analogy.

Hence, the correspondence between warning limit and focal element is as shown in Table 8.

Table 8 Correspondence between warning limit and focal element

Warning limit	Security	Mild warning	Medium warning	Serious warning	Insecurity
Focal element	m(X)	m(XY)	m(Y)	(YZ)	m(Z)

Table 9 Correspondence between grain-security indicators and focal elements during the period 1996 to 2011

Year	R_1	R_2	R_3	R_4	R_5	R_6	R_7	R_8
1996	m(X)	m(YZ)	m(X)	m(X)	m(X)	m(X)	m(Z)	m(XY)
1997	m(X)	m(Y)	m(Z)	m(X)	m(X)	m(X)	m(Z)	m(YZ)
1998	m(X)	m(XY)	m(X)	m(X)	m(X)	m(X)	m(YZ)	m(Y)
1999	m(X)	m(X)	m(X)	m(X)	m(X)	m(X)	m(YZ)	m(Y)
2000	m(YZ)	m(X)	m(YZ)	m(X)	m(YZ)	m(XY)	m(Y)	m(Z)
2001	m(Z)	m(X)	m(X)	m(X)	m(Z)	m(Y)	m(XY)	m(YZ)
2002	m(YZ)	m(XY)	m(X)	m(X)	m(YZ)	m(YZ)	m(XY)	m(Y)
2003	m(Z)	m(Y)	m(X)	m(X)	m(Z)	m(Z)	m(XY)	m(Z)
2004	m(YZ)	m(Z)	m(Z)	m(Y)	m(YZ)	m(Z)	m(XY)	m(X)
2005	m(YZ)	m(Z)	m(X)	m(XY)	m(Y)	m(YZ)	m(XY)	m(X)
2006	m(Y)	m(YZ)	m(X)	m(Y)	m(Y)	m(YZ)	m(X)	m(Y)
2007	m(Y)	m(YZ)	m(X)	m(Y)	m(Y)	m(YZ)	m(X)	m(Y)
2008	m(XY)	m(Y)	m(X)	m(Y)	m(X)	m(YZ)	m(X)	m(XY)
2009	m(XY)	m(XY)	m(Y)	m(YZ)	m(XY)	m(Y)	m(X)	m(XY)
2010	m(X)	m(X)	m(XY)	m(Z)	m(X)	m(Y	m(X)	m(X)
2011	m(X)	m(X)	m(Y)	m(Z)	m(X)	m(Y)	m(X)	m(X)

Second, the focal element that corresponds to each indicator of every year can be determined as shown in Table 8 based on the indicator data in Table 2, the warning-limit definition criteria in Table 4, and the correspondence between warning limit and focal element in Table 9.

Third, generate the basic probability-assignment function based on the indicator weight in Table 6 and the correspondence between indicators and focal elements of each year shown in Table 8. The generation method for basic probability-assignment function is described by taking year 1996 as an example:

$$
\begin{aligned}
m(X) &= 0.096 + 0.052 + 0.116 + 0.095 + 0.175 = 0.534 \\
m(XY) &= 0.131 \\
m(Y) &= 0 \\
m(YZ) &= 0.218 \\
m(Z) &= 0.117
\end{aligned}
$$

In this way, the basic probability assignment function for year 1996 should be: m(X) = 0.534; m(XY) = 0.131; m(Y) = 0; m(YZ) = 0.218; m(Z) = 0.117.

As shown in Table 10, the basic probability-assignment function can be generated for other years in the same way through computation.

(2) Fusion Method

The basic probability-assignment function is weighted based on the credibility of evidence to avoid the defect of the DS information-fusion method in processing highly conflicting data.

Because this chapter evaluates grain security based on the combination of data from the previous year and the current year, the weighted mean of these data should be determined. In consideration of the fact that grain security is principally affected by current-year data, the current-year data's degree of confidence is set to 0.8, whereas the previous-year data's degree of confidence is set to 0.2. The weighting method for basic probability assignment function is described by taking year 1997 as an example.

$$\text{Weight m(X)1997} = \text{m(X)1997} \times 0.8 + \text{m(X)1996} \times 0.2 = 0.4924.$$

As shown in Table 11, the weighted basic probability-assignment function can be generated for the period 1997 to 2011 in the same way through computation.

Second, fuse the data shown in Table 10 using the Dempster combination rule. Because there are two pieces of evidence in the system, the weighted basic probability-assignment function is autonomously fused for "n − 1" times (i.e., one time). The data fusion data are shown in Table 12.

Take the data fusion for the year 1997 for example. The fusion result for each focal element is m(X) = 0.5803, m(XY) = 0.0015, m(Y) = 0.2144, m(YZ) = 0.0476, and m(Z) = 0.1562; they respectively represent the probability of warning limit that

Table 10 Basic probability-assignment function of grain security for the period 1996 to 2011

Year	m(X)	m(XY)	m(Y)	m(YZ)	m(Z)
1996	0.534	0.131	0	0.218	0.117
1997	0.482	0	0.218	0.131	0.169
1998	0.534	0.218	0.131	0.117	0
1999	0.752	0	0.131	0.117	0
2000	0.334	0.175	0.117	0.243	0.131
2001	0.386	0.117	0.175	0.131	0.191
2002	0.168	0.335	0.131	0.366	0
2003	0.168	0.117	0.218	0	0.497
2004	0.131	0.117	0.116	0.191	0.445
2005	0.183	0.233	0.095	0.271	0.218
2006	0.169	0	0.438	0.393	0
2007	0.169	0	0.438	0.393	0
2008	0.264	0.227	0.334	0.175	0
2009	0.117	0.54	0.227	0.116	0
2010	0.657	0.052	0.175	0	0.116
2011	0.657	0	0.227	0	0.116

Table 11 Weighted basic probability-assignment function for the period 1996 to 2011

Year	m(a)	m(b)	m(c)	m(d)	m(e)
1997	0.4924	0.0262	0.1744	0.1484	0.1586
1998	0.5236	0.1744	0.1484	0.1198	0.0338
1999	0.7084	0.0436	0.131	0.117	0
2000	0.4176	0.14	0.1198	0.2178	0.1048
2001	0.3756	0.1286	0.1634	0.1534	0.179
2002	0.2116	0.2914	0.1398	0.319	0.0382
2003	0.168	0.1606	0.2006	0.0732	0.3976
2004	0.1384	0.117	0.1364	0.1528	0.4554
2005	0.1726	0.2098	0.0992	0.255	0.2634
2006	0.1718	0.0466	0.3694	0.3686	0.0436
2007	0.169	0	0.438	0.393	0
2008	0.245	0.1816	0.3548	0.2186	0
2009	0.1464	0.4774	0.2484	0.1278	0
2010	0.549	0.1496	0.1854	0.0232	0.0928
2011	0.657	0.0104	0.2166	0	0.116

Table 12 Results of autonomous fusion of weighted basic probability-assignment function for the period 1997 to 2011

Year	K value	m(X)	m (XY)	m(Y)	m(YZ)	m(Z)	Main focal element	Warning level determination
1997	0.4623	0.5803	0.0015	0.2144	0.0476	0.1562	m(X)	Security
1998	0.6619	0.6901	0.0459	0.2283	0.0217	0.0140	m(X)	Security
1999	0.6486	0.8689	0.0029	0.1071	0.0211	0.0000	m(X)	Security
2000	0.5761	0.5057	0.0340	0.2796	0.0823	0.0983	m(X)	Security
2001	0.5230	0.4544	0.0316	0.3027	0.0450	0.1663	m(X)	Security
2002	0.7567	0.2221	0.1122	0.4970	0.1345	0.0341	m(Y)	Medium warning
2003	0.4872	0.1687	0.0529	0.3234	0.0110	0.4440	m(Z)	Insecurity
2004	0.5631	0.0915	0.0243	0.2272	0.0415	0.6155	m(Z)	Insecurity
2005	0.6240	0.1638	0.0705	0.3350	0.1042	0.3265	m(Y)	Medium warning
2006	0.6952	0.0655	0.0031	0.6870	0.1954	0.0490	m(Y)	Medium warning
2007	0.7191	0.0397	0.0000	0.7455	0.2148	0.0000	m(Y)	Medium warning
2008	0.7190	0.2072	0.0459	0.6804	0.0665	0.0000	m(Y)	Medium warning
2009	0.8898	0.1812	0.2561	0.5444	0.0184	0.0000	m(Y)	Medium warning
2010	0.6069	0.7673	0.0369	0.1737	0.0009	0.0213	m(X)	Security
2011	0.5103	0.8727	0.0002	0.1008	0.0000	0.0264	m(X)	Security

corresponds to each of the five focal elements. M(X) was the largest value in year 1997 indicating the highest probability of grain security, so the main focal element is defined as m(X), based on which the level of warning should be "security." The rest may be deduced by analogy.

3 Evaluation Result Analysis

As a whole, the early warning analysis result in this chapter is in a position to reflect the grain security status in recent years, which indicates excellent operability of the early warning model. First, the key evaluation indicators are selected more objectively by summarizing relevant evaluation indicators based on references and expert opinions; moreover, the indicators with implicit correlation are removed to obtain a more accurate evaluation result. Second, this model enables more objective index weight calculation by taking into account the change of index weight for key affecting factors in different years. The indicator-fusion method used is in a position to determine changes in weight values of a certain indicator in different years, thereby identifying the key factors threatening grain security in different years to take appropriate measures. For instance, the decline in grain production was the major factor that led to the decline of the grain security level at the beginning of this century; the sharp increase in grain import volume and the decrease of grain self-sufficiency in China have attracted increasing attention in recent years; hence, it is important to make a difference between threats. Last, this early warning model exhibits certain dynamic characteristics. The model could be used for the dynamic prediction of the ensuing year's level of grain security by combining related data of the previous year or years. Therefore, this dynamic prediction technique is more rational compared with traditional static prediction models for grain security.

References

1. Bin X (2003) Research and system realization of food security assessment based on grey model. Comput Comput Technol Agric 1:703–711
2. Christiaensen LJ, Boisvert RN, Hoddinott J (2000) Validating operational food insecurity indicators against a dynamic benchmark: evidence from Mali. World Bank Policy Research Working Paper, p 2471
3. Chung K, Haddad L, Ramakrishna J et al (1997) Identifying the food insecure: the application of mixed method approaches in India. IFPRI, Washington
4. Doocy S, Teferra S, Norell D, Burnham G (2005) Credit program outcomes: coping capacity and nutritional status in the food insecure context of Ethiopia. Soc Sci Med 60:2371–2382

Research of China's Financial Safety Indicator

Xuemin Chen

Abstract We discuss China's finance security indicator. The term "finance security indicator methodology" refers to the factors in economy that affect finance security as well as their assessment and quantification method. The focus of the methodology is analyzing and investigating past domestic and foreign economic strength and operation, international balance of payment and external influence, relevant government administrative capability, fiscal policies, monetary policies, financial operation and developmental level, financial soundness and control power, etc. In the assessment system of China's finance security indicator, we highlight international experience. When analyzing the safety degree of each factor, we widely adopt internationally recognized threshold standards. The data used by this indicator system are mainly from China's official organization, but also included are those from the World Bank, IMF, and other reliable sources.

Keywords Finance safety · Finance risk · Economic policy

1 Introduction

The term "finance security" refers to the security of monetary capital financing and the stability of the financial system. Finance security is the basic issue studied in financial economics. With rapid economic globalization, finance security plays an increasingly important role in national economic security. The concepts of finance security and financial crisis are closely related. Financial crisis is the result of the large-scale accumulation of risks. It is a kind of extreme state of finance security. The concept of finance security is also the natural extension of such concepts as financial deepening and financial stability.

X. Chen (✉)
Institute for Culture Industrial Security Research, Beijing Institute of Graphic Communication, Beijing 102600, China
e-mail: chernxm@126.com

M. Li et al. (eds.), *Proceedings of 2015 2nd International Conference on Industrial Economics System and Industrial Security Engineering*,
DOI 10.1007/978-981-287-655-3_15

In the assessment system of China's finance security indicator, we highlight international experience. When analyzing the safety degree of each factor, we widely adopt internationally recognized threshold standards. The data used by this indicator system are mainly from China's official organization (China's National Bureau of Statistics), but also included are those from the World Bank, IMF, and other reliable sources.

We emphasize the definition of China's finance security indicator and also stress the interpretation of a finance-security indicator. Because the determining factors are complicated, we found that it is not enough to assess finance-security indicators merely through qualitative or quantitative analysis. Therefore, the method we used combines both qualitative and quantitative assessment and is based on objective facts and international experience.

The characteristics of assessment methods of China's finance security indicator are as follow:

(1) Respect China's stages and characteristics of economic development as well as study the features of China's finance security from an overall as well as systematic perspective. We attach importance to the policy systems related to industry and their developmental levels and even more importance to the matching of their degree with the stages of economic development.
(2) We chose highly sensitive and expressive indicators as well as adjust new risk factors with the change of environment. We will review the methodology of evaluation as well as make careful adjustments to adapt to world development as well as that of China' economy.

Jiang and Su [1], He and Lou [2], Gu [3], and Shen and Zhang [4] have studied China's finance security in their articles, and Edison [5] discussed whether or not an indicator can really predict financial crisis. Gu divides finance security into three factors, i.e., macro-financial risk, financial operation risk, and external financial risk, totaling 12 indexes, as well as uses analytical hierarchy process (AHP) for determining subjective weight. Jiang and Su divides finance security into three factors, i.e., microfinance prudent operation index, macroeconomic index, and international financial market index, totaling 17 indexes. He uses AHP for subjective weight as well as adopts principal component analysis (PCA) for dimensionality reduction and calculation in the safety evaluation of each factor. He and Lou divides finance security into four factors, i.e., microfinancial institution safety index, medium financial market safety index, macro-economic operation index, and international external risk medium index, totaling 22 indexes. He adopts PCA for dimensionality reduction and calculation.

In general, the basic starting points in the above-mentioned articles are correct. However, there are several problems with them: (1) the research is not systematic; the selection of indexes is separated; and the description of China's finance security is not comprehensive, systematic and profound enough; and (2) the methodology is not correct, nor has a solid theoretical foundation been laid for index weight. As far as the several articles mentioned above are concerned, because an insufficient number of factors are selected, the use of AHP is meaningless. Actually, it only plays the role of the title of weight. Second, PCA is not a weighing method. Rather, it is a descending-dimension method. The working principle of AHP is to select an index with large variation as the main analysis object as well as assign greater weight. However, the indexes with large variation may not be important in determining finance security. In contrast, some indexes with no large variation (e.g., inflation rate and nonperforming loans ratio of banks) are important factors used to judge finance security.

As far as indicator weighting is concerned, this chapter recognizes the sovereign credit rating of Standard & Poor and the practice of Xiang [6]. Xiang studied the measurement and analysis of China's economic imbalance; applied five factors (industrial structure, investment-consumption structure, financial structure, regional economic structure, and balance of payments structure) totaling 22 indexes; set imbalance thresholds for 22 indexes (scale of 1 to 6); and used simple arithmetic mean to obtain a final score for China's economic imbalance.

2 Finance-Security Indicator Methodology

Finance-security indicator methodology/criterion refers to the factors in economy that affect finance security as well as their method of assessment and quantification. The focus of the methodology is analyzing and investigating past domestic and foreign economic strength and operation, international balance of payment and external influence, relevant government administrative capability, fiscal policies, monetary policies, financial operation and developmental level, financial soundness and control power, etc.

The finance-security index system has 2 aspects, 5 key factors, and 32 indexes. The two aspects are "economic strength and policy environment" and "evaluation of financial sector."

"Economic strength and policy environment" reflects our assessment of national economic strength and operation, international balance of payment and external influence, relevant government administrative capacity, fiscal policies, and monetary policies. The score for this aspect is the average of the scores of such factors as "economic strength and operation," "international balance of payment and external influence," and "administrative capability and policy performance."

"Evaluation of financial sector" reflects our assessment of financial operation, financial developmental level, financial control power, and financial soundness. The score for this aspect is the average of the scores of "financial operation and developmental level" and "financial soundness and control power."

Our analysis includes two stages. In the first stage, the five key factors are graded. The rating scale is from 1 (the lowest degree of security) to 6 (the highest degree of security). The rating of each key factor is based on a series of quantitative and qualitative indexes. In the second stage, the average of the scores for the five key factors is calculated to obtain the finance-security indicator.

We use the simple arithmetic mean as the average of key factors in the second stage. There are three reasons for this: First, different measurement indexes constitute the measurement whole of the finance-security indicator. It is difficult to distinguish their degrees of importance. Second, some objective weighing methods, such as PCA, have a certain degree of irrationality because they determine weight merely according to variation and neglect the quality differences of an index or factor. Third, the measurement of any socioeconomic field, especially the measurement of complex phenomena, is not entirely accurate but rather is relatively objective. As long as the selection of index and factor at various levels is basically symmetrical, adoption of the simple arithmetic mean is feasible. This is also consistent with the calculation of many important indexes in the world. Either percentage method or text can be used for the assessment of finance security.

The followings constitute the five key factors for assessment. Those inside parentheses are corresponding quantifiable indexes. Unless marked "qualitative," the following data can be obtained on the website of China's National Bureau of Statistics.

Key factor 1 is economic strength and operation: (1) individuals' degree of affluence (i.e., the real per-capita GDP), (2) economic scale and developmental level (real GDP), (3) economic growth prospect (growth rate of real per-capita GDP), (4) economic stability (fluctuation of real per-capita GDP and unemployment rate), and (5) economic structure (each industry as a percentage of GDP).

Key factor 2 is administrative capability and policy performance: (6) effect of financial policy (change rate of government debt/GDP), (7) government debt level (debt burden/GDP), (8) financing cost (government debt cost), (9) financial revenue and expenditure (financial revenue or expenditure/GDP), (10) currency value stability and inflation (CPI), (11) monetary policy management experience (qualitative), (12) monetary policy tool and transmission (qualitative), and (13) performance of lend of last resort (qualitative).

Key factor 3 is international balance of payment and external influence: (14) degree of being export-oriented (i.e., dependence on foreign trade), (15) degree of balance of international payments (balance of current account/GDP), (16) capital and financial account (capital inflow or outflow/GDP), (17) foreign exchange reserve scale rationality and liquidity (total international reserve), (18) debt-servicing capacity (debt-servicing ratio), (19) debt risk (liability ratio), and (20) external financing capacity (relationship, cooperation, and agreement with economic powers and international systems).

Key factor 4 is financial operation and development: (21) financial growth and scale (financial interrelation ratio or monetization rate), (22) finance developmental level (the percentage of bank deposits or indirect financing in total household savings), (23) fixed-assets investment structure (financial investment/total investment), (24) financial assets price (real interest rate), and (25) exchange-rate marketization level (qualitative).

Key factor 5 is financial robustness and control power: (26) soundness of the banking industry (asset size, capital adequacy ratio, nonperforming loan ratio, degree of matching asset to liability, intermediate or off-balance-sheet business),

(27) soundness of the securities and futures industry (net capital of securities company, profitability of listed companies, risk-resistance capacity and profitability of securities companies, profitability and comprehensive strength of futures companies), (28) soundness of the insurance industry (asset size and number of institutions, premium income and growth, gains of capital use, liquidity, and solvency), (29) soundness of the financial market (credit risk of bond market, valuation [PE ratio] and volatility of stock market, and fluctuation in interest rate of interbank lending), (30) price stability (inflation rate), (31) influence of foreign banks (ratio of foreign shareholders, assets of foreign banks as a percentage of total bank assets), and (32) other influences of foreign capital (qualitative).

3 Key Factor and Grading Standard

We discuss the grading standards for the key factor "economic strength and operation" as an example. Our grading standard is based on industry experience as well as the latest research results.

The historical experience of economic security shows that A rich, large, diversified, resilient, and market-oriented economic entity that has always achieved sustainable economic growth can enable a country to be secure economically as well as improve its antirisk capacity. We find that market-oriented economic entities tend to be more affluent. They can more effectively allocate resources as well as promote long-term sustainable economic growth.

As the country and its people become richer, the country has more resources to meet challenges in the economic field. In "economic strength and operation," we take the index of real per-capita GDP as the measure of people's wealth and its basic score as well as adjust the basic score in light of other factors to obtain the final score of "economic strength and operation." This chapter compares the per-capital GDP of China with that of the international level to determine the corresponding scores and thresholds of indexes. Table 1 describes the grading process of "economic strength and operation."

In the past several decades, the world's per capita GDP has been increasing, and thus countries have become richer. However, more wealth does not necessarily mean more secure finance. Therefore, we will make adjustments to maintain the relativity and accuracy of assessment. In addition, we can judge the influences of global economic growth and change in exchange rates on countries' different development stages. We also find that regular adjustment will not necessarily lead to changes in scores.

Table 1 Grading of Economic Strength and Operation

<table>
<tr><td>Per-capita GDP (USD)</td><td>Less than 1000</td><td>1000–5000</td><td>5000–14,000</td><td>14,000–27,000</td><td>27,000–38,000</td><td>More than 38,000</td></tr>
<tr><td>Basic score</td><td>1</td><td>2</td><td>3</td><td>4</td><td>5</td><td>6</td></tr>
<tr><td colspan="4">Adjustment of items for an additional point</td><td colspan="3">Adjustment of items for deduction of a point</td></tr>
<tr><td colspan="4">High economic aggregate</td><td colspan="3">Economic growth lower than world-average level</td></tr>
<tr><td colspan="4" rowspan="2">Moderate economic growth exceeding world-average level</td><td colspan="3">Economic growth relies mainly on assets price after adjustment of inflation implying asset bubble caused by credit expansion.</td></tr>
<tr><td colspan="3">The economic fluctuation is too large or economy is too concentrated in industries, such as cyclical ones, where the GDP grow rate is more than 20 %.</td></tr>
<tr><td colspan="7">Final score = Basic score + adjusted score, and the net adjustment range does not exceed 2 grade points.</td></tr>
</table>

4 Conclusion

In light of the problems existing in the field of China's finance security as well as judging from international experience, we must study the solutions to finance security to draw on advantages and avoid disadvantages, set up professional finance-security research and decision-making organs, establish finance-security policies and standards, increase input to research and develop a financial computerizing software platform and core technologies of financial computerizing equipment, improve the domestic level of financial equipment, lay a solid foundation for finance security, strengthen the legal system construction of finance security, improve the regulatory level of financial supervisory authorities, eliminate hidden financial trouble in a timely manner, continuously deepen financial reforms, and establish orderly opening-up patterns suited to China's national conditions, to follow the world's trend of economic globalization and financial liberalization.

References

1. Jiang H, Su L (2009) An empirical analysis of financial safety index in China: 1998-2007. Contemp Financ Econ 10:47–53
2. He D, Lou F (2012) Construction and test of China financial security index. Chin Financ Rev 5:1–12
3. Gu H (2014) The national economic security capacity of finance. Chin Acad Res. 9:69–78
4. Shen Y, Zhang Z (2007) Establishment research on precaution system of Chinese financial security. J ShanXi Financ Econ Univ 10:89–94

5. Edison HJ (2003) Do indicators of financial crises work? An evaluation of an early warning system. Int J Financ Econ 8:11–53
6. Xiang J (2008) Measurement and analysis of China's economic structural imbalances. Chin Manage World 7:1–11

Research on the RMB Index from the Perspective of Financial Security

Zhanzhong Cao and Xuezhi Qin

Abstract The RMB index was calculated with July 21, 2005, as the base period; the weight of both the broad- and narrow-basket RMB indexes was determined separately; and a functional test was performed. The research indicated that the broad-basket RMB index was more stable and can better reflect the trends of the RMB against a basket of currencies both prior to and as well as in financial crisis. The broad-basket RMB index has better stability compared with the USD index and the Shanghai composite (security) index because the broad-basket RMB index fluctuates less.

Keywords RMB index · A basket of currencies · Currency weight

1 Introduction

Since the United States subprime crisis in 2007, the international financial order dominated by the US dollar has increasingly been blamed for the crisis. Global and local financial security has been threatened to varying degrees. At present, the total economy of China ranks second in the world. The internationalization of the RMB pushed by China has not only considered its own interests to some extent, but it is also necessary for the construction of a new international financial order, stability of the regional financial security, and the protection of interests of related fields.

Z. Cao (✉) · X. Qin
School of Economics, Dalian University of Technology, Dalian 116024, China
e-mail: caozhanzhongxh@163.com

X. Qin
e-mail: qinxz@dlut.edu.cn

M. Li et al. (eds.), *Proceedings of 2015 2nd International Conference on Industrial Economics System and Industrial Security Engineering*,
DOI 10.1007/978-981-287-655-3_16

2 Literature Review

McGuirk applied a convex combination operation to the bilateral import and export weights of basket currency countries using trade- and consumption-related data to obtain the competitive weights of different currencies [1]. Turnovsky calculated the weights of different currencies in a sample basket using a general equilibrium model of a small open economy, but the calculation is too complex to be operated [2]. Maxwell determined the exchange-rate index of a basket using the GDP deflator index as price conversion indicator, which is better than the CPI index [3].

In recent years, Chinese researchers have begun to pay more attention to the RMB index. Jianbin and Cuihong pointed out that the monetary-condition index is the weighted sum of the change of interest and exchange rates compared with the base period, and it is a comprehensive index that incorporates the impact of interest and exchange rates on the monetary environment [4]. Zhigang dissected the theoretical basis and feasibility of different methods [5]. Universities and research institutions launched the RMB-related index, and the RMB exchange-rate index was compiled by Fudan University [6].

The selection of sample pool reflected in this analysis is mostly focused on currency that is directly traded with the RMB while taking into consideration the currency of those who have a close business relationship with China.

3 Construction of the RMB Index

After the collapse of the Bretton Woods system, a floating exchange rate replaced the fixed exchange rate. The dollar is the strongest currency in the world, but the global economy as well as investors are concerned about the fluctuation of the US dollar. The US dollar index emerged over time to reflect the strength of the dollar. The US dollar index d can be divided into two categories: the exchange-traded US dollar index and the non exchange traded US dollar index.

The exchange-traded index includes the ICE dollar index in the Intercontinental Exchange (the New York Cotton Exchange combined with the coffee, sugar, and cocoa exchange in 199; the combination exchange became the New York Board of Trade; NYBOT was acquired by the Intercontinental Exchange in 2007 and renamed the ICE Future US) and the CME dollar index launched by the Chicago Mercantile Exchange. The non exchange traded index includes the Fed's FRB index, Morgan Guaranty index (MG index), which is guaranteed by the Morgan exchange and the index of Special Drawing Right designed by the International Monetary Fund.

3.1 Selection of Base Period

Before July 21, 2005, the exchange rate of the RMB against the US dollar remained fixed for a long time. Later, the revolution of the RMB-exchange rate officially started with reference to a basket of currencies and a managed floating exchange-rate system. This day was chosen as the calculation base period of the RMB index. In the future, the RMB will be recalculated according to the base period based on free convertibility of the RMB. This will assess the true value of the RMB more accurately.

3.2 Sample and Weight

During the process of the design of weight, after processing the national trade data we considered the importance of trade to different countries or regions in China and, combined with the international pay ability of a particular currency, finally decided on a two-basket currency. The first basket is the narrow-sense currency basket, which only includes the currency of four important trade partners/regions: the US States dollar, the Euro, yen, and the Won. The second basket is the broad-sense currency basket. In addition to the four currencies just mentioned, it also includes Dollar A, ringgit (Malaysia), ruble (Russia), rupee (India), and the Singapore dollar and pound (in total 10 currencies) considering the ranking of trade and stability of the currency and choosing trade partners within a wider range.

Calculation of the currency index weight can be divided into unilateral, bilateral, and multilateral trade weight through previous research. Among all three, multilateral trade weight is more suitable for RMB index construction. According to the calculation method of multilateral trade competition weight, taking narrow-basket currency weight as an example (1), we calculated the import weight of each country/region:

$$W_i^m = \frac{m^i}{m_c} = \frac{m^i}{\sum_{k=1}^{n} m^i} \tag{1}$$

The result shows that the import weight of the Euro is 30.76 %, whereas that of US dollar, yen, and won is 17.36, 28.32, and 23.56 % respectively

Bilateral export competition weight is calculated as follows:

$$W_i^{bx} = \left(\frac{x^i}{x_c}\right)\left(\frac{y_i}{y_i + \sum_{h \neq China} x_h^i}\right) \tag{2}$$

The calculation needs the internal supply y_i of country/region i. We used the 2011 GDP of the country/region minus the export volume to approximately represent y_i according to the data availability. We kicked out 250 billion re-export trade

Table 1 Result of calculation of comprehensive weight of basket currency

	EUR (%)	USD (%)	JPY (%)	KRW (%)	AUD (%)	MYR (%)	SUR (%)	INR (%)	SGD (%)	GBP (%)
Narrow	39.38	35.48	16.33	8.28						
Broad	27.51	21.72	16.97	11.25	5.81	3.85	4.08	3.65	2.37	2.96

volume from total export volume of Singapore because huge re-export trade exists in Singapore. W_i^{bx} equals the export weight of each country multiply by a bilateral competition coefficient: The greatest W_i^{bx} was in Europe and the lowest in Korea after adjustment using the bilateral competition coefficient.

Now, the export competition weight represents the third market:

$$W_i^{bx} = \sum\nolimits_{k \neq j} \left(\frac{x_c^k}{x_c}\right) \left(\frac{x_i^k}{y_k + \sum\nolimits_{h \neq China} x_h^k}\right) \tag{3}$$

Finally, the comprehensive weight of basket currency is calculated as follows:

$$W_i = \left(\frac{m_c}{x_c + m_c}\right) W_i^m + \left(\frac{x_c}{x_c + m_c}\right) W_i^x \tag{4}$$

The import and export weight coefficients were 0.45 and 0.55, respectively, according to the trade situation in 2011 in China. After the calculation and standardization of four kinds of basket currency weight, in consideration of the third competition relationship, euro weight will always be the first whether in narrow or broad basket, but the weight gap between the USD and the Euro decreased compared with using the other two methods. The proportion of the Australian dollar is more than 5 %, which is the highest within noncore currency, and the ringgit weight dropped to lower than 5 %, which showed that the trade between China and Australia still demonstrates strong growth even though third-market competition exists. When calculating the broad-basket currency weight, the order of the currency weight changed when using different methods. But the weight of the four currencies, i.e., the Euro, USD, JPY, and KRW, are significantly greater than that of the other four kinds of currency in a narrow basket regardless of the method used. Therefore, it is scientific to select these four currencies as a narrow basket when constructing the RMB index (Table 1).

4 RMB Index Simulation and Function Test

After calculating the sample currency weight of the RMB index, on the basis of the exchange rate data of the RMB, against sample currency since reform of the RMB-exchange rate, the simulation and function tests of both the narrow- and broad-basket RMB index started, and the science and stability of the RMB index was checked.

The data in this chapter came from the RMB mid-rate data announced by the State Administration of Foreign Exchange and from the exchange rate of member-country currency against SDR released by IMF every workday since the data were available. Among those, some of the direct transaction data of the RMB to foreign currency was lost due to differences in the base period; thus, we used SDR data instead.

4.1 Broad- and Narrow-Basket RMB Indexes

The empirical results showed that the trend of broad- and narrow-basket RMB indexes was similar because the correlation coefficient of the two indexes was 0.9947. This showed that selection of select key currency is important when choosing a sample currency of the RMB index; currencies with low trade weight had little effect on the final result. According to the volatility (Table 2) of both the broad- and narrow-basket RMB index, the broad-basket RMB index was more stable both prior to and in financial crisis.

Table 2 RMB index and related index function test

Title		Yield per day	SD
Before financial crisis (2006.12.01–2007.11.30)	Broad-basket RMB index	0.000064	0.001496
	Narrow-basket RMB index	0.000073	0.001623
	Shanghai composite index	0.003666	0.022459
	USD index	−0.000302	0.003109
In financial crisis (2007.12.01–2009.11.30)	Broad-basket RMB index	0.000387	0.002614
	Narrow-basket RMB index	0.000357	0.002695
	Shanghai composite index	−0.003333	0.028257
	USD index	−0.000010	0.006956
After financial crisis (2009.12.01–2011.05.06)	Broad-basket RMB index	0.000096	0.002373
	Narrow-basket RMB index	0.000113	0.002440
	Shanghai composite index	−0.000206	0.013429
	USD index	0.000019	0.005509

4.2 Comparison of the RMB Index with USD Index and Shanghai (Security) Composite Index

The term "United States dollar," which refers to the dollar index launched by the American intercontinental exchange, fluctuated dramatically due to the America subprime mortgage crisis in 2007. The US dollar depreciated at first and then significantly appreciated at the end of second quarter in 2008, when the US dollar fell to its lowest point. Meanwhile, the RMB index increased. In January 2009, the JPY and the EUR against the USD and the RMB greatly appreciated, whereas the RMB index and USD index dropped sharply because JPY and Euro are the most important sample currencies of the RMB and USD indexes.

According to the basic analysis frame of economics, capital flows freely between the stock index and the currency index. If we do not consider other factors that may influence the stock index, the profit-driven capital will be invested in RMB assets under the continuous appreciation of the RMB. For the external environment, the China stock index underwent a depression of valuation in the global capital market, but theoretically the trend of the RMB index should be consistent with that of stock index. The stock index should have declined when the RMB depreciated dramatically.

The empirical results show that the track of the RMB index deviated from that of the Shanghai composite index, but this did not match the theoretical assumptions. To consider the theoretical assumptions correct, we can explain the result based on the logic presented below: Because the system and the reform process of the exchange rate in China is strictly under government control, the RMB index described in this chapter cannot yet completely reflect the real strength of the RMB. This means that the inherent value of the RMB was distorted to some extent under the circumstances that the capital account was not completely open in China; the RMB was the currency in circulation, but it was not through the inspection of an open-market system. The situation will change in the future because at present China actively promotes the process of RMB internationalization.

4.3 RMB Index and Related Index Function Test

We compared the yield per day and the SD of the broad-basket RMB index with that of the narrow-basket RMB index, the Shanghai composite index, and the USD index (Fig. 1).

Judging from yield per day, the yield per day of both the broad- and narrow-basket RMB indexes was always positive during three periods; even in the serious crisis phase, it is still relatively strong, and this played a large role in keeping the RMB, the financial market, and the macroeconomy stable. It also served as a buffer while the international monetary system was coping with the crisis. In contrast, the US dollar index had already been weak before the crisis, and

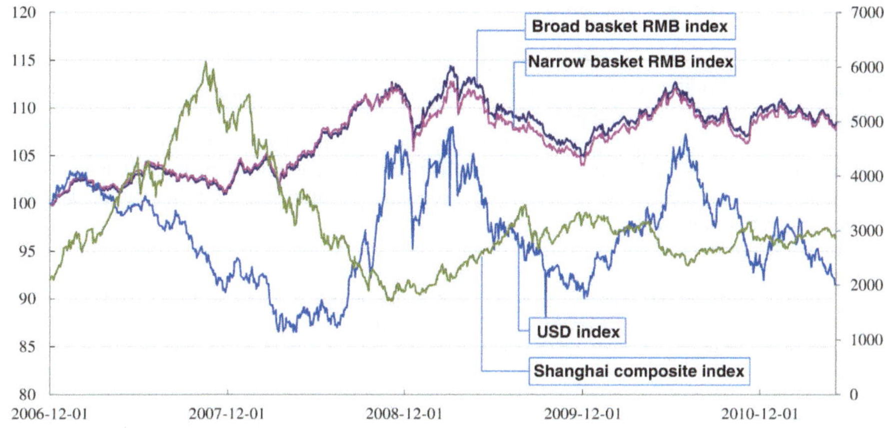

Fig. 1 RMB index and related index function test

the yield per day can still be negative after the crisis. The yield per day of the Shanghai composite index was greater than that of the USD index and the RMB index before the crisis, but it decreased quickly after the crisis due to the impact from the world's economic situation.

Judging from volatility (i.e., the SD), the SD of the broad-basket RMB index was the smallest indication relative stability. The crisis had a huge impact on the US dollar, and the volatility of the USD index in far beyond that of both the broad- and narrow-basket RMB indexes, which indicates poor stability. The volatility of the Shanghai composite index was the greatest due to sensitivity to the global economy; thus, the Shanghai composite index is not as stable as the RMB index and the USD index.

5 Conclusion and Future Directions

First, the broad-basket RMB index is better than the narrow-basket RMB index in terms of stability both before and during the financial crisis. Second, the SD and the volatility of the broad-basket RMB index are smaller than those of the USD index. Third, the stability of the RMB index had positive influences on the stability of the economy and financial market in China. It will also stabilize the regional financial market that revolves around China and will promote sustainable development of the global economy.

Disadvantages and future of the analysis include the following. The RMB index method is based only on static timing data with poor performance of real-time data. The weight coefficient must be adjusted dynamically according to the continuous opening of capital; therefore, the transaction index will be the new direction of the future in RMB index research.

References

1. McGuirk AK (1987) Measuring price competitiveness for industrial country trade in manufactures. IMF working paper
2. Turnovsky SJ (2000) A determination of the optimal currency basket. Am Econ Rev 5:1–16
3. Maxwell O-A (2004) Measuring the real effective exchange rate (REER) in Ghana.Credit research paper of centre for research in economic development and international trade, University of Nottingham
4. Jianbin C, Cuihong L (2006) Construction of China currency situation index and test the effect of monetary policy. Contemp Financ Econ 5:31–37
5. Zhigang Z (2007) RMB effective exchange rate index method selection. Stat Edu 12:14–15
6. FuDan University (2010) The RMB exchange rate index. http://www.fudan.edu.cn/fudannews/2010/1025/25829.html

Research of Industrial Upgrading Basing on Industrial Technology Chain

Ruimin Xu

Abstract Basically, technological upgrading is the foundation of industrial upgrading. Analyzing the upgrading method of a technology chain is of benefit for developing countries to shorten the gap of industrial technology and increase industrial competitiveness. This chapter shows a research method of industrial upgrading, explains the concept of an industrial technology chain, and analyses the process of industrial upgrading based on the industrial technology chain through four perspectives. Finally, this chapter makes some recommendations for the industrial upgrading of our country based on the industrial technology chain.

Keywords Industrial technology chain · Industrial upgrading · Technology chain

1 Introduction

With the technology diffusion and permeation and the development of global economic integration, developed countries have extended their own industrial chain to different countries and regions, thus realizing technology resources integration and utilization based on the global platform. However, because the high-value link in the industrial chain link is usually occupied by developing countries that block technology and market-channel monopolies from developed countries, developing countries can only complete with low-value link, which results in industrial upgrading being locked into a limited range. Development of an industrial technology chain is the foundation of industrial upgrading; furthermore, upgrading the

R. Xu (✉)
Faculty of Management and Economics, Dalian University of Technology,
100044 Beijing, China
e-mail: xrm@sinoauchan.com

M. Li et al. (eds.), *Proceedings of 2015 2nd International Conference on Industrial Economics System and Industrial Security Engineering*,
DOI 10.1007/978-981-287-655-3_17

industrial technology chain is the key to industrial upgrading. Therefore, the influence of the industrial technology chain on industrial upgrading is worthy of further discussion.

2 Theoretic Research Achievements of Industrial Technology Chain and Upgrading

There are two main overviews of domestic and foreign research on industrial upgrading: One is industrial restructuring theory, and the other is upgrading of the value chain. The former concentrated on the rule of varying pattern of industry hierarchy, such as the transformation from labor-intensive industry to capital- and technology-intensive ones. The latter mainly focused on choice among all aspects of links in the industrial chain. Both of the two above-mentioned factors are beneficial to industrial restructure optimization and improving industrial efficiency and finally driving industrial upgrading. And at the same time, the level of economic and technological development restricts these two factors. Therefore, industrial technology is the key to industrial upgrading. Although foreign literature rarely covers the concept of the technology chain, many scholars have considered the idea of the technology chain when conducting research of production or industrial technology. In China, Deyu yuan [1] proposed the term "industrial technology chain," first considering it as an industrial cluster that contained an inner link and both a forward and backward link.

3 The Mechanism by Which the Industrial Technology Chain Affects Industry Upgrading

3.1 Coordination Mechanism of Technological Innovation and Industrial Development

Industrial technology innovation can be divided into two types based on its influence on industry: breakthrough (or fundamental) and evolutionary technology innovation [2]. Industry productivity will increase rapidly when the industry system obtains huge uptakes from innovation. Consequently, high benefit is brought by high industrial productivity, which makes the industry rapidly grow and expand. Breakthrough technology innovation can promote the rise of new industry. When the industry matures, breakthrough technology innovation has been replaced by the concept of improved technology innovation and productivity growth rate, and thus the cost-reduction rate tends to slow down. Therefore, the essence of industrial upgrading is the process of the transition of the industry from a low-skilled, low value added state to new technology and a high value added state.

3.2 *The Fusion of the Stretch of Industry Technique and Development*

According to the industry growth curve, the time-line of industry developing to the mature stage is long. The reason is that after the first generation of products and technology innovation, the industry must go through a new round of technological innovation to achieve a second- and third-generation product to promote vigorous development of the industry. The development of industrial technology actually "stretches" the industrial technology chain. Sometimes the industrial technology in developing countries creates a system of it own because of regional segmentation and poor information, which makes local industry dissociate from world-leading technology. In contrast, developed countries integrate those branched and short-chain outward technological innovations into the global technological chain so that they can take advantage of global processing and technical resources [3]. Therefore, developing countries can learn how to integrate technological innovation into the mainstream technology system of the global technological chain to improve the capability of industrial technology.

3.3 *Industrial Technology Chain and Upgrading*

Upgrading of the industrial technology chain can be divided into two types: progressive upgrade and fundamental upgrade. The term "progressive upgrade" refers to the progressive innovation of technology, product design, and manufacturing technology, thus resulting in an improvement of production and efficiency with no change of the technology paradigm and track [4]. In contrast, the term "fundamental upgrade" means there is technology reformation, or even the appearance of a new industry, along with changes of the technology paradigm and track. Fundamental upgrade is classified into interchain upgrading and across-chain upgrading depending on whether the middle technological chain has been crossed and directly reaches the most advanced technological chain. In addition, upgrading within the chain refers to overall industrial technology improvement that does not belong to the category of fundamental innovation, e.g., computer upgrade from model 286 −386, while model 286 did not change in terms of basic design thought, logic, technology principle, and technical paradigm didn't change (refers to Fig. 1).

3.3.1 Technological Disruptions

According to the industry growth curve, the technology experience develops to maturity and then is eliminated by the generation process, which is sown as an S-shaped curve being determined by the technology development rule [5]. At the

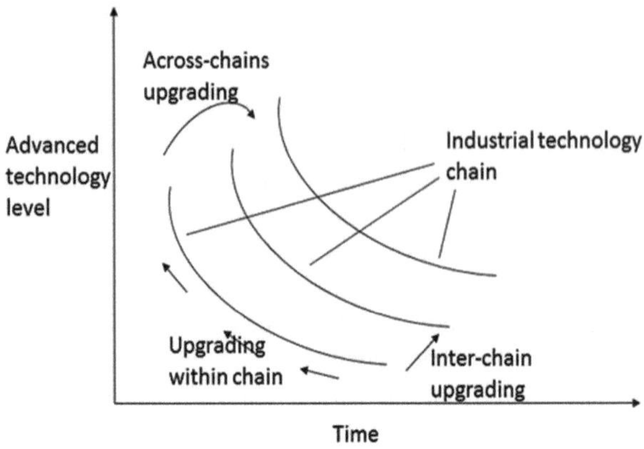

Fig. 1 Upgrading model of the industrial technology chain

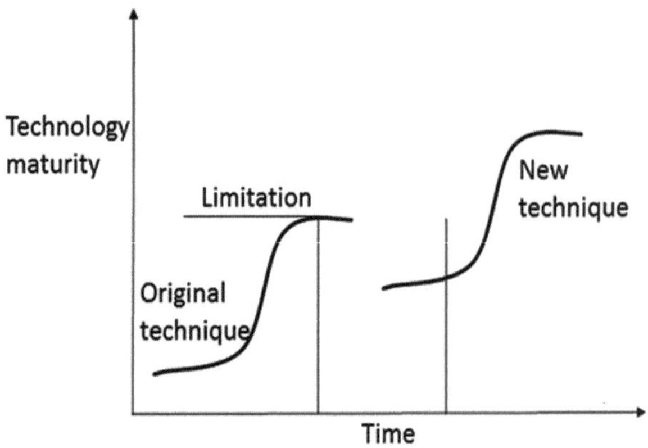

Fig. 2 Limits of industry technology and disruptions of the technological chain

start-up stage of technology development, the speed of technology development is slow because there are many problems to overcome. As the technology is gradually perfected, the growth and spread of the technology becomes fast. However, when the technology matures, it is difficult to create significant improvements, and thus the technology reaches its limit. Therefore, industrial technology chain will appear to have disruptions in the process of alternating old with new technology, which also signals the right timing for industrial upgrading (Fig. 2).

3.3.2 Across-Chain Upgrading

Technological "leapfrogging" was first described by the Dutch scholar Soete in [6]; Lee and Lim [7] considered industrial technology leapfrogging as the transition of the overall industrial chain based on analysis of the industrial technology progress of D-RAM chips, PCs, and machine tools [8]. Industry development includes the appearance of new industrial technology that arises from fundamental change in the

Table 1 The characteristics of industrial upgrading

Advanced technology level	Market growth rate		
	Low	Middle	High
Low	Industry blank Introduction of products as main characteristics	Industrial inserting Low-end technology input	Industry upgrading Industry university institute cooperation development
Middle	Industrial breakthrough Product differentiation	Industry vigor Technology licensing cooperation	Industry alliance Mature technology and production scale
High	Develop industry Seek cooperative production	Perfect industrial structure Self-dependent innovation	Industry consolidation Leading innovation

technology paradigm or is related to basic a technology innovation (Table 1). People can take advantage industry development by the leaps-and-bounds development rule, making an enterprise pass across some of the industry technology development phase, achieve direct access to the forefront of the industrial technology system, and become involved in the new industry competition. This process can also make a country or a region with technical innovation and development of new technology bypass the level of industrialization of developed countries in some intermediate process of industry development, in effect shortening the process of industrialization and entering directly into the stage of industrial modernization.

4 Suggestions for Industry Upgrading Based on the Development of the Industrial Technology Chain

4.1 Scientifically Position and Select the Appropriate Mode of Technology Upgrading

Some industrial markets in China have entered into a high degree of development. The government should enforce Industry–university–institute cooperation as well as cooperation with foreign advanced-technology enterprises to build an industry alliance and realize large-scale production. Enterprises with advanced technology should introduce PE with related background and realize innovative technology, finally realizing the breakthrough technology leap between chain and node.

4.2 Take Measures to Break Down Technology Barriers

The situation of technical reserves and technology governance of the world's industry shows regular variation. The Chinese industry must be aware of the reserves and governance so that industrial upgrading and leapfrogging development can be realized. The main methods are as follows: (1) improve R&D capability by bringing in a technology core team; (2) acquire technology-leading enterprises; and (3) cooperate by way of patent licensing while avoiding technological dependence.

4.3 Government Supports, Guides, and Promotion of Industrial Upgrading

Industrial upgrading needs both a leading role in the market mechanism and supporting role in government public policy. The government can take measures such as (1) establishing a public service platform including opening the platform of R&D, information exchange, and international trade; (2) set up a foundation of industrial development as a "technology-driven" model of industrial development adopted by developing countries; and (3) establish industrial preference policy on tax, financing, and land and labor policies to guide the social foundation toward industrial development.

Acknowledgments I extend sincere gratitude to my professor for instructive advice and useful suggestions on my thesis. Special thanks go to my dear family members and friends for their support.

References

1. Deyu yuan (2000) The definition of industrial technology. J Northeast Univ (1):22–25
2. Lee K, Lim C (2001) Technological regimes, catching-up and leapfrogging: Findings from the Korean industries. Res Policy 30(3):459–483
3. Humphrey J, Schmitz H (2002) How does inserition in global value chains affect upgrading in industrial clusters. Reg Stud 9(36):1017–1027
4. Gao R, Ji Y (2006). Systematic analysis of technological chain and industrial selection. R&D Management, vol 58−64
5. Mao Jianqi (2007) Preliminary analysis of global technological chain. Sci Res management 23 (11):85–92
6. Soete L (1985) International diffusion of technology, industrial development and technological leapfrogging. World Dev 13(3):409–422
7. Lee K, Lim C (1999) Technological regimes, catching-up and leapfrogging: findings from the Korean industries. Res Policy (30):459–483
8. Sharif N (1999) Strategic role of technological self-reliance in development management. Technol Forecast Soc Chang 62(3):219–238

Analysis of Formation Mechanism of Industrial Hollowing-Out in China

Zhengquan Li

Abstract Since the financial crisis in 2008, China's manufacturing industry began to move operations to Southeast Asia, South America, and inland China, thus deepening the "industrial hollowing-out" even further. This chapter analyzes the phenomenon of industrial hollowing-out in China from four production perspectives: hollowing-out, appreciation of RMB, excessive expansion of the virtual economy, and foreign direct investment. It also proposes coping strategies for industrial hollowing-out such as speeding up industry upgrading and strengthening supervision of the virtual economy.

Keywords Industrial hollowing-out · Formation mechanism · Coping strategy

1 Introduction

Industrial hollowing-out first appeared at the second half of the 19th century in Britain. Since then, it has changed forms and caused trouble to countries around the world involved in economic globalization. Industrial hollowing-out can affect some industry factors or products depending on the external market by making a country (or area) lose market control temporarily or forever. In addition, it results in real-economy transformation or recession, excessive virtual-economy expansion, unbalance of international trade and capital flows, etc. Based on such global experiences as those of American, Japan, and Latin America, industrial hollowing-out is the main factor influencing industry security. As a result, it is of practical significance to perform further research and analyze the formation mechanism of industrial hollowing-out so that we can develop and adopt corresponding methods of prevention.

Z. Li (✉)
CCISR, Beijing Jiaotong University, Beijing 100098, China
e-mail: sunran0414@163.com

M. Li et al. (eds.), *Proceedings of 2015 2nd International Conference on Industrial Economics System and Industrial Security Engineering*,
DOI 10.1007/978-981-287-655-3_18

Table 1 Net amount and growth rate of manufacturing FDI by China

Year	Net amount of manufacturing FDI (million dollars)	Growth rate (%)
2003	62404	
2004	75555	21.07
2005	228040	201.85
2006	90661	–60.24
2007	212650	134.56
2008	176603	–16.95
2009	224097	26.89
2010	466417	108.13
2011	704118	50.96
2012	866741	23.10

Data source Past years China statistical yearbook

2 Theoretic Research Achievements of Analyzing Factors of Industrial Hollowing-Out

Relevant academic analysis of the factors of industrial hollowing-out has not yet reached an agreement; most of the existing literature show results from the following perspective. Most scholars agree that FDI (foreign direct investment) is the fundamental cause of industrial hollowing-out, and this view has infuenced the judgment of political and academic circles for a very long time. Cowling and Tomlinson [1] researched multinational companies' foreign investment in Japan, finding that the FDI was apparently higher than that of Group of Seven, which lead to the industrial hollowing-out in Japan [2]. Kim represented a industrial hollowing-out model showing that industrial transfer will have a significant influence on the economy of resource-export countries in the formation of industrial hollowing-out [3]. Many other scholars researched factors of industrial hollowing-out from other perspectives, verifying that FDI only caused the industrial transfer and not the whole reason for industrial hollowing-out. This perspective includes factors such as technology hollowing-out and unbalance of input and output (Table 1).

3 Current Situation and Analysis of the Formation Mechanism of Industrial Hollowing-Out in China

3.1 Current Situation of Industrial Hollowing-Out in China

Industrial hollowing-out appeared in the late 1990s in China and has become more prominent in recent years. During the past 20 years, great changes have taken place in the economic structure of China. The proportion of primary industry in the national economy fell sharply from 27.1 % in 1990 fell to 10.1 % in 2012; the proportion of second industry basically remained stable; and the proportion of tertiary industry had a large increase from 31.6 % in 1990 to 43.1 % in 2012. Apparently, the increase the third industry in the national economy has been far greater than that of the second industry. The industrial contribution to national economic growth is gradually being replaced by contributions from tertiary industry and the virtual economy is expanding excessively, thus leading to a worse situation of industrial hollowing-out.

3.2 *Factors of Industry Hollowing-Out*

3.2.1 Technology Hollowing-Out (Market for Technology)

In 1992, the Chinese government established the Joint Venture Law, which allowed foreign holdings and marketing as well as chairman. "Market for technology" was developed as a basic strategy to use foreign capital aiming at changing the domestic market for foreign developing technology. However, the strategy of market for technology did not achieve the expected effect, and industrial hollowing-out has become increasingly serious because of the following problems caused by the strategy:

1. Multinational companies blocked core technology to maintain their technological monopoly advantages, e.g., equipment-manufacturing and other industries.
2. Before the 1990s, foreign investment was mainly concentrated on medium-sized project in low-level technology or labor-intensive industries and rarely in new and high-technological industries.
3. The international competitiveness of our brand product and exported-oriented industry did not come to fruition [4].

3.2.2 Labor Hollowing-Out

With the speeding up of industrialization, more domestic economic regions (such as the Yangtze River delta) and developing countries, such as Vietnam, have chosen labor-intensive export as the main development path. In the process of domestic economic development in our country, the shortage of migrant workers m is a concentrated reflection of the negative effects of low wages. Specifically, "migrant worker shortage" refers to workers who have almost economic basis for continuous education and training investment after paying for the higher cost of city life and migration. Therefore, it is difficult to improve workers' qualities and skills, thus leading to low wages making workers trapped in a vicious cycle of "low income → lack of human capital → low income accumulation → low wages."

In addition, it is not possible to improve labor resources to be of higher quality and thus optimize the demand for improved industrial structure.

3.3 Imbalance of Industrial Profits

3.3.1 Virtual-Economy Profits Induce Flight of Manufacturing Capital

After the outbreak of the international financial crisis in 2008, China introduced a series of economic stimulus plans to make huge amounts of fund flow into the virtual economy of the real estate and financial markets. On one hand, the high profits from virtual capital induced funds in the real economy turned to the virtual economy. In contrast, the flowing funds increased the expansion of the virtual economy, and now the false prosperity of the virtual economy and industrial hollowing-out are in a vicious cycle of mutual promotion. According to the 2012 list of China's top 500 enterprises published by China Enterprise Confederation, 14 commercial banks achieved a total revenue of 3.25 trillion RMB, an increase of 30.32 % over the year before. The total operating revenue of the largest five commercial banks accounted for 32.2 %. Meanwhile, although operating revenue of 272 manufacturing enterprises reached a proportion of 42.7 % among the top 500 enterprises, the profit only accounted for 25.04 %.

3.3.2 Appreciation of RMB Shrinks the Processing Industry

The value of the RMB has increased quickly under the mutual interaction of the expectations of appreciation and external pressure. On one hand, the price advantage of China's manufacturing products in the international market is gradually weakening. In contrast, there has not been enough domestic demand to make up the gaps caused by weak international market demand. Compared with 2005, the RMB-to-USD exchange rate was 2.68, 7.17, 15.21, 16.6, and 17.4 %, respectively. As of October 2012, the exchange rate broke through 6.30 %. Based on the relevant experience of Japan, the yen increased at an average rate of 20 % in a few years since 1985, which resulted in an export crisis and industrial hollowing-out.

Contrary to the expansionary monetary policy in Japan, China's government implemented prudent monetary policy when facing the situation of RMB appreciation, which included increasing the reserve ratio and raising interest rates. However, the interest rates growth is also am cause of industrial hollowing-out by causing continuous currency appreciation and serious shortages of the domestic manufacturing sector [5].

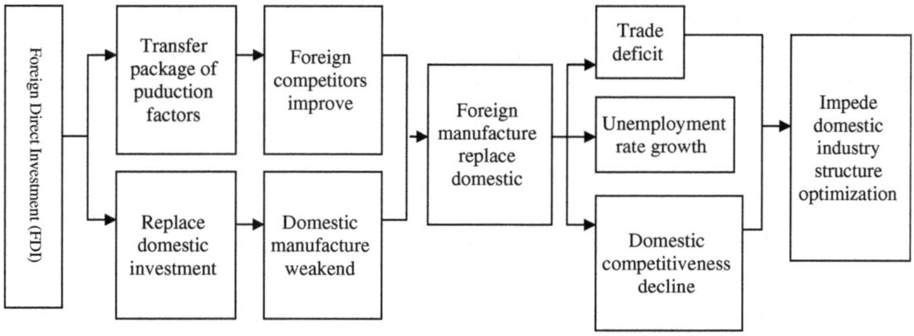

Fig. 1 FDI leads to industrial hollowing-out

3.4 Foreign Direct Investment (FDI)

Generally speaking, many scholars believe that foreign direct investment can lead to industrial hollowing-out and cause a series of adverse impacts on the macroeconomy of the investment country as seen in Fig. 1. There are analyses from three perspectives:

1. FDI will lead to an international trade deficit, which includes the reverse impact of export and import through moving the manufacturing base to abroad and importing overseas production [6].
2. Reduce international competitiveness of foreign investors by transferring the package of manufacturing factors, such as capital, technology, and management experience, thus making the competitor country increase its competitive ability.
3. There will be shrinkage of domestic manufacturing industry, along with a reduction of employment opportunity, due to transfer of the manufacturing base to foreign countries,.

Combined with the data and situation in China, although there is rapid growth of FDI by China in the recent 10 years, foreign investment in China still involves far more than investment overseas by China to other countries. Theoretically, FDI by China is not enough to result in domestic industrial hollowing-out so far. But according to sheet 3.1, except for 2006 and 2008, the net amount and growth rate is increasing yearly.

4 Coping Strategy for Industrial Hollowing-Out in China

4.1 Positive Response to Appreciation of RMB

To deal with appreciation of the RMB, the government needs to establish a reasonable exchange rate policy that is suitable to our enterprises benefits. From another point of view, corporations that are influenced heavily by fluctuation of the exchange rate must adapt to the change. The corporation can take measures such as expanding the domestic market, opening the international market through various trading methods, and using capital operation to increase risk-response ability.

4.2 Speeding up Industry Upgrading

Industry upgrading is a complicated systematic project that requires the cooperation of the government, enterprises, and all aspects of society in developing a different industrial strategy for various industries and regions. The government should pay more attention to supporting the innovation ability of corporations' R&D departments. In addition, both the government and individual enterprises should strengthen the training of qualified human recourses and provide a guarantee of the talent needed for industrial upgrading [7].

4.3 Strengthen Supervision of the Virtual Economy

Supervision of the virtual economy mainly includes financial securities and the real estate industry. Supervision should include establishing a legal system to control unreasonable profits to prevent severe fluctuations in the financial market and enforcing the reasonable use of land recourses to so that excessive expansion of the real estate market can be avoided.

Acknowledgments I extend sincere regards to my supervisor for instructive advice on my thesis. In addition, grateful thanks are given to my dear family and friends for their help.

References

1. Cowling K, Tomlinson PR (2000) The Japanese Crisis-A case of strategic failure? Econ J 110:358–381
2. Cowling K, Tomlinson PR (2001) The problem of regional "hollowing-out" in Japan: lessons for regional industrial policy. Working paper, University of Warwick
3. Kim YJ (2007) A model of industrial hollowing-out of neighboring countries by the economic growth of China. China Econ Rev 2:122–138
4. Xinqiao P (2000) Did the market change for technology. Beijing Int Econ Rev 5:33−36
5. Tejima S (2000) Japanese FDI, the implications of hollowing-out on the technological development of host countries. Int Bus Rev 06:555–570
6. Jiyong C (1996) Research of foreign direct investment in America. Wuhan University Press, Wuhan
7. Sisi T (2012) Enhancing the vitality of the real economy to avoid industrial hollowing out. Macroecon Manag 9:40–42

Analysis of the Changes of Drug Control Law in the People's Republic of China from the View of Medical Industry Security

Lu Yu

Abstract The Drug Control Law of the People's Republic of China, which the basic law to adjust the relationships of the medical industry in China, was released in 1984 and revised twice, respectively, in 2001 and 2013. This chapter analyzes the characteristics of the Chinese medical industry as well as future problems that may arise the industry's development using changes in the content of the law as the point of penetration. It also analyzes and estimates the future stability and security of policies in terms of security laws of the Chinese medical industry.

Keywords Industrial security · Pharmaceutical enterprises · Laws · Regulations

1 Introduction

Currently the most fundamental and most important law for the Chinese medical industry is the Drug Control Law of the People's Republic of China. It is related to the medical industry because it expressly regulates the establishment conditions and relevant responsibilities for "medicine-production enterprises" and "medicine-operation enterprises" as industrial organizations.

The Drug Control Law of the People's Republic of China was first released in 1984 (hereafter referred to as the 1984 Control Law), and it has been revised for twice in the 3 decades since then, i.e., the 2001 revision (hereafter referred to as the 2001 Control Law) and the 2013 revision (hereafter referred to as the 2013 Control Law). This chapter analyzes and evaluates policy security in terms of laws in the Chinese medical industry and estimates the future variation trend in relevant laws related to the Chinese medical industry from the aspect of industrial security by comparing the initial Drug Control Law 1984 with the revisions in 2001 and 2013.

L. Yu (✉)
School of Economics and Management, Beijing Jiaotong Univesity, Beijing 100044, China
e-mail: 304238591@qq.com; asmodeus108@163.com

M. Li et al. (eds.), *Proceedings of 2015 2nd International Conference on Industrial Economics System and Industrial Security Engineering*,
DOI 10.1007/978-981-287-655-3_19

The industrial economic data in this article consists of three types of data: rate of capital accumulation, quantity of medium-sized enterprises of the Chinese medical industry, and total assets value of the Chinese medical industry. The reason for selecting the three types of data lies in the fact that the Drug Control Law of the People's Republic of China constitutes legal measures to regulate and control the entire medicine industry as well as pharmaceutical enterprises in China from the state level upward. Changes in departmental or regional data cannot reflect the execution results of the law or its influence of the medicine industry.

2 Changes of the Content of the Drug Control Law of the People's Republic of China

Except for general rules and supplementary provisions, the 1984 Control Law of China mainly covers medicine production operation, quality, packaging, and instructions; special medicine; medicine for medical departments; medicine advertisement and trademark; and medicine supervision and relevant responsibilities, etc. The legal relationships adjusted by this law are retained in the 2001 and 2013 revisions. Although the 2001 revision made a great deal of large-scale alterations, the overall framework of the 1984 law has not been broken: Only "special medicine management" was omitted and the content incorporated into the chapter on "medicine management." Compared with 1984 Control Law, the 2001 Control Law mainly focused on the following revisions:

First, it confirmed the legal status and responsibilities of State Drug Administration (based on which the SFDA—State Food and Drug Administration—was established in 2003), and then provided the unified administrative authority of county, municipal, and provincial drug administration.

Second, it confirmed the legal status of the Administrative Regulations on Medicine Production and Quality (Article 1.9 of the 2001 revision) and Administrative Regulations on Medicine Operation and Quality (Article 2.16 of the 2001 revision), in particular the new articles in the 2001 revision of the Administrative Regulations on Medicine Operation and Quality.

Third, it clarified the effect on medicine planning and industrial policies and fourth, it emphasized responsibilities corresponding to various illegal acts. Compared with the 2001 revision, only a new article on "entrusted production of medicine" was added in 2013 revision. Therefore, it can be seen that from 1984 to 2001, China had experienced the most obvious development and fierce changes in the Chinese medicine industry, which resulted in the significant revisions to the law. In addition, two special legislative trends, in addition to general clauses on the quality and prices of medicine, appearing in the original as well as the 2001 and 2013 revisions of the Control Law should be noted: (1) support and encourage technological innovation in the research and development of new medicine; and (2) regulate the Chinese medicine industry.

3 Analysis—Interaction Between the Drug Control Law and the Chinese Medicine Industry

From the above-mentioned descriptions, we can see great changes to the medicine industry of China from the end of the 20th century to the beginning of the 21st century while major changes emerge regarding the establishment and supervision of medicine production and operation enterprises. The law of 1984 stated that only the "hygiene administration authority under the State council" was responsible for supervision, while the 2001 law directly stated that the "China Drug Administration" was responsible for supervision, which was then transferred to the State Food and Drug Administration in 2003 and then to the China Food and Drug Administration (CFDA) in 2013. Before the establishment of the above-mentioned supervision institutions, especially in the 1984 Control Law, the establishment of medicine-production enterprises was approved by the "drug production and operation administration authority of the provinces, autonomous regions and municipalities directly under the central government", while the operation enterprises of the medicine industry only needed approval from the "local medicine production and operation administration authority." The inaccurate written expression created a huge vague space for the establishment and development of production and operation enterprises in the medicine industry. Although both production and operation enterprises need corresponding professional qualifications according to the law, procedures and results approval by local governments and medicine departments were in chaos at the end of the previous century in China due to relatively backward medicine technologies, low industry professionalism, and vague legal regulations. In particular, medicine-operation enterprises only needed a permit from the hygiene administration authority at a county level, so an enterprise in the medicine industry that met the conditions for provincial and municipal policy encouragement would find it very possible to obtain the permit.

Meanwhile, because the 1984 Control Law created extensive regulation of legal responsibilities, mostly adopted administrative punishment as the measure, and applied the option of he word "may" in many provisions regarding relevant responsibilities (note: using "may" for administrative punishment in laws indicates that the administrative authority may or may not make the decision to punish), some enterprises that engaged in illegal production or operation may be released or exempted from punishment due to their potential influence in some regions or the government's consideration of local interests (such as taxes and employment) and thus could continue with production and operation. Even when they were cancelled, they did not bear many relevant responsibilities. Therefore, when some enterprises were closed due to illegal production and operation, the same managers would establish a new enterprise in a short time to continue with the illegal production and operation.

For the above-mentioned two reasons, from the end of the previous century to the beginning of the current century, all regions established different kinds of pharmaceutical enterprises and operation enterprises for the sake of their interests

(GDP, taxes, employment rate). This resulted in production redundancy and duplicate construction, ignored the overall demands of the state and relevant industrial structure and layout, and led to insufficient innovation of medicines and low production technologies, etc., due to the dispersed investment. Put this way, until the 1990s the 1984 Control Law fell behind the need for development of the medicine industry and was unfavorable for industrial security. This phenomenon was ameliorated only by the release of the 2001 Control Law. As already mentioned, the 2001 Drug Control Law expressly specified that "(The establishment of medicine production enterprises needs to) meet the development planning and industry policies of the medicine industry formulated by the state, to avoid duplicate construction" (Article 2.7 of the 2001 Drug Control Law). This independent provision proved from the reserve side that before issuance of the 2001 Control Law, local duplicate construction and redundant production had reached an extreme level in the medicine industry.

This provision also matches the diagram of the changes of the average capital growth rate for 3 years in the medicine industry of China in Fig. 1. Due to the above-mentioned duplicate construction and redundant production, the capital growth rate of such enterprises during 3 years of the medicine industry of China had seen a decline, while the development potential of the whole industry was not optimistic. It was not until 2002 the situation started to improve. With respect to the legal aspect explaining the phenomenon, strict professional supervision led to a decline of the profitability of medicine production and enterprise operation, but the situation rebounded after low-quality enterprises were weeded out. This shows the great significance of the release and execution of the 2001 Control Law and reflects the previous chaos of China's medicine industry.

The total quantity of medicine- production enterprises in China also supports the above analysis. From Fig. 2, we can see that the number of medicine-production

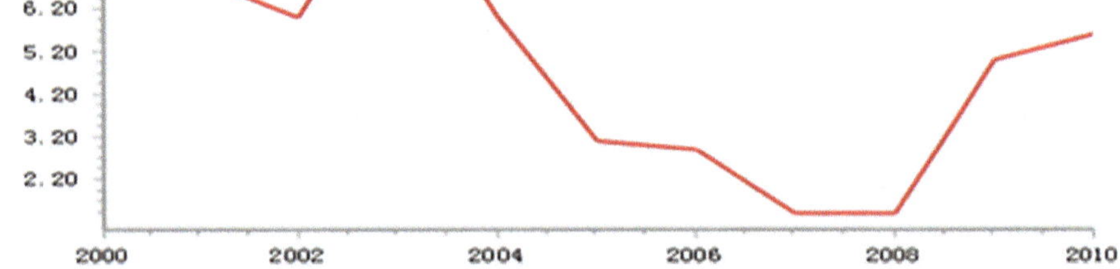

Fig. 1 Chinese pharmaceutical industry: the average growth rate of capital in 3 years

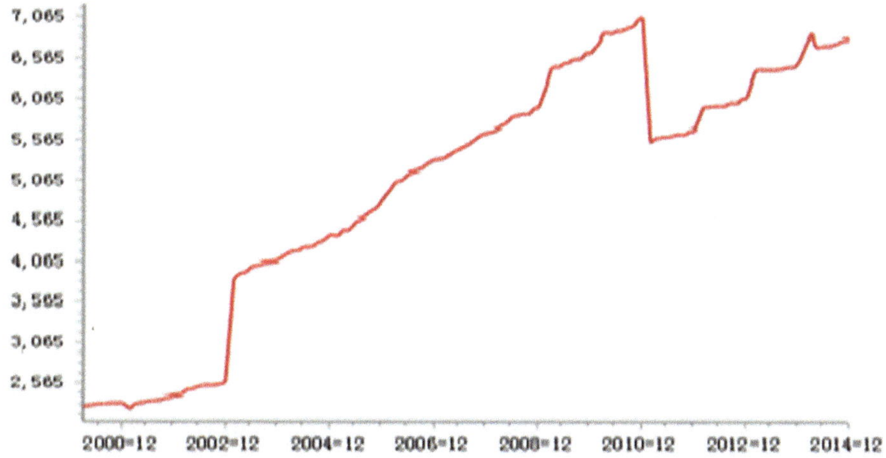

Fig. 2 Chinese medicine manufacturing: enterprise quantity

enterprises of China had sharply increased between December 2002 and February 2003 from 2572 to 3850. This did notoccur because of poor execution of the 2001 Drug Control Law. In contrast, the law was issued in February 2001 and came into effect in December 2001. In the 10 months between the time of the law's release and effect, large numbers of local capitals, especially those with potential advantages or policy support, seized this time to seek establishment and obtain administrative approval. Therefore, by the end of 2001 when the law came into effect, medicine production enterprises had completed administrative approval of the medicine and hygiene system and started other work related to establishment, such as registration at commercial and industrial departments, factory construction, etc. Therefore, by February 2003, a large number of local medium- and small-sized enterprises had completed construction and been incorporated into the total quantity of industrial organizations, thus leading to the sharp rise. It is the same with Fig. 3: Total assets of China's medicine industry showed a significant increase between December 2002 and February 2003.

Finally, it should be clarified that from Figs. 2 and 3, we can see the decline of the quantity of medicine enterprises from 2009 to 2010 as well as small fluctuations in total assets. From a legal perspective, several governmental entities—including the Ministry of Health and State Medicine Supervision—issued documents to require local medicine-supervision departments to strengthen supervision, reduce the checkpoints of medicine circulation, strictly approve new medicine-production enterprises, and strictly control the quantity of medicine production and wholesale enterprises. In 2011, the above-mentioned departments issued notices again making "reduction of quantity of medicine production enterprises" the evaluation standards for work in 2009.

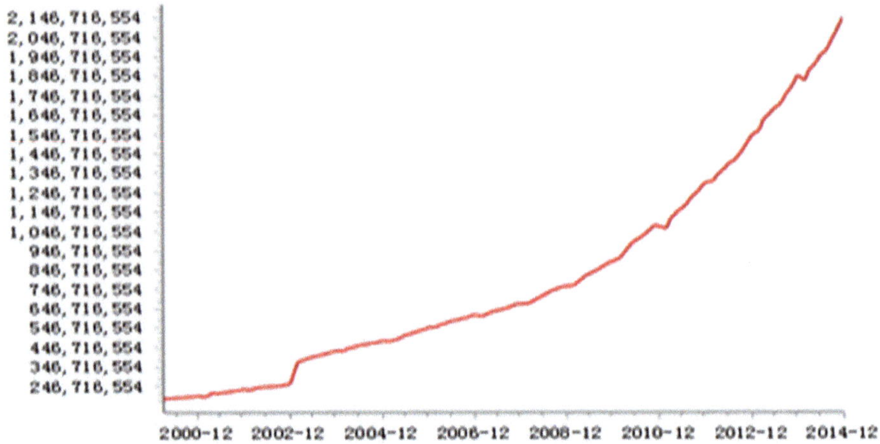

Fig. 3 Chinese medicine manufacturing: accumulated assets

4 Conclusion

Through the above analysis, it can be seen that China's medicine industry has maintained the backward features of low innovation capability, lack of high-technology talent, imbalanced organizational structure, and relative monopoly. Due to these features, the twice-revised Drug Control Law adhered to the legislative trend of "encouraging innovation" and "strengthening supervision." However, owing to the high risk, high investment, and high profits of medicine industry, it can still attract a great number of capitals. With relative monopoly as the backdrop, under nonlegal factors such as the interest consideration of local governments, various types of capitals, especially local capitals with policy support, are still inclined to invest in the industry even without high-technology competitiveness. The result is that despite the fluctuations in the number of medicine enterprises along with changes of the laws and policies, the total quantity of enterprises and industrial assets are rising steadily, but the development potential shows a significant fluctuation along with the changes of supervision. This indicates that China's medicine industry is still in the development stage in terms of both technological competitiveness and management competitiveness. From the aspect of industrial security, it still lacks the ability to resist the entry of foreign advantageous capitals and competition of high-end enterprises. Therefore, the policy protection and support will remain as the focus in recent future to protect the industrial security.

Correspondingly, the law-based adjustment will relatively affect China's medicine industry According to our analysis, the release of a law will reach an obvious effect within 1 or 2 years. The regulations and administrative rules of relevant departments will also be released rapidly. Based on the release and execution of relevant laws, the legislation and jurisdiction institutions can understand and forecast economic trends. Therefore, from the aspect of security of China's medicine industry, the stability of laws and policies can be expected. As can be seen from the figures, the growth of the medicine industry has been relatively steady within a certain time period after major adjustments have been to some policies. In this way, there is no urgent need to change the current industrial policies for China's medicine industry. In addition to domestic supervision, other policy features will remain at a high level including strengthening medicine-quality supervision, encouraging research and development of new medicine and high technologies, protecting Chinese medicine, and attracting foreign capitals. According to the analysis of current laws, regulations, and policy trends, China's medicine industry is expecting the entry of foreign capitals and having to new skills and cultivating talent from foreign enterprises. Therefore, in a predictably short time, the laws and relevant policies for China's medicine industry of China will show no major changes.

References

1. Li M (2012) Research on industry security theories [M], 3rd edn. Economic Science Press, Beijing
2. Li M, Gao X (2014) Report on security of Chinese medical care industry [M]. Social Sciences Academic Press, Beijing
3. Xing Y (2006) Research on medicine industry policies, explorations on economic issues [J], 3rd issue
4. Kong X (2009) Development and integration direction of medicine industry under ackground of new medical reform, medicine and philosophy (Humanistic Society Medicine Edition) [J], 4th issue
5. Yu H (1998) Administrative regulations on Chinese government, reform [J], 3rd issue

Impact of Industrial Development on the Spatial Structures in the Danube Region in Serbia

Slavka Zeković

Abstract This chapter summarizes the key problems of development changes as well as industrial development and its degradation, and provides an analysis of industrial development in the Danube and Belgrade regions (level of NUTS 2). The analysis includes two components: (1) economic growth and developmental changes; and (2) territorial concentration. In the chapter are identified the results of development changes and economic growth by applying shift-share analysis. The chapter shows that the consideration of national share, industrial mix, and regional share in total shift-share employment growth of the regions indicate a strong process of deindustrialization. An allocative component of regional economic growth has a positive value reflecting above-average industrial productivity. Comparative analysis of the regional industrial territorial concentration is based on location quotient. The results indicate a decrease of the territorial industrial concentration in the two Serbian regions.

Keywords Belgrade and danube region · Industrial growth · Development changes

1 Introduction

In the Republic of Serbia, like in other ex centrally-planned economies, long-term development strategies have been performed based on industrialization. Industry played the most important role in contribution in GDP growth, employment, and export. This chapter presents the analysis of key problems related to developmental changes involved in industrial restructuring, decrease of industrial growth, decrease in industrial employment and competitiveness, deep deindustrialization, industrial

S. Zeković (✉)
Institute of Architecture and Urban & Spatial Planning of Serbia, 11000 Belgrade, Serbia
e-mail: slavka@iaus.ac.rs

M. Li et al. (eds.), *Proceedings of 2015 2nd International Conference on Industrial Economics System and Industrial Security Engineering*,
DOI 10.1007/978-981-287-655-3_20

degradation, and weakening of the territorial industrial concentration in the two developed regions during the post-socialist period.

In 2008, GDP growth in Serbia was 3 % with a negative GDP rate in some years, a decrease in competitiveness, a delay in economic reforms, high unemployment, and an increase in regional-development discrepancies. The collapse of industrial development in the Serbian economy, as a key consequence of the transitional development policy, was induced by different factors and contexts. The main contextual factors can be identified in both the transition process as well as the global economic and financial crisis and its repercussions on territorial development. During the post-socialist transition recession in Serbia, 700,000 jobs were lost in industry with almost 1 million unemployed, 1.3 million (or 20 %) poor inhabitants, and bankruptcy of enterprises [1]. In the Belgrade and Danube regions 27.1 % of the total population and 41.6 % of the total employees [1] are concentrated, and they realizing 60 % of the GDP with the allocation of approximately 65 % SMEs of the Serbia.

2 Applied Approach and Methods

Quantitative approaches have been applied using the tools of comprehensive and comparative analysis. Identification of the main effects of territorial development in the two regions is based on the comprehensive development framework approach. The methodological approach involves correlation between the national and regional levels. The suggested approach implies mobilization of strengths and resources in development under conditions of prolonged global economic and financial crisis with emphasizing regional responsibility for structural change and spatial components. The applied comprehensive analysis of the regional industrial development includes (1) the dynamics and components of regional growth and development changes; and (2) the spatial concentration. Comprehensive evaluation of industrial development has an “imperialistic” and “hybrid” character [2]. Evaluation depends on contextual factors and indicators that play an important role at the national and regional levels. Many *quantitative methods* exist for the analysis of regional industrial development including the techniques of economic base, production functions, shift-share analysis, input-output analysis, location quotient, optimization techniques, cost-benefit methods, and qualitative research [3]. The quantitative approaches applied in the comprehensive analysis of industrial development in the two regions included shift-share analysis, Spider method, and location quotient.

Shift-share analysis is a widely applied analytical technique used for retrospective decomposing of changes in employment in different regions. The aim of this analysis is to identify changes in industry with consideration of comparative advantages in particular areas regardless of whether or not they show growth or decline of employment and inhabitants. According to the general form of analysis, total employment in the regional area is e, while employment is the activity ith of the region e_i (e_i^t at the beginning of the period and e_i^{t+n} at the end). Calculation includes the framework of the reference area (country) along with the total number of employees E (E^t at the beginning of the period and E^{t+n} at the end) with employment in the ith activity E_i (E_i^t at the beginning of the period and E_i^{t+n} at the end). The shift-share model is based on employment dynamics in the ith activity of the regional area, which is a function of three components [4]: (1) regional share in national growth; (2) mix of changes in the activities themselves; and (3) shift and change of activities in the region. Changes in employment in the ith activity of the region from the time t to time $t + n$ can be measured by the share, mix, and shift changes due to the following formula [4]:

$$e_i^{t+n} - e_i^t = e_i^t\left[\frac{E^{t+n}}{E^t} - 1\right] + e_i^t\left[\frac{E_i^{t+n}}{E_i^t} - \frac{E^{t+n}}{E^t}\right] + e_i^t\left[\frac{e_i^{t+n}}{e_i^t} - \frac{E_i^{t+n}}{E_i^t}\right]$$

Shift-share analysis is used to determine the contribution of each component to regional growth using the formula:

$$\mathrm{SS} = \mathrm{NS} + \mathrm{IM} + \mathrm{RS}$$

where SS = shift-share, i.e., the share of changes; NS = the proportion of changes at the national level; IM = the share of industrial mix/structure; and RS = regional allocation changes. The component "*national share*" (NS) is measured as the increase of total employment in the regional area due to growth of national economy in the analyzed period. The component "*industry (structural) mix*" (IM) identifies the growth rate of the industry in the region based on the national growth rate for this sector. The allocative component "regional change" (RS), or the competitive effect, is perhaps the most important among the components. It points to the potential and role of leading and lagging industries in the regional/ metropolitan area. The competitive effect compares the growth of the regional/ metropolitan area in the industrial sector with a growth rate of the same sector at the state level (or e.g., labor productivity). A leading industry is the one in which a local area has a higher growth rate compared with that of industry in the state.

The advantage of the shift-share method is that it uses a simple way to decompose the territorial differences in economic or sectoral growth by analysis of the three growth components: structural, competitive/differential, and allocative growth [5].

Location quotient (LQ) is widely used analysis of the economic base as well as measure for determination of the spatial distribution of industry, i.e., the level of spatial concentration of activity in an area compared with that in a larger area. Industry development is measured regarding to the number of inhabitants or total number of employed in an area. The numerical value of this indicator is used as a basis for typology and classification of region compared with the national level, whereas the average value LQ = 1 indicates average industry development. Values greater than LQ > 1 imply a more developed region with production specialization.

LQ 1 implies week industrial development. The formula for calculating LQ (the so-called “Balassa index”) is **as follows** [6]:

$$(\mathrm{LQ}) = \frac{e_i/e}{E_i/E}, \text{or } \mathrm{LQ} = \frac{e_i/e}{s/S}$$

where e_i is the number of people employed in regional industry; e is the total number of people employed in the region; E_i is the number of people employed in industry; E is total employment on a national level, s is the number of inhabitants in the region; and S is number of population in the country [7, 8].

3 Results and Discussion

By the application of shift-share analysis in decomposition of the regional economic growth, we offered in this chapter a way of general assessment to determine which part of the regional differences in average employment can be attributed to specific regional employment and which part to the effects of certain sectoral structures.

Comparative evaluation of the dynamic of development of the two regions was implemented during the period 1990 to 2012. The analysis shows that the greatest contribution to the decline of industrial employment was weak and inappropriate competitive industry structure, then the impact of factors, and then components of national economic growth trends. Due to intensive deindustrialization, both regions are characterized by a considerable volume of adverse impacts of industrial structure with fewer adverse impacts due to national components. The empirical results show that although it shows some negative values, the structural component of the shift-share analysis of the regions shows a slightly better effect of regional economic decline than the national average. Industrial structure and the factors that determine it led to the decline of industrial and overall employment in both areas. This is a consequence of favorable regional conditions (Table 1).

Results of the empirical analysis indicate that the process of deindustrialization, as measured by a drastic drop in employment has been very intense in the Belgrade and Danube regions. Increased employee productivity and favorable regional conditions, as well as better management arrangements, have contributed to

Table 1 Relative values of shift-share analysis of industrial growth in two regions during the period 1990 to 2012

	Belgrade	Danube region
Shift-share (SS)	−0.1010732	−0.200592
National share (NS)	−0.0521483	−0.104350
Industrial mix (IM)	−0.0602689	−0.120599
Regional share (RS)	+0.0113440	+0.024357

Table 2 Dynamics of spatial industrial concentration by LQ in the Belgrade and Danube regions during the period 1990 to 2012

Indicator	Belgrade region	Danube region
LQ 1990	1.31	0.85
LQ 2012	0.47	0.34
Change LQ	−0.84	−0.51
Difference (%)	−278.7	−250.0

alleviation of the overall decline of industrial employment in this area compared with the Serbian average. An allocative component of decomposed growth of the regions has a positive value (+0.0113440 in the Belgrade region and +0.024357 in the Danube region). This shows that the both regions are specialized in the industrial sector, the productivity of which is above the national average.

A strong process of deindustrialization, the concentration of economic activities and productive forces in the both regions, along with increasing geographical differences in the overall level of industrial development, are the consequence of transitional recession and reflect the lack of adequate regional policy, policy of industrial innovation, strong impact of global economic and financial crisis, the use of available territorial capital, and spatial directing of activities.

For analysis of *the spatial industrial concentration* in the Belgrade and Danube regions regarding industry distribution, we used quantitative LQ values expressed by relation of industrial and total employment in these regions and on a national level according to the number of population in these territories. Based on LQ values in both regions during the period 1990–2012, we estimate that there is a significant drop of LQ role in industry development and spatial concentration of industry (Table 2). Spatial concentration of industry in the Belgrade is 2.8 times less than it was in 1990.

Territorial disposition of the location-development potential and resources and trend of growth of the Belgrade and Danube regions (as the most developed) could intensify the increase of regional differences in Serbia [9]. This is a consequence of attractive and competitive conditions for development along the Danube corridor.

4 Conclusions

The empirical results of the comprehensive analysis and evaluation of the three components of industrial development in the Belgrade and Danube regions in Serbia show that economic growth and competitiveness are almost entirely explained by differences in its specificities in terms of employment. The results show that although having negative values, the structural component of the shift-share analysis of the regions indicates a slightly better effect of regional economic decline than the national average. The allocative component of decomposed economic and industrial growth of the regions has a positive value as a reflection of specialization in the sectors of the regions, the productivity of which is above the national average. Shift-share analysis indicates that deindustrialization was very intensive in both regions. Comprehensive analysis of the economic development shows substantive development changes, decreased competitiveness, strong deindustrialization, and higher labor productivity in both regions. Zeković [9] There is a need to support better competition and territorial cohesion of industry in accord with European commitments [10] and the Spatial Plan of the Republic of Serbia [11].

Acknowledgments This chapter is a result of the project "The role and implementation of the national spatial plan and regional development documents in renewal of strategic research, thinking and governance in Serbia" (no. 47014) financed by Ministry of Education, Science and Technological Development of Serbia.

References

1. Vujošević M, Zeković S, Maričić T (2012) Post-socialist transition in Serbia and its unsustainable path. Eur Plan Stud 20(10):1707–1727
2. Dogan M (1996) Political science and other social sciences. In: Goodin R, Klingemann H (eds) A new handbook of political sciences. Oxford University Press, Oxford, pp 97–130
3. Guba EG, Lincoln YS (2005) Paradigmatic controversies, contradictions, and emerging influences. In: Denzin NK, Lincoln YS (eds) The sage handbook of qualitative research, 3rd edn. Sage, Thousand Oaks, pp 191–215
4. Stevens BH, Moore LC (1980) A critical review of the literature on shift-share as a forecasting technique. J Reg Sci 20(4):419–437
5. Esteban-Marquillas J (1972) A reinterpretation of shift-share analysis. Reg Urban Econ 2(3):249–255
6. Balassa B (1965) Trade liberalization and revealed comparative advantage. Manch Sch 33:99–123
7. Vresk M (1980) Osnove urbane geografije. Školska knjiga, Zagreb
8. Krešić I (1982) Prostorna ekonomija - osnove teorije, lokacije, razmeštaja i organizacije u prostoru. Zagreb, Informator
9. Zeković S (2009) Regional competitiveness and territorial development of industry in Serbia. Spatium 21:27–38
10. Zeković S, Vujošević M (2014) Some global challenges for socioeconomic growth in the South-Eastern Europe: the role of industrial policy. Second conference of LSEE's Research Network on Social Cohesion in South East Europe Post-crisis recovery in Southeast Europe and beyond: policy challenges for social and economic inclusion, London
11. Spatial plan of the Republic of Serbia/Prostorni plan Republike Srbije 2020 (2010) Službeni glasnik Republike Srbije: Belgrade, no. 88/10

Value Exploration of Introducing the Theory and Technology of Economic Psychology into Research on Industry Security

Guangli Luo

Abstract Industry security is a component of economic and national security of a country or region. It is an important channel through which the government obtains reliable and valuable data to formulate policies and is the basic parameter essential for national and regional development. It is also an important part of macroeconomics. Research on economic industry security in various countries and regions is long-standing, but most studies have established a theoretical system by relying on policy regulations and economic behavior as a presupposition along with logical reasoning, a lack of economic activity, psychology empirical induction, and empirical analysis research. At the same time, economics began to reflect on postexperience single logical reasoning limitations. It was found that there are important reasons of human economic behavior that affects people's economic behavior: It is a form of psychological motivation. It has been found through long-term study that human economic behavior has its own psychological motivation. To understand the motivation behind the behavior can help us grasp the nature of the behavior more clearly and accurately. The architecture method of psychology's samples method of testing has the advantage of theory confirmation and construction, which the logical deductive method in economics does not have. It is in this context that psychology has a dialogue and fusion with economics resulting in the independent cross-discipline of economic psychology. Economic psychology not only broke the constraints of using purely logical reasoning in economics, it has also produced rich empirical technology; therefore, both early and future theory of economics can use these methods to perform empirical analysis research so that the research results and parameters are more scientific and realistic.

G. Luo (✉)
General Manager of Shanghai Dinglan Business Consulting Co Ltd,
Shanghai 201999, China
e-mail: lgl_0525@sina.com

G. Luo
China Industrial Safety Research Center, Beijing Jiaotong University Postdoctoral
Research Institute, Beijing, China

M. Li et al. (eds.), *Proceedings of 2015 2nd International Conference on Industrial Economics System and Industrial Security Engineering*,
DOI 10.1007/978-981-287-655-3_21

Industry security in the economic field is naturally influenced by the development of economics: It introduces the theory and empirical technology of economic psychology to different degrees and in different ways to make studies more scientific and empirical. This is also the topic of this chapter. Currently evolutionary psychology and neural economic psychology have become hot topics [1]. The effect of psychology on economics has been widely recognized and applied. This chapter uses qualitative, comparative analysis, empirical method, etc.; discusses personal insights on the introduction of the theory and technology of economic psychology; clarifies respective performance effects, advantages, and practical significance of economic psychology and economic security; and discusses the orientation and focus of economic psychology in the industry security system. The innovation of this chapter is as follows: Question the limit of industry security fully established in subjective logic thinking; discuss the importance of introducing psychology to theoretical empirical analysis and technical guidance at every stage of industry security establishment and operation; put forward the author's view of economics from the perspective of psychology; intentionally arouse more voice and special thinking on this subject from different fields; and jointly promote the development of China's industry security research center and international industry security research system. The term "economic psychology" in this chapter refers to the common-concept category of narrow economic psychology with general economic psychology.

Keywords Economic industry security · Economic psychology · Empirical

1 Overview of Industry Security

Security is a basic requirement for both people and countries. Any subject who loses security falls into potential crisis. Industry security is the extension and supplement based on the field of economic security as well as a component of security factors of a country or region. British classical economist Adam Smith put forward the problem of national economy protection from the level of national security. Smith's argument was summarized as a theory of national-defense needs. The view of implementing protectionism for national defense was widely accepted and thus became the basis on which many countries formulate an industry-protection policy [2]. With the development of economic globalization, especially international procurement trade, foreign investment, technology and talent, and other economic behaviors and connotations have become more complicated, economic security and industry security systems have also become more complicated, which poses higher requirements and new challenges for economic security and industry security research. Countries start to explore security research suitable for the formation of theory and policy in terms of realistic significance as well as seek a more realistic and predictable method of establishing a security-value system.

Industry security is closely linked with economic security, defense security, people's happiness, and comprehensive national strength. Determining how to explore a corresponding security system synchronously in the dynamic development of the society and economy is the first priority of a country or region. It is of great social and practical significance that a group working for industry security can explore the results of a "security fortress" with realistic significance. Factors in the formation of security threats are in many ways environmental, economic, cultural, and behavioral factors, etc. The prerequisite to reducing security risk factors is understanding the essence of related factors. The important point of reducing industry security risk and establishing an industry-security system is to understand the factors related to industry security including economic behavior related to industry security, factors related to the establishment of an index system, and the psychology and behavior empirical analysis of relevant people involved in industry security. At present the mathematical model can logically deduce an assumption, but it is isolated from the empirical study of human behavior. Theoretical circles developed economic psychology when they increasingly encountered the limitations of economic logical reasoning. Industry security is a branch and theoretical supplement in the field of economics; thus, it is also suitable for the introduction of economic psychology. To make the industry-security theory system more empirical and scientific by introducing the methodology and epistemology of economic psychology in the whole process of industry security research has great practical significance.

1.1 The Current Situation of the Industry-Security System

Countries have attached great importance to industry-security problems, but there has been a lack of theoretical study mostly in the area of macroanalysis of problems and policy formulation. Thus far, the research on industry security has mainly run through the relevant theories of international trade, especially trade protectionism theory and theory of national industry protection. No independent theory system has been formed, so research in this area has not been developed fully and independently [2]. For example, It has not been linked with research on microeconomic behavior, nor has it formed practical technical departments. Due to its late start of economic development, lack of experience in economic behavior with other developed countries, and an imperfect security system, China's industry-security system has many hidden dangers that must be improved. In 2007, under the efforts of Chinese economists with a broader perspective, Beijing Jiaotong University China Industry Security Research Center was set up. It is currently the largest and most authoritative specialized agency of industry-security research and promotes China's industry-security research to the forefront of the international academic arena, especially the Model and Index Research Institute at the center, which provides information parameters and research models for the development of local industry security, academic construction, and government economic policy; it also plays a positive role in promoting local economic and social development.

1.2 Limitation of the Industrial-Safety System Based on Logical Reasoning

At present, whether in the world or in China, achievements in industry security have been studied based on a certain hypothetical premise in the logical deduction model. However, logical derivation with certain subjectivity and assumption, along with limitations in economic science, may lead to unsustainable results. How can this problem be solved? Modern scientific research shows that single-subject research cannot solve all of the problems in the academic field, and with the appearance of new problems in social development, this limitation will become even more evident. Economic psychology was developed to solve this long-standing weakness in the field of economics.

2 Economic Psychology

2.1 Overview on Economic Psychology

The publication of La Psychologie Economique, which was written by French psychologist G. Tarde in 1902, signalled the birth of economic psychology. After a series of developmental and subject definitions, economic psychology is currently defined in the following way: "Economic psychology is a science to describe, explain and predict the economic behavior of individual small groups (such as family) and large human range (such as all consumers)" [3], and this definition is the basis on which the following thesis is explored.

2.2 Past and Present Development of Economic Psychology

Economic psychology is an applied science to study human economic behavior; explore psychological mechanism and process combined with political economy, psychology, and sociology; and explain and predict human economic behavior. The scope of study covers all aspects from the micro level of the family and individual consumption to the macro level of the behavior of the whole country or region. Interdisciplinary study and the empirical research method of economic psychology intersect with the empirical requirements of the establishment of economics and the industrial-safety system: Make the establishment of the system more practical; and make the system always be dynamic. The method of interdisciplinary combined with empirical research methods can also perform analysis of mental behavior and identify early warning system settings for external factors that can cause potential safety hazards. The method can also determine a model of a self-assessment psychology system, which can reduce safety invasion occurring in a certain geographic range, self-maintain in a perfect dynamic perfect state, and take the trends in industry-security systems to a higher level. The increase of industry-security safety can avoid errors due to personal factors of the system's creators as well as deviations caused by the logical reasoning method while at the same time studying economic behavior and policy to make the system more rigorous and empirical as well as improve the reliability of industrial economic security.

With the integrative development of the world, our understanding of the meaning of the industry security index, as well as the integration of politics and real communication caused by cultural and mental behavior factors in different countries and regions, provides reliable data for the establishment of a more complex and extensive security system. Although psychology is not the main parameter in building the model, it is a necessary one. With the development of economy and science, the special uses and effects of psychology have increasingly been shown: No investigator whose subject of study is human behavior can ignore the research and understanding of the psychological mechanism to achieve benign infiltration of a human population and make positive contributions to it.

Economic psychology is experienced as the dynamic process of "production of concept, subject definition, reach content and objectives, the academic status quo." Economic psychology uses human economic behaviors as the research object to directly observe human economic activities as well as the inner psychological activity factors that can affect economic activities corresponding to external motivation, attitude, judgment, and choice. It takes the psychological variables in the economic field as the main object and studies human psychological activity and the various economic behaviors dominated by such. It especially focuses on discussing the effects of psychological variables on economic activities within a psychology theory framework [3]. Economic psychology's theoretical premise is "the utility theory, the expectation theory, uncertainty and risk theory, game theory," etc. The models under these theories well measure and predict behaviors, as well as the psychological motivation to produce the behaviors, by combining the "psychophysical physical concept," which is the result of consumption and gratification, with empirical study on the internal conversion to materials. In economic psychology, "expected" refers to the ex ante estimation made by the economic parties or behaviors on economic variables (e.g., prices, interest rates, profit, or income) in terms of the direction and fluctuation of future changes. As the characteristic and premise of economic parties' economic activities, uncertainty undoubtedly dominates real human economic behavior. The element of uncertainty is never out of the economist's view for very long. The term "uncertainty" is usually associated with "risk" and "probability." In economic psychology, the reach on uncertainty and risk, such as the "prospect theory" put forward by Daniel

Kahneman, explains the judgments and decision-making behaviors of humans under uncertain conditions. "Prospect theory" can better describe or explain human judgment and decision-making behaviors under uncertain conditions. Especially in financial markets, such as the constantly changing stock exchange market, the prospect theory's explanation and persuasion are stronger than the those of the expected utility theory. Game theory studies the conflict and cooperation between tactful and rational decision-making bodies.

Economic psychology has not established a perfect system, and of course is related to the developmental history of the subject. Because economic psychology has stronger applicability, scholars are not always eager to construct a framework of the subject's system; instead they prefer to focus on hot topics and difficult problems that economics cannot explain in the context of social and economic activities of the period. China's economic psychologist Yuanming [4], using the economic behaviors as a clue and following the usual pattern in domestic teaching materials and the way of thinking more suitable for Chinese scholars, put forward a framework of economic psychology whose center is psychological variables, which better meets the requirements of subject content review and lays the foundation of and provides space for further development of the subject.

2.3 Functions and Advantages of Economic Psychology

According to sociologists' prediction, economic psychology and economics shall be linked with the leading psychology of school subjects in the twenty-first century and learn from each other to form a new "subject for the centuries." They will be the milestones of guidance on the economic forecast now and in the future. Industry-safety reach belonging to the same branch of industrial economics will benefit from achievements made based on interdisciplinary studies. It is also inevitable to optimize its research field.

The emergence and development of economic psychology is based on economics. Through psychology, it explains unaccountable phenomenon that cannot be explained by itself. Finally, through development, it forms an independent interdiscipline whose function is irreplaceable. The main roles are as follows: (1) economic psychology introduces psychological factors and subjective variables (such as motivation, attitude, emotion, expectation, etc.) into the research on economic phenomena, which can better analyze economic phenomenon such as investment, unemployment, stagflation, etc. to grasp the rule and mechanism, and improve the understanding of economic behavior and phenomenon; (2) economic psychology involves the conduct of psychological applied research on hot topics and difficult problems in the field of social economy to help businesses and economics operate better and enhance the scientific basis and foreseeability of decision-making; (3) economic psychology analyzes practical problems arising during the development of China's economy, as well as the regional economic status in other counties in the new stage, new normal, and new background, in the context of corresponding theories in economic psychology; and economic psychology combines with the scientific appraisal made by economics on all economic policies and plans to provide theoretical support and reference information for the development of the macroeconomic policy formulated by the government; (4) through tracking and studying the irrational behaviors in markets with import problems in social and economic activities, such as group violence, parade, blind investment behavior, and population characteristic, etc., economic psychology determines potential dangerous problems related to human social life, provides early warning, and suggests preventive measures that contribute to the stability of the socialist economy as well as harmonious social development.

The system of economic psychology is still in the process of development, but it is indeed its own independent subject. Based on the traditional research methods of psychology and economics, economic psychology has gradually formed a set of unique research methods such as experimental, case study, sampling, market test, scaling (such as that used for consumer sentiment index), direct measurement, and in-depth conversation. It has a mode of three factors (icon prediction, etc.). It can use its own theoretical model to give reasonable explanations for various kinds of economic behaviors, and it shows the psychological activity routines of people who conduct economic behavior and analyzes the routine effect on economic behaviors. At the same time, because of the reality, applicability, maneuverability, positivity, effectiveness, and other characteristics of economic psychology, it has become a highly practical subject. The logical starting point of economic psychology's research is psychological and subjective factors, which are psychological variables that can affect human economic activities. Its central logic is also psychological variables. If we gather the main subjects of economic psychology research in both China and abroad and run them through the whole body of economic psychology research, Research shows that psychological variables can be found in all economic behavior and that psychological variables play a specific role in all economic behaviors.

3 Introduce the Value of Economic Psychology into the Study of Industry Security

3.1 Description of Introducing the Value of Economic Psychology into the Study of Industry Security

The Resolution Study of Economics Limitation (i.e., the problem that cannot explain the mental motivation behind economic behavior) is proceeding under efforts of both economists and psychologists. A series of subsystems in the field of economics, such as industrial economy and industry security, will also benefit from the exploratory development of economic psychology. Introducing economic psychology into the systematic study of industry security will have a significant impact on the study of industry security: A subversive new viewpoint will bring new enlightenment and results to researchers as well as to the subject as a whole. The values based on present study are summarized as follows:

1. According to results by deduction or calculation of logical thought and the actual presentation of how they complement each other, economic psychology can further adjust the inherent limitations of subjective and logical reasoning and reduce deviation.
2. Economic psychology makes study execution and success closer to reality.
3. Economic psychology test the psychology and behavior of the researcher and mode maker and avoid connatural defects and mistakes that are caused by nonrational reasoning.
4. By forecasting and planning for the professional psychological technology of external security risks, economic psychology gains the source path and the real reason for the risk and protects safety objectively. Studies such as prospect theory, framing effect, anchoring effect, people's judgment, and reasoning of the in psychology of decision making all create hitherto unknown and new concepts for the economic field. It allows the study be more scientific and the results be more quantifiable, appropriate, clear, and easy to understand, operate, copy, and evaluate.

3.2 Methods of Introducing the Value of Economic Psychology into the Study of Industry Security

Introduction of the value of economic psychology into the study of industry security must proceed with purpose, steps, and a plan. First, according to different purposes, identify the goals and size of the study and then select a relevant psychological theory and a corresponding index; second, mix and adjust it with the index system selected by the security model of economic industry; third, use the empirical method to obtain the index and determine the effect of the model, compare it with the plan data, adjust the index data and effect, and finally, after several corrections have been made, identify a new model that is closest to reality.

To improve the effectiveness of the final model, at the initial stage of selecting a design the model must identify the psychological samples and behavioral indicator of the study's research objectives, conduct psychology analysis, and determine sample collection for each step of the changes in economic behavior as well as the participators; meanwhile, it must consider the psychological behavior factors of the researchers at each step (i.e., pay attention to the index data for changes of their emotion and condition) and let the subjective emotion affect it as best as possible. Find out the unusual thing and show and correct it according to the standard in time, and fully exclude the unscientific data formed under abnormal conditions. Then further study the data, and adjust and compare it with those found in the last step.

At the end, form a scientific index based on rational logic and irrational demonstration. Meanwhile, remember to build a timeliness mechanism (let every model index only work in its valid time; if it exceeds the valid time, correct it and develop an appropriate new index), and make the reliability of the model index always be optimal under dynamic conditions.

There are many methods of economic psychology study and demonstration. Due to the length limit here, I have described only a few of them. Methods based on the reference and demonstration of the industrial-security model under economic behavior include the prospect theory of decision-making psychology in the context of economic psychology, the cognitive mechanism of cooperation behavior in gambling psychics, and the brain-imaging study of an individual's emotional experience in neuropsychology. They all make the study and production process of the model index more close to that of the study with more psychological motivation factors and thus closer to actual human behavior and decision-making. The psychologist, for example, regarding the study of industrial security, believes that uncertainty factors of decision-making in the work is something one must experience, but the unknown situation and risk caused by the uncertainty factors of decision-making cannot be obtained by logical reasoning, We can use the prospect theory of economic psychology, as well as the empirical method of economic psychology, if the uncertainty event occurs in the industrial security's outer model and submodel. Of course, more technical psychology methods can be used in the study of industrial security, but they require researchers to continually discover and use them. During the sampling and proving process, perform psychology research evaluations of the illogicality of the study object of industrial security, minimize the influence of the sample and the study object by subjectivity, and determine the sample that is closest to the psychological model. According to this process, including that of further empirical study, the whole process dynamically integrates the mental motivation factors of economic behavior, thus making the sure study always combines logical and empiric elements. Two powers from two different subjects make the study of economic psychology more scientific and practical and avoids the downfalls occurring from the deviation of single logical derivation from material reality.

3.3 Some Advice for Industrial Security Study Based on the Empirical Study of Economic Psychology

This study of economic psychology stands at the angle of economic psychology. My present personal opinion and hope is that I can give some information and advice to this field with a different study perspective as well as lend give more theory and technology that can be practically introduced into economic psychology in the study of industrial security. Because the study mode considers two important subjects—psychology and economics—it should make the study of industrial security be more practical and valuable. Try to avoid the abuse caused by the limit of the single staff (logical reasoning) and the limit of the event itself (economics).

4 Analyze the Future of the Economic Psychology and Industrial Economics Combined with Economic Development

With further development of economic globalization, complication of the economic situation will pose challenges to industries in each country, area, and field. It will become more complicated to create an industry security policy. To understand recent economic behavior and better forecast the future, the models and methods formed using only mathematical models and logical reasoning will not be adequate. Detailed study of the economic behavior and psychology of relevant events as well as personal psychological patterns will be fully used. The fusion of these two subjects not only push development itself, but it also improves the subject of artificial intelligence. Meanwhile it can also make the subject of improved artificial intelligence serve the subjects psychology and the industrial economics. In the end, each subject will gain new improvement during their mutual penetration and serve the community's economy and industry security in a better way.

5 Conclusion

This chapter is not limited to the ideas and suggestions put forth by the author for his country, it also proposes an opinion based on the academic sector of industry security. The writer believes that based on the importance of industry security and its advantage for economics as well as the industrial economy and its security, economic psychology will become more valued by society, and these two subjects will be more improve and develop together. Industrial economics and psychology are two absolutely necessary powers in future socioeconomic development, and together they will maximize each other's value in the development of the social economy.

References

1. Ziqiang X (2014) Economic psychology P8, Beijing
2. Li M (2006) CNKI.net research on industry security theory
3. Guangrong W, Fanyu Z (2010) CNKI.net economic psychology: integration an innovation
4. Yuanming (1999) Pay attention to knowledge innovation, develop high-tech industry. Sci Tech Prog Policy 4(16):34–35

The Impact of Safety Climate on Safety Behavior for Drivers in High-Speed Railway Industry Using Safety Control as Mediator

Li Zhang and Yingqi Liu

Abstract The chapter is based on drivers'-safety management of China's high-speed railway. It shows the empirical process of developing a safety climate to promote safety behavior. Data were collected by means of questionnaire from 161 drivers in 7 workshops of the high-speed railway. Data were analyzed using two approaches: liner regression and mediating-effect test. Results showed that individual perceptions of safety climate exerted a causal effect on individual safety behavior. All facets of safety climate were significantly associated with safety behavior. Furthermore, we found that the relationship was mediated by the individual's control of safety. These findings highlight the importance of high-speed driver characteristics that improve safety performance.

Keywords High-speed railway · Safety climate · Safety behavior · Safety control · Multiple analysis

1 Introduction

In China, the total mileage of the high-speed Railway has exceeded 11028 km since 2008. However, the consequences of trains operating at speeds >300 km/h present some concerns in terms of operation. Safety, stability, and punctuality of high-speed railway. Meanwhile, some strict requirements regarding hardware facilities of the railway, working environment, and skilled employees have been put forward. The status of drivers, currently a hot topic, impacts the safety and sustainability of the traffic system (Wang and Guo [1]). Currently, railways rely almost exclusively on

L. Zhang (✉) · Y. Liu
School of Economics and Management, Beijing Jiaotong University, 100044 Beijing, China
e-mail: 12120603@bjtu.edu.cn

Y. Liu
e-mail: liuyq@bjtu.edu.cn

M. Li et al. (eds.), *Proceedings of 2015 2nd International Conference on Industrial Economics System and Industrial Security Engineering*,
DOI 10.1007/978-981-287-655-3_22

retrospective data, to gauge levels of organizational safety. Yet, increasingly, railways are looking at the practices of other high-reliability industries (i.e.,, industries with low rates of incidents but potentially high associated costs (e.g., nuclear power plants)) and using leading indicators of safety to aid in identifying unsafe conditions before incidents occur [2]. Safety climate is a prime example of a leading safety indicator because it provides evidence of the psychosocial conditions that may encourage or discourage safety behavior. For high-speed railways, the safety environment impacts on the perception of the drivers. Improving safety requires a continuous refinement of practices and a strong commitment to prevention, especially when the risk involves human lives.

The relation between safety climate and safety behavior is clear, but the mechanism is unclear. To better understand the causal relationships between safety climate and safety behavior, more studies based on multiple measurement points are needed. Huang et al. [3] found that a plausible effect of implementing safety policies on safety outcomes is likely mediated by employees' perceived control on safety, which builds on the results of their earlier study. They hold that the existence of a safety climate is an objective of a safety culture. What precautions individuals choose to take depends on the degree of control of safety risks and the ability to improve or eliminate unsafe conditions in the workplace. With safety control, a safety climate will influence employees' safety behaviors.

2 Literature Review

2.1 *Safety Climate and Safety Behavior*

Zohar [4] defined safety climate as the coherent set of perceptions and expectations that employees have regarding safety in their organization. Among the research, Christian et al. [5] categorized safety-climate perceptions into two levels of analysis resulting in shared group-level safety climate and individual/psychological safety climate. They defined psychological the term "safety climate" as an individual's perception of his or her working environment that arises from his or her cognition and sense-making [6]. A shared group-level safety climate emerges when individual workers' perceptions are shared and consensus forms within a particular work environment. Due to the fact that this study involved high-speed drivers who rarely interact with other coworkers, we analyzed just the individual level.

Zohar and coworkers suggest that, based on shared perceptions of management safety commitment, that employees infer the relative value of safety performance in the organization. Neal and Griffin [7] and Griffin and Neal [8] identified two types of safety outcome: compliance and participation. Safety compliance is defined as adhering to safety procedures and carrying out work in a safe manner, whereas safety participation is safety-oriented behavior that involves the individual participating in safety meetings, setting safety goals, and providing safety suggestions within the organization.

A psychological safety climate is a valuable addition to the safety literature in that it can help explain workers' safety behaviors. Recent meta-analyses suggest a positive relation between safety climate and safety outcomes (Beus et al. [9], Christian et al. [5], [10]). For instance, research using samples of workers from the fields of construction, manufacturing, and other modes of transportation generally show that a positive safety climate is associated with greater compliance with safety standards [11, 8] and a lower incidence of workplace accidents (Mattila et al. 1994), [12] Railway-specific safety-climate research is sparse, so it is expected that the relationship between safety climate and safety behavior will hold for railways drivers. However, there are also noteworthy exceptions suggesting that this relationship is not always so straightforward. Meanwhile, research suggests that differences in culture, industry, or job position may partially account for the failure to replicate factor loadings across studies.

2.2 *Safety Control*

Karasek and Theorell [13] contended that the job environment alone does not lead to psychological or physical strain (e.g., distraction or fatigue). Instead, job strains resulting from demands at work can be exacerbated if workers lack control in their jobs. It is not difficult to infer that an antecedent of behavior may be control, i.e., behavior is affected by one's control. Safety control is a person's perception of the ability or opportunity to manage work situations to avoid injuries and accidents [14]. However, little research has been performed to evaluate evaluated the potential effects of employee safety control on safety behavior and accidents. Harris (1998) reported that safety control was negatively related to unsafe behaviors but was ineffective in predicting injuries. Simard and Marchand [15] also showed workers' control as a significant predictor of employee involvement in promoting safety.

3 Method

3.1 *Participants*

A total of 161 participants were railway drivers from 7 different workshops. Of these, 158 returned completed surveys (98.1 % response rate). The majority of the sample (73.4 %) reported having received safety training between 6 and 11 times per year. In addition, 61.5 % reported working >5 years with the railroad, and 53.2 % of the sample was age 35 years or older. In terms of educational level,

55.7 % had a high school diploma, 35.4 % had a college degree, and 7.6 % had a university degree or higher. All workers in this organization belonged to the different unions. Participants were asked to indicate agreement using a five-point Likert scale.

3.2 Measures

The safety-climate facet scales and items included in those scales were selected based on those of similar transport industries. The results of this work indicated that safety climate (i.e., commitment to safety, previous safety, work-safety tension, safety communication) was measured with 17 items. The coefficient alpha was 0.820.

Safety behavior was assessed using two items from Neal and Griffin [7]: compliance and participation. Thirteen items were included in the current study. The coefficient alpha was 0.668.

The job-stress literature has conceptualized control as operating in conjunction with two related variables: understanding why and how stress-causing events occur (Snyder et al. [16]). Snyder et al. [16] developed a safety-control scale on the basis of the definition of safety control. The scale consists of seven items. The coefficient alpha was 0.488.

4 Results

In terms of the safety awareness and perception of high-speed drivers, this study designated X as the safety climate to present, which includes three different aspects: previous safety, work-safety tension, and safety attitudes of leaders. Meanwhile, safety behavior was designated Y, which consists of safety compliance and safety participation. Safety control, designated M, is a mediating variable involving five questions that represent the control of risk by high-speed drivers. Furthermore, all of the variables are processed by a significance approach. Table 1 illustrates the regression result. Obviously, the coefficient is significant ($P < 0.05$), and the coefficient of c is 0.517.

Table 1 Coefficients of safety behavior on safety climate

Model		Unstandardized coefficients		Standardized coefficients	t	Sig.
		B	Std. error	Beta		
1	(Constant)	0.000	0.027		−0.006	0.995
	X	0.444	0.059	0.517	7.543	0.000

Table 2 Coefficients of safety control on safety climate

Model		Unstandardized coefficients		Standardized coefficients	t	Sig.
		B	Std. error	Beta		
1	(Constant)	0.002	0.053		0.031	0.975
	X	0.342	0.114	0.234	3.000	0.003

Table 3 Coefficients of safety behavior on safety climate and safety control

Model		Unstandardized coefficients		Standardized coefficients	t	Sig.
		B	Std. error	Beta		
1	(Constant)	0.000	0.027		−0.011	0.991
	X	0.417	0.060	0.485	6.940	0.000
	M	0.081	0.041	0.138	1.983	0.049

Respectively, this result is not consistent with those of Kuenzi and Schminke [10]. Beus et al. [9] actually found that occupational injuries were more predictive of safety climate than safety climate was of injuries. Indeed, the organizational climate theory supports the view that changes in safety behavior will promote safety policy changes, thus resulting in safety-climate changes. However, although safety policy and management come from an organizational point of view, this study took the perspective of individual psychological control to examine the impact of perceived climate on individual behavior. The result suggests that the safety behaviors of drivers in China's high-speed railway are influenced to a large extent by the railways safety climate. This may be due to the Chinese situation of railways being the country's economic lifeline, which are strictly limited by administrative personnel. Compared with safety-policy changes, changes in the behavior of drivers occur more frequently. To further explore the mechanism of organizational climate on driver's behavior, i.e., determine the mediating variable, this chapter will continue to explore and analyze the connection between them.

Table 2 shows that the coefficient is not insignificant ($P < 0.05$); at a significance level of 5 %, the coefficient of a is 0.234. Consequently, Table 3 shows the coefficients of safety behavior on safety control.

It is significant as well. The coefficient of b is 0.138. Analyzing the result of regression 3, the coefficient of c' is 0.485, which is significant.

Above the theory of Baron and Keriny [17] is satisfied. They suppose that if the coefficient is less than the critical value and is furthermore significant as well, it can be concluded that the effect of the mediating variable is partial. Therefore, safety climate will impact safety behavior either directly or indirectly through the mediating variable of safety control.

In conclusion, a, b, and c are significance. Thus, the mediating effect of safety control is obvious. Safety control is a partially mediating effect. Safety climate does not completely impact safety behavior by way of safety control. The effect of the mediator is 0.0624, which is calculated as follows: (Effect m = ab/c = 0.234 × 0.138/0.517 = 0.0624). The variance of the dependent variable explained by the effect of the mediator is 11.4 % (sqrt(0.276–0.263) = 0.114).

5 Conclusions

This study, based on China's high-speed railway, conducted empirical research on the safety climate and safety behavior of high speed railway drivers. It was found that safety control is a mediator between safety climate and safety behavior, which depends to some extent on the level of drivers' individual control. Generally, the higher the level of safety climate, the more frequent occurred safety behavior of drivers. However, if the degree of safety control is weak, e.g., weak ability to deal with emergencies and incorrect drive behavior, the accident and injure will occur. Drivers should also improve their personal safety control because this can play a role even if the safety climate is not strong. Conversely, the influence will be more powerful.

This study still has some limitations. It is evidence and refining based on the results previous studies and further showed the important role of safety control as a mediator between safety climate and safety behavior. However, these conclusions rely on cross-sectional studies and a lack of longitudinal studies. In current studies, researchers have recognized the positive role of safety climate for employees, but it they have still been performed within the outcome variables of safety climate. In fact, exploring the antecedents of safety climate may be more meaningful. How to improve safety climate resulting in safe employee behaviors will be an important direction for future research.

References

1. Guo H, Wang W, Guo W, Jiang X, Bubb H (2012) Reliability analysis of pedestrian safety crossing in urban traffic environment. Saf Sci 5(4):968–973
2. Flin R, Mearns K, O'Connor P, Bryden R (2000) Measuring safety climate: identifying the common features. Saf Sci 34(1):177–192
3. Huang YH, Chen PY, Krauss AD, Rogers DA (2004) Quality of the execution of corporate safety policies and employee safety outcomes: assessing the moderating role of supervisor safety support and the mediating role of employee safety control. J Bus Psychol 18(4):403–426
4. Zohar D (1980) Safety climate in industrial organizations: theoretical and applied implications. J Appl Psychol 65(1):96
5. Christian et al. (2009) WASP-10b: a 3MJ, gas-giant planet transiting a late-type K star. MNRAS 392(4):1585–1590
6. James LA, James LR (1989) Integrating work environment perceptions: explorations into the meaning of meaning. J Appl Psychol 74:739–751

7. Neal A, Griffin MA (1997) Perceptions of safety at work: developing a model to link organizational safety climate and individual behavior. The 12th annual conference of the society for industrial and organizational psychology, St. Louis, MO
8. Griffin MA, Neal A (2000) Percetions of safety at work: a framework for linking safety climate to safety performance, knowledge, and motivation. J Occup Health Psychol 5(3):347–358
9. Beus J, Payne S, Bergman M, Arthur W (2010) Safety climate and injuries: an examination of theoretical and empirical relationships. J Appl Psychol 95(4):713–727
10. Kuenzi M, Schminke M (2009) Assembling framgents into a lens: a review, critique, and proposed research agenda for the organisational work climate literature. J Manag 35:634–717
11. Goldenhar LM, Williams LJ, Swanson NG (2003) Modeling relationships between job stressors and injury and near-miss outcomes for construction laborers. Work Stress 7(3):218–240
12. Clarke S (2006) Safety climate in an automobile manufacturing plant: the effects of work environment, job communication and safety attitudes on accidents and unsafe behavior. Pers Rev 35:413–430
13. Karasek R, Theorell T (1990) Healthy work: Stress, productivity, and the reconstruction of working life. Basic Books, New York
14. Anderson L, Chen P, Finlinson S, Krauss AD, Huang YH (2004) Roles of safety control and supervisory support in work safety. Society of Industrial/Organizational Psychology (SIOP), Chicago
15. Simard M, Marchand A (1995) A multilevel analysis of organizational factors related to the taking of safety initiatives by work groups. Saf Sci 21:113–129
16. Snyder et al. (2001) Internalization of ionotropic glutamate receptors in response to mGluR activation. Nat Neurosci 4:1079–1085
17. Baron RM, Kenny DA (1986) The moderator-mediator variable distinction in social psychological research Conceptual, strategic, and statistical considerations. J Personal Soc Psychol 51:1173–1182

Part III
Empirical Studies

Synergy Mechanism Research Based on the Integration of Beijing-Tianjin-Hebei Region

Shaoqing Tang, Yating Sui, Jianling Li and Jianbin Chen

Abstract Integration of the Beijing-Tianjin-Hebei (B-T-H for short) region development has raised the level of to national strategy, and industry synergy is the essences of integration of the Beijing-Tianjin-Hebei region. It is also the power of Beijing-Tianjin-Hebei regional integration, so research on synergy mechanism is very important. In this article, we starts with B-T-H regional industrial comparative analysis, determine the breakthrough point of collaborative mechanisms, select the operational strategies, and finally explore an industry-coordinated development path.

Keywords Regional integration · Metropolitan circle · Industry synergy

1 Introduction

Regional integration is the advanced stage of rural-urban integration, and it is also the product of the development of rural-urban integration. The key to its realization is industry-coordinated development. The function of the industry-coordination mechanism is achieving joint development and a mutually beneficial situation by industrial collaboration in the B-T-H metropolitan area.

S. Tang (✉)
Business College of Beijing Union University, Beijing 100025, China
e-mail: 13910144495@139.com; shaoqing.tang@buu.edu.cn

Y. Sui · J. Li
Department of International Business, Business College of Beijing Union University, Beijing 100025, China

J. Chen
Research Administration Office, Business College of Beijing Union University, Beijing 100025, China

M. Li et al. (eds.), *Proceedings of 2015 2nd International Conference on Industrial Economics System and Industrial Security Engineering*,
DOI 10.1007/978-981-287-655-3_23

Xi Jinping, the general secretary, specifically presented on February 26, 2014, at the Coordinated Development of B-T-H Work Symposium. To realize the strategy for the coordinated development of B-T-H, we must first solve the coordinated development of industry; therefore, the key to realizing the integration of the B-T-H region is industry coordination. Realizing the coordinated development of the B-T-H industry will ensure the development of B-T-H in terms of sustainable and stable growth.

2 Comparative Analysis of the Regional Industry

2.1 Primary Industry of the B-T-H Region

The primary industry of Hebei accounts for a significant proportion of primary industry. The primary industry of Beijing and Tianjin, however, accounts for smaller proportion. The added value of primary industry of Beijing has increased, but its proportion has been declining. The situation of Tianjin is similar to that of Beijing: Only the added value of the primary industry of Hebei is increasing and the proportion remains unchanged. That is to say, the problems of agriculture, rural areas, and peasants still exist in Hebei. To solve the problem of peasants, the population must be transferred from the primary industry to secondary and tertiary industries, and the population transfer should be achieved in an orderly fashion among the industries.

2.2 Secondary Industry of the B-T-H Region

In B-T-H region, the secondary industry of Tianjin and Hebei accounts for a large proportion and has similar structure. From 2009 to 2013, although the added value of the secondary industry of Beijing continued to grow, the proportion continues to decline; the added value of the secondary industry of Tianjin also continues to grow, but again the proportion continues to decline; and the added value of Hebei increases considerably, but the proportion remained unchanged.

2.3 Tertiary Industry of the B-T-H Region

From 2009 to 2013, the absolute value and the proportion of the B-T-H tertiary industry continue to grow indicating that the structural adjustment of B-T-H has been optimized. The tertiary industry of Beijing has the obvious advantages: It maintains an overwhelming superiority with regard to both quantity and quality;

the industrial structure of Beijing approaches that of developed countries; and while Tianjin and Hebei proportions are lower by 30 and 40 %, respectively, than Beijing, this still shows that the development of Tianjin and Hebei is at an typical industrialization stage.

3 Analysis of Industrial Synergy Development

The B-T-H region has the special historical bond. Economic and social development of the three regions include gradient, complementary, and mutuality so there is a good economic cooperative foundation in the B-T-H region. We can analyze the synergy development of integration of the B-T-H region based on the following four dimensions.

3.1 Role of Government and Market

Coordinated development means that different areas within and outside of these regions carry out cooperation among local governments for mutual benefit. The role of government in the process of regional cooperation is mainly promoting the rational flow of resources and correcting market failure to realize regional synergy development. The role of the market is for industry convergence and collaboration development in the B-T-H region.

3.2 Gradient Development

Due to the economic and social development gap between different regions, gradient development becomes the periodical characteristics and endogenous power of regional cooperation. Each region can make a reasonable division of labor and technology transfer to achieve regional harmonious development based on the theory of gradient development. According to different country income group standards of the World Bank, Beijing-Tianjin regional development has reached the level of wealthy countries; however, Hebei is at the middle-income level. In 2013, the proportion of tertiary industry reached 76.9 % in Beijing, while in Tianjin it is 48.1 % and in Hebei is 35.5 %. There is a large gap in economic and social development. The character of gradient provides theoretical preconditions to the B-T-H regional cooperation and promotes orderly industry transformation to realize a mutually beneficial situation among these three regions.

3.3 Complementary Development

Complementary development confers the possibility to benefit from trade based on comparative advantages in various areas. According to the theory of comparative advantage, each region should focus on industries having a comparative advantage and reap mutual benefits through regional trade. The factor endowments of B-T-H are different. Beijing is rich in education, science, and cultural resources. Tianjin has advantages in shipping, logistics, and manufacturing. Hebei has a cost advantage in labor force and land. The regions should complement each other effectively to enlarge their own local advantages. Complementarily has created the conditions for leverage development in different regions. The development of a single region tends to be limited by local resources at a certain stage. If different regions can complement each other, the constraints could be alleviated effectively; Beijing and Tianjin both face many problems of cities such as excessive population, traffic jams, and air pollution. Hebei province has a large space-bearing capacity for industry, but because of a shortage of developmental resources the pressure for industrial transformation and upgrading is high. Regional cooperation can achieve regional complementary advantages and promote regional sustainable development.

3.4 Symbiotic Development

Symbiosis refers to the character of interdependence among all the areas within the region. The reason for the existence of the symbiosis lies in the externality of regional public goods. Air pollution, water-resources exploitation and use, water pollution, and food safety all have the characteristic of cross-domain flow, but they cannot be solved within a single region. Only a systemic solution can be put forward through regional synergy in the integration framework. At present, B-T-H is facing common problems of serious air pollution and a shortage of water resources, so it urgently needs to strengthen regional cooperation and promote orderly regional overall integrated development.

4 Breakthrough Point of Synergy Mechanism

4.1 Circular Economy Strategy

The largest disagreement about industrial docking in the three regions is this: Beijing is unwilling to give up high-end industries, whereas Tianjin and Hebei seem unlikely to accept low-end industries. Undertaking Beijing and Tianjin industry should not pollute the environment! This is what Beijing-Tianjin-Hebei should insist along its road to sustainable development.

If the traditional industries could reuse waste as a resource and reduce pollution through technological innovation, transformation, and upgrading, then it can realize a recycling and green economy.

For example, the byproduct of heat from China Resources Power can be supplied to the surrounding businesses and residents, and the afterheat can be used for the desalination of seawater. Likewise, the waste heat of Shougang can be used for desalination as well, and the concentrated salt produced by desalination can be supplied to Sanyou Chemical. The Caofeidian desalination project will become an important water source in Beijing, and soon one million tons of Bohai seawater will be supplied to Beijing every day after desalinization.

4.2 Scientific and Technological Innovation Driven Strategy

First, the international development of Beijing should be synchronized with the coordinated development of the Beijing-Tianjin-Hebei region. On one hand, by introducing the latest international technology and a cutting-edge innovation team, industrial parks such as Zhongguancun will obtain better and faster development to accelerate the innovation function in the capital city. The effect of spillover and radiation must contribute to industry consolidation and integration in the coordinated development of the B-T-H region.

Second, industrial parks such as Zhongguancun should do more work to nurture and guide services. If there is a well-developed infrastructure in the B-T-H region, it could obtain greater achievements in terms of industrial cooperation and technology synergy.

5 Conclusion

First and foremost, to achieve the industrial coordinated development of the B-T-H region, the government should formulate policies for B-T-H industrial development policy guidance and reasonably allocate the industrial division of the network cities to form an orderly industrial chain. Second, resources should be fully used relying on new technologies, new processes, and producer service to achieve a circular economy. Finally, development of the B-T-H region should be sustainable so that balanced economic development between the regions can be achieved.

Empirical Analysis of Agricultural Exports Competitive of Henan Province

Mei Feng and Yafen Yan

Abstract Great changes have taken place in the agricultural environment in Henan province since China's entering the WTO. In this chapter, the present situation of agricultural-product export trade in Henan was chosen as the breakthrough point and indexes such as export market share (MS) and revealed comparative advantage (RCA) were selected to analyze and evaluate the agricultural-product export competitiveness of Henan. Accordingly countermeasures and suggestions to enhance the competitiveness of its exports were put forward.

Keywords Henan province · Agricultural products' export · Export competitiveness

1 Introduction

Export competitiveness, namely the international competitiveness of export products in foreign markets, is the main topic of this study, the aim of which is to explore the advantage of a country's goods or services in the overseas market as well as the country's ability to capture the overseas market. The earlier research on international competitiveness can be traced back to the 18th century with Adam Smith's theory of absolute cost to David Ricardo's theory of comparative advantage and then to the subsequent factor endowment theory of Eli F. Heckscher and the following human capital theory. All of these theories have had an important influence on the formation of market competitiveness. Hongjian Zhang used the methods of revealed comparative advantage (RCA), trade competitiveness index (TC), and international market share to calculate the level of the international competitiveness of China's exported agricultural products and concluded that

M. Feng · Y. Yan (✉)
Donlinks School of Economic and Management, University of Science and Technology Beijing, Beijing 100083, China
e-mail: bianmumian@126.com

M. Li et al. (eds.), *Proceedings of 2015 2nd International Conference on Industrial Economics System and Industrial Security Engineering*,
DOI 10.1007/978-981-287-655-3_24

combined efforts from the government, enterprises, and agriculture are needed to improve our the international competitiveness of China's agricultural products [1]. According to the above analysis, the domestic and foreign literature on the export competitiveness of agricultural products is greater, and the research direction and emphasis also varies. Most research has been analyzed from the macroscopic level, whereas research on the regional export competitiveness of agricultural products export is less.

Henan province is an important food provider in China, and the agricultural production base has long undertaken the task of supplying our country's food and promoting the steady development of agriculture. After joining the WTO, the domestic and international markets gradually became integrated, and competition in the agricultural market has given priority to domestic and international competition and the new trend of the integration of international and domestic agricultural markets and domestic competition [2]. Therefore, the export competitiveness of agricultural products in Henan province was analyzed using the method of empirical analysis, and countermeasures and suggestions to enhance the competitiveness of Henan agricultural exports are put forward.

2 Current Situation of Henan Agricultural Products' Export

2.1 The Overall Analysis of Agricultural Products' Exports

In recent years, exported agricultural products in Henan province have showed an increasing status. Agricultural exports sales increased from $411 million in 2005 to $966 million in 2012, but the agricultural product export trade accounted for a tiny proportion of agricultural output; meanwhile the proportion of export trade in the province's total trade declined gradually from 8.10 % in 2005 to 2.96 % in 2011. From the perspective of the agricultural-product export trade of the country, the proportion of agricultural exports sales of Henan province across the country witnessed an increased trend, but the proportion is still beyond the national tenth, thus indicting a relatively weak competitiveness of Henan agricultural-products export, which must be improved.

2.2 Structure Analysis of Export Varieties

The agriculture of Henan province has shown continuous development due to the strategic adjustment of the agricultural structure. In terms of the varieties of agricultural-product import and export, exports of pork, pig, garlic, tobacco, and natural honey occupied a top position in Henan province and exports of wheat, casing, and apple juice also [3]. However, the variety of exported agricultural

Table 1 Main varieties of agricultural products exported by Henan province in the period 2006 to 2010

Year	Honey	Casing	Apple juice	Pork	Chicken	Garlic
2006	1089.0	1158.8	5273.6	4060.9	2841.7	5034.2
2007	894.2	1019.7	8484.3	2481.7	3696.7	3849.5
2008	926.8	1251.1	4644.6	1410.7	2394.6	2357.8
2009	984.5	1274.3	3194.0	1752.3	2865.1	5058.4
2010	1420.9	1854.9	2785.1	3667.4	3985.0	13431.9

products is singular; the proportion of primary products is greater; the proportion of processed agricultural products is low; and export was mainly concentrated in the area of labor-intensive agricultural products. The main agricultural products exported by Henan province are shown in Table 1.

2.3 Structure Analysis of Export Market

The export markets of Henan agricultural products were very concentrated. From the intercontinental perspective, the main export destinations of Henan agricultural products were Asia, Europe, and North America, and export has further expanded to Africa, Oceania, and Latin America in recent years. From the viewpoint of countries and regions, the export markets of Henan province have expanded to >120 countries and regions including the United States as well as in Hong Kong, Japan, South Korea, Indonesia, Russia, the Association of South-East Asian nations (ASEAN), and the European Union market. The proportion of agricultural products to these countries and the area has been maintained at approximately 70 %.

3 Empirical Analysis and Evaluation

3.1 Structure Analysis of Export Varieties

Export market share (MS), also known as the international market share, refers to the proportion of which the export volume of a country (region) accounts for the world's total exports, thus reflecting the overall competitiveness of the country (region) of export; a larger proportion represents an enhanced competitive power. The formula is as follows:

$$MS_{ij} = (X_{ij}/X_{iw}) \times 100\%. \qquad (1)$$

where M_{ij} is the export market share index of i product and j country; X_{ij} is the total product exports i of one country or region; and X_{iw} is the total export of product i.

Generally, the higher the MS_{ij}, the stronger the export competitiveness and the higher the level of profitability and vice versa.

The trade competitiveness index (TC index) represents the proportion of which the import export trade balance of a country (or region) accounts for total import and export trade [4]. Its formula is as follows:

$$TC_{ij} = (X_{ij} - M_{ij})/(X_{ij} + M_{ij}). \tag{2}$$

where X_{ij} is the export sales of i product and j country; M_{ij} is the import sales of i product and j country; and TC_{ij} is the trade competitiveness of i product and j country. It varies from −1 to 1.

The revealed comparative advantage index is the ratio between the proportion the export sales of one product in one country that accounts for the overall export sales of that country as well as the overall proportion for which this product accounts for overall export sales of the whole world. The formula is as follows:

$$RCA_{ij} = \left(X_{ij}\Big/\sum iX_{ij}\right)\Big/(X_{iw}/iX_{iw}). \tag{3}$$

In detail, $RCA \geq 2.5$ indicates very strong export competitiveness of the product or industry; $1.25 \leqq RCA < 2.5$ indicates relatively strong export competitiveness; $0.8 \leqq RCA < 1.25$ indicates middle-strength export competitiveness; and $RCA < 0.8$ indicates weak export competitiveness.

3.2 *Empirical Analysis of Henan Agricultural Export Competitiveness*

From the point of export market share index. In 2007 and 2008, a decrease of export market share showed weak export competitiveness. After 2009, due to the adjustment of production structure in Henan province and the increased supervision and support of agricultural-products export, export market share and competitiveness increased (Table 2).

From the point of index indicators of trade competitiveness, the average trade competitiveness index of Henan province in 2008 to 2010 was −2.161; in 2011, 2012 the agricultural-products trade competitiveness index was 0.174 and 0.165, respectively. The negative trade competitiveness index in recent years showed that Henan's agricultural export competitiveness is weak. However, the aggrandize trade competitiveness index shows the development trend of the agricultural-products export scale and earnings after 2008 as well as the export competitiveness of Henan province.

From the point of the export revealed comparative advantage index, the RCA index of exported agricultural products in Henan province was <1 from 2007 to 2012, showing a strong comparative advantage in the international market.

Table 2 Export market share of agricultural products in Henan province from 2007 to 2012

	2007	2008	2009	2010	2011	2012
Agricultural exports sales of Henan province	4.89	4.53	5.07	7.92	9.66	10.10
Agricultural exports of the world	11328.8	13401.5	11688.5	13664.7	16749.3	16571.1
Export market share	0.0432	0.0338	0.0434	0.0580	0.0577	0.0609

Table 3 RCA index of agricultural products in Henan province from 2007 to 2012

	2007	2008	2009	2010	2011	2012
Agricultural export sales of Henan province	4.89	4.53	5.07	7.92	10.75	10.10
Goods export sales of Henan province	83.91	107.19	73.46	105.34	192.4	296.8
Agricultural export sales of the world	11328.8	13401.5	11688.5	13664.7	16749.3	16571.1
Goods export sale of the world	139870	160700	125420	152740	182550	184010
RCA index	0.72	0.51	0.74	0.84	0.61	0.38

The RCA index of agricultural products in Henan province in 2009 to 2010 increased gradually, but the numbers are still low and >1. In 2011 and 2012, the RCA index continued to decline, and in 2012 it reached a low of 0.38. Clearly that Henan's agricultural-product export competitiveness is relatively inferior and has much room for development and increase (Table 3).

4 Suggestions to Enhance the Competitiveness of Henan Agricultural Exports

4.1 Strengthen the Quality of Agricultural Products

In recent years, blocked export of agricultural products due to quality occurred occasionally. The coordination of government, enterprises, and farmers is needed to strengthen the quality of agricultural [5]. First, the government must establish a unified system of improving the quality of agricultural-product quality and safety and strictly implement the international standard. Second, agricultural producers and enterprises must set up quality consciousness and improve their capability to test and monitor the quality of agricultural products. Finally, great importance should be attached to the establishment of the industry association with clear and effective responsibility. With the help of industry association, agricultural products can be supervised and the quality and safety of the export of agricultural products be ensured while at the same time increasing competitiveness.

4.2 Implement Brand Strategy

To improve the export competitiveness of agricultural products in Henan province, it is necessary to strengthen the cultivation of leading enterprises in the export and implementation of brand strategy. The first thing is to establish brand awareness, actively develop local advantages and characteristic agricultural products, and cultivate a well-known agricultural-products brand. Second, it is advised to establish production base of agricultural products, introduce advanced technology, expand the scale of production, promote the industrialization of agricultural production and operation, and cultivate leading enterprises in foreign-exchange earnings [6]. Again, cultural connotation should be attached to the role agricultural products play in the advantages of geography and culture in Henan province to improve brand value, build a green brand image, and improve the competitiveness of the agricultural-products brand.

4.3 Optimize the Export Structure of Agricultural Products

Based to the international competitiveness of Henan's agricultural products, it is important to know the traditional export market share, understand consumer demand as well as policies and regulations regarding time, ensure the quality of the export products, and actively expand the export potential of emerging markets. At the same time, export-growth potential should be nurtured while strengthening the competitive-product edge [7]. Items having the largest existing export advantage in Henan province are apple juice, sesame, honey, tea, and pigs and chicken and products made of such. This requires the promotion of localized production; construction of production bases of high-quality, large-scale, and low cost commodities; and realization of large-scale production, standardized management, and industrialized operation.

References

1. Zhang H (2010) Countermeasure research of the improving of international competitiveness of characteristic agricultural in Hunan Province. J Cent. S Univ For Technol 6(3):61–63
2. Xue X (2012) Empirical analysis of Henan agricultural international competitiveness. Res Dev 6(3):112–114
3. Ma X (2008) Research of present situation and the development strategy of Henan agricultural products import and export trade. Value Eng 12(4):28–30
4. Chen C (2007) The study of improving agricultural products' export competitiveness of Sichuan Province. Mark Mod 36(1):37–38
5. Gan X, Zhou S, Piao Y (2012) The empirical study of export competitiveness of agricultural product in Shandong Province. North Econ 24(6):63–64

6. Zhao Y (2013) Export competitiveness research of agricultural product in our country. J Henan Bus Coll 6(3):11–13
7. Chen Q (2013) Analysis research of agricultural export competitiveness in Hubei Province. Logist Eng Manag 12(10):169–173

Research of the Technology–Organization–Service Synergetic Evolution Model in the Process of Service-Enhancement—A Case Study of the Haier Group

Qingtao Meng and Yusen Xu

Abstract Investigation of the synergetic evolution mechanism of manufacturing enterprises' service enhancement is performed through a case study of the Haier Group. The main findings of this study are as follows. First, this paper presents a model of synergetic evolution of technology–organization–service (TOS) model in the process of service enhancement for Chinese manufacturing enterprises. Second, technology and service innovation interact with each other. The technology innovation appears to be gradually user-oriented, and a new services project provides a platform for the new technology. Finally, organization innovation is the synergistic support tool for technology and service innovations.

Keywords Technology innovation · Organization innovation · Service innovation · Synergy

1 Introduction

During the past 20 years, service and manufacturing have gradually integrated, which bring up opportunities and challenges of there own. This phenomenon has concerned by academia and industry widely [1]. Increasingly more scholars have started researching manufacturing service enhancement. They explored the influence of service enhancement from a variety of angles on manufacturing enterprises

Q. Meng (✉) · Y. Xu
Faculty of Management and Economics, Dalian University of Technology, Dalian 116024, China
e-mail: qingtaomeng@126.com

Y. Xu
e-mail: ipmp@dlut.edu.cn

Q. Meng
School of Business Administration, Qilu University of Technology, Jinan 250353, China

M. Li et al. (eds.), *Proceedings of 2015 2nd International Conference on Industrial Economics System and Industrial Security Engineering*,
DOI 10.1007/978-981-287-655-3_25

in China faced with globalized competition. The research of service enhancement can be said to deepen the research of service innovation.

In the study of service innovation, researchers have paid more attention to the one-way effect of technology on service and less attention to the mutual influence between them [2, 3]. The former show neither the essence of service innovation, nor do they explain the process and mechanism of service enhancement. There have been some investigations of the relationship between technology, service, and organization [4]. However, it still focuses on the combination of technology, organization, culture and market, etc. Thus, it is necessary to study the combination effect of technology, service, and organization on manufacturing enterprises' service enhancement.

Based on the theories of synergy and combinatorial innovation, this paper presents research of the process of service enhancement in manufacturing firms from the perspective of the synergy of technology and service and attempts to portray the organization as the collaborative supporting tool and the bridge of the two. Furthermore, combined with the specific case of the Haier Group, this paper presents a construction of the synergistic evolution model of "technology–organization–service" in the process of service enhancement for manufacturing firms.

2 Synergetic Evolution Process of the Haier Group

The Haier Group is one of the world's largest white-goods manufacturers: It has established 5 R&D centers, 21 industrial parks, and 66 trading companies in the world. Currently the Haier Group has become a large-scale multinational business conglomerate. Its revenue reached RMB 200.7 billion in 2014. Because of its rapid growth and impressive performance, Haier has attracted wide attention in the world, especially for its innovation practice, which includes the implementation of overall control and a clear (OEC) management mode, a market chain, and a combination model of person and order.

2.1 Technology Innovation of the Haier Group

Through the construction of technology innovation network system, the Haier Group has continuously upgraded its technical innovation ability. Haier's inner technology innovation system is composed of Central Research Institute, Product Research Institute, Haier Center Laboratory, State Key Laboratory of digital home appliances, etc. Its huge external innovation network includes an international technology development network, global talent network, a global information network, and so on. Its design branches around the world facilitate full use of global design resources, which enables it to design and develop localized products.

2.2 Service Innovation of the Haier Group

Along with the development of industrial standards, the difference of technology level in the household electrical-appliance industry has gradually vanished. Improving service and quality are the two most effective measures to enhance competitiveness. Haier became aware of this need very early and chose to change its identity from home-appliance manufacturer to lifestyle-service provider. Haier takes service to be part of the product and provides service to customer during the entire product process of

The Haier Group has constructed a design system, a flexible manufacturing system, an e-commerce platform to meet the needs of users. By way of the e-commerce platform, Haier states that the product can be designed by users and then produced by its factories. In this way, Haier reduces the distance between the product and market and makes the most of external market resources.

2.3 Organization Innovation of the Haier Group

To innovate R&D personnel based on market demands, the Haier Group created a model-manager system, which requires the research staff not only to design the product but also to take charge of sales, costs, profits, and after-sales. It forces the developer to pay close attention to the needs of users to develop valuable product such as cartoon television, double-power washing machines, and so on.

The Haier Group has also transformed the strategic business unit (SBU). Taking market competition as the criterion, SBU is the unit production or service, which is used to evaluate employees. In this way, employees are forced to innovate and compete for customers.

Another organization innovation of Haier is the post debt–funded mechanism, which is a further extension of the OEC management mechanism market chain–management mechanisms. Haier has established an incentive mechanism for employee innovation including a job-performance evaluation system, an interest-distribution system, etc.

3 Case Analysis and Discussion

3.1 The Model Accords with General Rules

Berger and Lester [5] were the first to put forward the concept of service enhancement for manufacturing enterprises. They suggested that enterprises take advantage of service enhancement to improve the competitiveness of their products through a combination of manufacture and service. Many manufacturing enterprises are changing from production-manufacturing to service-oriented manufacturing. This is in accord with the general rules of service enhancement in manufacturing firms, in which the boundary of technology and service has been steadily been eroded. Meanwhile, the innovative activities of any organization cannot do without the support of an organizational system.

The success of the Haeir Group depends on the synergy of technology, organization, and service, which is in accord with the general law of manufacturing enterprises' service enhancement. In this way we draw a synergetic evolution model of technology–organization–service (TOS) for Chinese manufacturing enterprises based on the case of the Haier Group.

3.2 *Technology Innovation Promotes Service Innovation*

Panizzolo [6] proposed that manufacturing enterprises change their product strategy into a product-and service-strategy and offer the personalized, flexible, and more extensive solutions to meet the needs of customers. In that sense, the ever-changing needs of the customer are an important incentive for manufacturing enterprises to develop service enhancement. Due to this, the Haier Group takes customer demand as the starting point of technology innovation. The clean installation service of air conditioners is a new service program developed according to customer demand, enabling users to say good-bye to the old installation mode, which involved much dust. As a result, this program has becomes a major competitive advantage of Haier.

Marceau and Martinez [7] put forward that service differentiation is an effective strategic tool for manufacturing enterprises to obtain a potential competitive advantage. However, one of the most important ways for manufacturing enterprises to achieve service differentiation is through technology innovation to develop new markets. The Haier Group has many successful products: the "little prodigy" mini washing machine, a color refrigerator, and the desert air conditioner, which adapts to the development of the western region.

3.3 *Service Innovation Integrated with Technological Elements*

Service innovation may be technological as well as nontechnological innovation. Overall, technology is an important dimension of service innovation. The essence of service innovation is that noncodified knowledge is transformed into codified knowledge. The modern technology of information and communication plays an important role in service innovation. From the perspective of knowledge transformation, this research reveals that technology plays a very important role in service innovation.

Technology, especially information and communication technology (ICT), is a key factor to promote service innovation. It can be used not only as the bridge of communication among enterprises, suppliers and users, it can also become an important marketing channel. For example, the Haier Group has developed the Enterprise Resource Planning management system (ERP), which uses information technology. Based on this system, Haier has established a B2B platform for online purchase, online bidding, and online payment, which collects internal data such as production, inventory, sales, finance, and human resources, along with shared with external suppliers and distributors, to shorten the procurement cycle.

3.4 Organization Innovation of the Supporting Utilities

Davies [8] pointed out that enterprises are gradually creating new departments that are responsible for the integration of products and services. Volberda et al. [9] pointed out that a flexible organization can ensure the enterprise the competitive advantage in a fiercely competitive market environment. They also believe that the flexibility of an organization depends on the variety of the enterprises' management ability and the ability to respond to different circumstances.

Along with the Haier Group adopting an internationalized service-oriented manufacturing model, its internal organization has changed from divisional structure to interior market-chain structure. The flexible and flat organizational structure is an open system, not a closed one. The open system can meet the needs of customers through the integration of various resources. As an example, take the success of the MGM color TV due to the coordination of various departments. In this way, Haier can shorten its development cycle from 6 to 2 months; thus, the speed and quality of its innovation has been greatly enhanced.

Technology and service innovation together are essentially a kind of reform. Innovative implementation often means the negation of the old organization, old methods, and old systems. Because organization plays an important role throughout the whole process of technical and service innovation, creating a means of organizational innovation is the first thing to do to complete the process of technology and service innovation.

From this case-study analysis, we can see that the series of organizational innovations of the Haier Group—such as the establishment of a model-manager system agers, the transformation of the SBU, the combination of person and order, and development of a post debt–funded mechanism—are meeting the need to merge technology and service innovation.

4 Conclusions

Based on case study of the Haier Group, this chapter indicates that theirs is a feasible way for manufacturing enterprises to obtain a competitive advantage through the synergy of technology, organization, and service. First, China manufacturing enterprise face the specter of a relatively backward level of technology level as well as fierce homogenous competition, so a differentiated-services strategy is a suitable way to gain the competitive advantage. China's enterprises should take service innovation as a strategy and try heir best to seek an innovative opportunity through the merging of service, service delivery, and technology innovation.

Second, when domestic manufacturing enterprises obtain the competitive advantage in the use of technology and service innovation, they should pay attention to improve its organization system. If the organization system cannot keep up, it will hinder the long-term development of technology and service innovation.

Last, government and relevant industry associations should actively set up appropriate measures to guide manufacturing enterprises to carry out innovation based on product differentiation competition, and regulate market management to create a favorable external environment for China's enterprises.

Acknowledgments This research was financially supported by the National Science Foundation (71272095G0203).

References

1. Guisheng W, Lei L (2011) An empirical study on service-enhancement in China's manufacturing sectors and policy implication. J Ind Eng Manag 25:87–95
2. Zhanghong JJ (2012) Research on distributor capability development of manufacturing enterprises for service-enhancement-a case study of Guangzhou Toyota Motor co., Ltd. Soft Sci 26:110–113
3. Hauknes J (1996) Innovation in the service economy. Access Download Stat 27:96–117
4. Qingrui X, Bin G, Yi W (2000) Development of technological innovation in China core competence-based innovation portfolio. J Ind Eng Manag 14:1–9
5. Berger S, Lester R (1997) Made by Hong Kong. Oxford University Press, Oxford
6. Panizzolo R (2008) A methodology to measure the value of services provided to customers in manufacturing firms. Meas Bus Excell 12:3–15
7. Marceau J, Martinez C (2002) Selling solutions: product-service packages as links between new and old economics, the DRUID summer conference, Copenhagen/Elsinore
8. Davies A (2003) Integrated solutions: the changing business of systems integration. In: Principe A, Davies A, Hobday M (eds) The business of systems integration. Oxford University Press, Oxford
9. Michiel DB, Van DBFAJ, Volberda HW (1999) Managing organizational knowledge integration in the emerging multimedia complex. J Manag Stud 36:379–398

The Three Successful Metropolitan Area Rail Transit Modes—the Revelation for the Yangtze River Delta Region

Yingying Gai

Abstract Today, with the rapid development of globalization, regional economic integration has been an inevitable trend. In recent years, Yangtze River Delta integration has been strengthened, which the most comprehensive strength of China. During its development, transportation played a crucial role as a bridge and a hub. By analyzing the international metropolitan area's rail transit system in Tokyo, Paris, and London, especially the rail between urban and suburban areas, we summarize the characteristic of each metropolitan area's rail transit system, and, finally, make suggestions for the development of Yangtze River Delta regional rail transport.

Keywords Yangtze River Delta · Metropolitan · Rail transit

1 Introduction

With the rapid development of globalization, implementing the strategy of regional integration has played an essential role for the development of countries and regions. Clearly, regional economic integration is irreversible and accelerating. In our country, Yangtze River Delta, Pearl River Delta, and Bohai Rim Region are still in the exploratory stage; the process of Yangtze River Delta is the fastest in terms of development. In addition, the Yangtze River Delta integration process was started earliest.

In recent years, domestic and foreign scholars have published a large number of studies about the Yangtze River Delta and its metropolitan areas that are worthy of reference and learning. Zou and Li analyzed the main bottleneck in the Yangtze River Delta region: the imperfect high-grade infrastructure network and the intercity railways [1]. By comparatively analyzing the rail-transit modes of three international metropolis circles—i.e., Paris, New York, and Tokyo—Zong (2011) summed

Y. Gai (✉)
School of Economics and Management, Beijing Jiaotong University, Beijing 100044, China
e-mail: 14120506@bjtu.edu.cn

M. Li et al. (eds.), *Proceedings of 2015 2nd International Conference on Industrial Economics System and Industrial Security Engineering*,
DOI 10.1007/978-981-287-655-3_26

up the similarities and differences of rail-transit modes of modern international metropolis circles [2]. Liu discussed the development of Tokyo rail transit and the metropolitan transformation of spatial structure and analyzed the development of the metropolitan spatial structure corresponding to construction of the railway system [3]. Xie pointed out that the competitive advantages make metro areas the engines of the United States' economic growth. Obviously, the international metropolitan area's rail-transit system is crucial, especially that between urban and suburban areas [4].

The remainder of this chapter is organized as follows. The next section discusses transportation development that emphasizes the important role of suburban railway, and this is followed by a summary table (Sect. 3) of the Yangtze River Delta's overview, development process, present situation, and transportation infrastructure. Section 4 describes the three successful metropolitan-area rail-transit modes. Finally (in Sect. 5), the critical role of rail transit mode in influencing the development of the Yangtze River Delta is discussed.

2 Transportation Development

As the science and technology have progressed, transportation options have increased. Modern high-tech traffic tools, such as high-speed rail, show us the absolutely unexpected speed and capacity compared with traditional modes of transport. Let us take look at our current options for urban transport :

According to the statistics, from the operation area to the divide, suburban railway has the largest single capacity. It is important to note that the suburban railway plays a large role in any successful track traffic system, especially in the more successful metropolitan areas.

3 Yangtze River Regional Integration

3.1 Overview of Yangtze River Integration

The Yangtze River Delta lies in China's eastern coastal area, and the core city of Shanghai is one of the largest cities in the world. It has the highest level of urbanization, the most dense population in terms of urban distribution, and the highest economic development level in China. Shanghai is its center and Nanjing, Hangzhou, and Hefei form the center of service. The urban agglomeration includes Shanghai, Nanjing, Jiang Su province, Suzhou, Wuxi, Changzhou, Xuzhou, Zhenjiang, Yangzhou, Nantong, Taizhou, Huaian, Yancheng, Lianyungang, Suqian, Hangzhou, Ningbo, Jiaxing, Huzhou in Zhejiang province, Shaoxing, Taizhou, Jinhua, Wenzhou, Lishui, Quzhou, Zhoushan, Hefei, Wuhu, Chuzhou, Maanshan, Huainan, Shanghai–Hangzhou Expressway, Shanghai–Nanjing Expressway, and all railways.

3.2 Forming the Process of the Yangtze River Delta Integration

The process of Yangtze River Delta integration process can be said to be "progressive," and the urban construction goes step-by-step from extensive to intensive. The transformation process is roughly divided into four stages:

(1) small town construction stage, namely, the stage of urbanization
(2) small- and medium-sized city-construction stage
(3) large city and megalopolis stage
(4) formation of urban agglomeration.

3.3 Current Situation of the Development of Yangtze River Delta Integration

To simplify data collection and analysis, this chapter uses Shanghai, Jiangsu, and Zhejiang to represent the Yangtze River Delta economic circle data. It analyzes the current situation of the Yangtze River Delta in terms of GDP and industrial structure.

(1) GDP: In 2013, the Yangtze River Delta region's population accounted for <12 % of China's population and its area for approximately 2 %. From 2009 to 2013, the region's GDP amounted to 7.24941 trillion, 7.24941 trillion, 10.062481 trillion, 10.890527 trillion, and 11.833236 trillion Yuan, respectively. Its GDP accounts for a proportion of the total, respectively, of 21.3, 21.5, 21.3, 21, and 21.3 %. Obviously, the Yangtze River Delta region has great development potential and has embarked on a stable "fast track" of economic development.
(2) Industrial structure: In 2013, primary, secondary and tertiary industry the Yangtze River Delta region was, respectively, 555.998 billion, 5.5568445 trillion, and 5.720393 trillion Yuan. These industries accounted for 4.7, 47.0, and 48.3 % respectively. The Yangtze River Delta region plays the role of city with comprehensive advantages, emphasizing its modern service industry. The structure of the three major industries in Shanghai is 0.6:0.6:62,which is similar to that of Zhejiang. Shanghai sets finance, trade, shipping, and information service as the development object of its modern service industry, and it is the comprehensive service center of the Yaghtze River Delta region. The degree of regional integration in the Yangtze River Delta economic circle is very high; the production technology level gap between provinces is not large; and industry division of labor is relatively mature.

3.4 Transportation Facilities in Yangtze River Delta

In recent years, with the increasingly process of perfecting infrastructure construction in the Yangtze River Delta region, the structure of urban agglomeration spatial layout has basically been formed, and the "city effect" has been highlighted.

By 2010, the three highways have formed a 3-h rapid traffic circle connecting the three central cities of Yangtze River Delta region: Shanghai, Nanjing, and Hangzhou. The rapid advance of traffic infrastructure not only accelerated the pace of integration of the Yangtze River Delta urban agglomeration, but it also played a role in the development of the entire country.

Driving the Yangtze River Delta region by urban agglomeration strategy is of great significance to China's economic and social development. Although the infrastructure in the Yangtze River Delta region has been particularly successful, it still has deficiencies such as imperfect rail transport [5]. The rail transit is still in construction and stops in every train station. Transporting passengers from People Square to Qingpu requires >1 h, but transporting from Kunshan to Hongqiao station or Xinke district requires only 20 min by high-speed railway. Thus, transporting passengers from a large suburban residential area to the city takes more time than transporting those from Jiangsu to the same destination.

4 Three Successful Metropolitan-Area Rail-Transit Systems

From the viewpoint of GDP and industrial structure in the Yangtze River Delta region, we can see that the development of Yangtze River Delta city clusters has made stepwise progress; however, compared with the world-famous city group, the gap still exists especially in the field of rail transit. By analyzing the international metropolitan-area rail-transit systems in Tokyo, Paris, and London, especially the railways between urban and suburban areas, we summarize the characteristic of each metropolitan-area rail-transit systems and give suggestions for the development of railways of the Yangtze River Delta region.

4.1 Tokyo Metropolitan

Tokyo is recognized by the world as a successful international metropolis, and its urbanization reaches 80 %. The government's focus on urban transportation infrastructure projects of orbit-transportation construction not only emphasizes the government's responsibility in rail-transit construction, it also attracts a consortium to participate in rail-transit construction by a variety of preferential policies. In Tokyo, the railway can be roughly divided into four groups: JR line, subway, iron, and other private railway. Its passengers mostly depend on the rail transits. According to the data of the Tokyo Metropolitan Urban Development Bureau, passengers who go to work or school every day account for 86 % of total passengers. During the rush hour, this proportion is as high as 91 % and holds first place in the world.

4.2 *Paris Metropolitan*

The Paris metropolitan area is located in northern France, surrounded by Paris and seven provinces, with a total area of approximately 12,012 km^2 (approximately 2.2 % of the French global area) and serves a population of approximately 11.49 million (approximately 18.8 % of the total population). Paris maintains fast-transport links by road, rail, and a rapid rail traffic network in the region (RER). To promote contact between new and old cities, Paris makes full use of the existing idle network and developed the suburb loop line to build 13 urban rail lines (RER + suburban railways) and the public transportation network.

4.3 *London Metropolitan*

The London metropolitan area—with a total area of approximately 4.5 km^2 and a population of 36.5 million (which should be approximately 60 % of the urban population)—is divided to 6 downtown areas by circular. It has 12 underground lines, whose length is 425 km and a suburban railway whose length 3071 km. Nearly 80 % of travel into the city center during the peak hours involves rail transit (Tables 1 and 2).

Table 1 Traffic volume off our country and the Yangtze River Delta region

Index	Unit	National	Yangtze River Delta region
Passenger capacity	One hundred million people	208.8	181
Passenger-turnover capacity	One hundred million/km	27,571.64	15867
Freight capacity	One hundred million tons	417	179
Freight turnover capacity	One hundred million ton/km	168,165	68203

Table 2 Comparison of different urban rail-transit systems

Spatial hierarchy	Metropolitan			Travel speed (km/h)	Station spacing (km)
	Tokyo	Paris	London		
Center city	Subway	Subway	Subway	25–35	0.5–1.2
Faubourg	JR	RER	Suburban railway	40–70	2–3
Outer suburb	JR	RER	Suburban railway	40–80	2–5
Homeland	Shinkansen	TGV high-speed rail	National railway	200–320	30–50

4.4 *Comparison Between the Urban Rail-Transit System*

Three metropolis circles depend on rail transit to carry passengers between economic and social active areas, and the intercity rail transportation network is highly developed. Thus, construction of an integrated suburban railway in the Yangtze River Delta region has much positive effect to gain from studying the successful cases of railways abroad (Figs. 1 and 2).

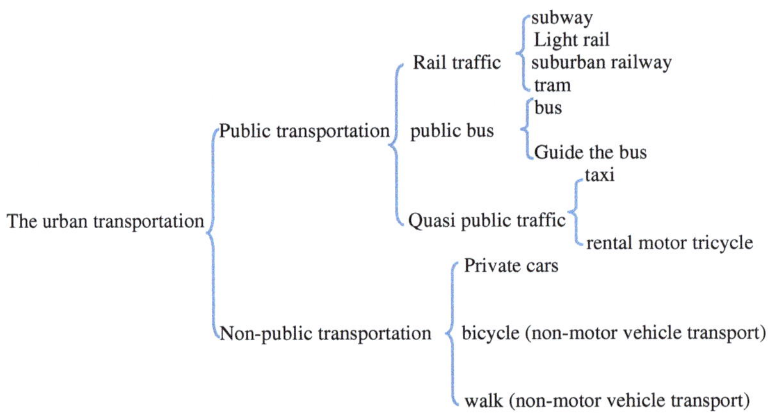

Fig. 1 The urban transportation

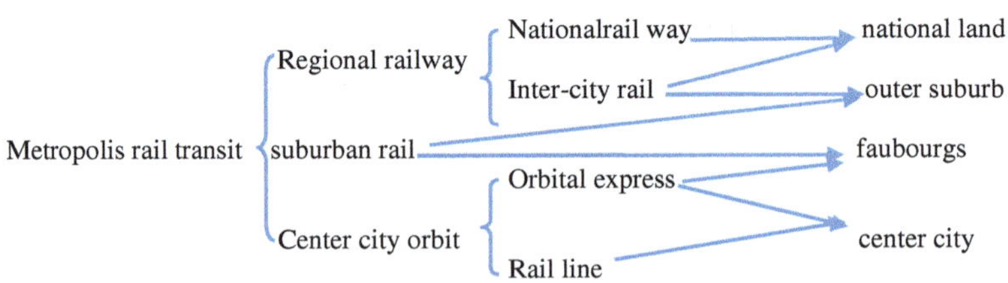

Fig. 2 Rail-transit network in the metropolitan area

5 Enlightenment for Yangtze River Delta Integration

Learning from generalization of the rail-transit system model in the above-mentioned three metropolitan area (Tokyo, Paris, and London), it is important to realize integration of the Yangtze River Delta railway region in China. World-class city groups tend to have a quick and convenient transportation network. Although integration of the Yangtze River Delta region railway system has overall planning and construction, it is still lacks a more convenient and effective multiform transport system. The city subway miles in Tokyo and Paris are not far from those of Shanghai, but these cities have a lot of regional railway, i.e., 23,000 km mileage, and three modes: fast, slow mode, and direct operating. Therefore, different areas should have different modes of rail transit. The Yangtze River Delta region should have various rail transit modes such as fast, slow, etc.

Yangtze River Delta regional integration should take the initiative to draw lessons from the successful experience of the world's urban agglomerations and believe in the near future that the Yangtze River Delta regional railway can be world-class.

Acknowledgments I express gratitude to my tutor. He has supported me since I entered the School of Economics and Management in 2014, whether it is in the usual form learning or his review of my chapters, which are full of my teacher's sweat and effort. I am also grateful to my family, especially my parents, who support me silently. Thanks go to experts and scholars who, despite being busy, have reviewed and ventured valuable opinions about this chapter.

References

1. Zou W, Li Q (2010) Analysis of regional economic integration: Yangtze River Delta, pearl River Delta and Bohai Rim region. Reform 10:86–93
2. Zong J (2011) Rail transit modes of three cities at abroad. Transp Stand 17:73–76
3. Liu X (2010) Tokyo rail transit development and the metropolitan transformation of spatial structure. Urban Mass Transit 10:6–12
4. Xie F (2007) Metropolitan areas and the economic growth of USA. The North Forum. 01: 144–147
5. Shen P (2005) Traffic need be first in Yangtze river delta integration. Compr Transp 11:01

Integration of a Value Chain on Choice of Alliance or Acquisition: An Empirical Study of the Fast-Moving Consumer Goods Industry

Weike Zhou and Hai Wang

Abstract For firms seeking to strategically combine their resources with those of other firms, two popular alternative governance structures emerge: alliance or acquisition. In this chapter, we use an integrated value chain to examine the how and why configurations of two firms' resources and knowledge affect the costs and benefits associated with each governance structure. It is assumed that the integrated value chain and market growth are factors that will affect the likelihood of the focal pair of firms forming an alliance versus engaging in an acquisition. We also predict a possible transition from alliances to acquisition base on the alliance's integration of the value chain.

Keywords Consumer goods · Value chain · Integration

1 Introduction

Acquisitions, alliances, and divestitures are strategic alternatives along a continuum of governance modes [1–3]; however, relatively little is known as to when firms should pursue one versus the other. Furthermore, few studies have focused on the comparative choice between alliances versus acquisitions. Kogut and Singh [4], Hennart [5], and Balakrishnan and Koza [6] have typically emphasized why a focal firm may choose an alliance versus an acquisition, leaving us with a limited

W. Zhou (✉)
College of Business Administration, Dalian University of Technology,
Dalian 116024, China
e-mail: zhouwk@dlut.edu.cn

H. Wang
Dalian University of Technology Library, Dalian University of Technology,
Dalian 116024, China
e-mail: wanghai@dlut.edu.cn

M. Li et al. (eds.), *Proceedings of 2015 2nd International Conference on Industrial Economics System and Industrial Security Engineering*,
DOI 10.1007/978-981-287-655-3_27

understanding on the factors that motivate two specific firms to choose an alliance versus an acquisition.

Balakrishnan and Koza [6] suggested that a focal firm prefers an alliance to an acquisition if the focal firm does not have enough information about the potential target. Hennart [5] found that if the potential target possesses some assets that are not separable from the firm's assets, then the focal firm would like to acquire the target. The studies used different units of analysis such as the transaction. These scholars also analyzed the research question from different perspectives such as information economics, inseparation of resources, macro views, and, more importantly, knowledge-based view. These perspectives each confer somewhat different viewpoints on issues related to the choice of acquisitions and alliances.

In this chapter, we examine the choice between alliance and acquisition based on the perspectives of knowledge-based view and value chain in the context of a fast-moving consumer goods industry. We try to recognize the knowledge in the whole value chain by linking the two concepts together. At the same time, we will try to explain the choice between alliance and acquisition in a dyadic way, i.e., we will propose a dyadic perspective by considering how firms' ally choice is influenced by comparison of the firms' value chain integration.

2 Value Chain Integration, Knowledge, and Alliance Formation

A value chain is a chain within the activities of a firm operating in a specific industry. The business unit is the appropriate level for construction of a value chain. Products pass through all activities of the chain in order, and on each activity the product gains some value. The chain of activities gives the products more added value than the sum of the independent activities' values [7].

However, in this chapter we concentrate on the knowledge in the value chain, not the value brought by activities. We also argue that the knowledge in the whole value chain has much more effect on a firm's competitive advantage than does independent knowledge in a single part of the value chain. As Kogut and Zander [8] suggested, firms having combinative capabilities can learn from both internal and external sources such as alliances and acquisitions. However, as some scholars [9] have suggested, it is difficult for firms to learn quickly if they lack a knowledge base or absorptive capacity. When firms have an incomplete value chain, they cannot understand the key activities and knowledge in some critical part of the value chain, which means they have no or little absorptive capacity in these parts of the chain, which may lead to the firms' failure.

By engaging in strategic alliances, firms may expect to learn knowledge from their partners, which can fill their knowledge gap in the value chain. But that holds true only if the partners have complementary knowledge that fits the focal firm's knowledge gap well.

Ever since the work of Penrose [10], much strategic management literature has proposed that firms tend to create value if they access complementary resources and knowledge. Now, as we said previously, this argument can be extended to the realm of alliances.

Several studies have illustrated the importance of complementarity in strategic alliances. Gulati [11] found that firms occupying complementary niches have higher chances of alliance formation.

Complementarity of knowledge in the value chain between a focal firm and a target is important for the formation of alliances. The complementarity of knowledge in a target firm's value chain will fill the knowledge gap of the focal firm and complete the focal firm's value chain.

However, if a firm with an incomplete value chain tries to acquire another firm to complement its value chain, what will happen? We argue that this firm will face greater risk than forming an alliance. Without related knowledge, firms have much difficulty integrating value chains of acquired firms with its own value chain. At first, this firm cannot prevent persons with specific knowledge from opportunistic behavior.

For Pepsi carbonated beverage, the value chain is incomplete, and its amounts of producing, channel, and sale knowledge are not strong enough to support Pepsi competing in the market. Therefore, we can say that Pepsi's value chain is not complete, and some critical knowledge is lacking in terms of the producing, channel, and sales knowledge. If Pepsi wants to compete with other firms that produce carbonated beverages in the Chinese market, it must acquire the knowledge absent from the value chain so it can cultivate a healthy and perfect knowledge cycle. As we can see from Fig. 1, Master Kong possesses the knowledge missing from Pepsi's value chain (white area), which is complementary to Pepsi's knowledge cycle. Together with the knowledge owned by Pepsi (yellow area), Pepsi can build up a complete value chain with sufficient knowledge (Fig. 2).

For Coca Cola and Huiyuan, the situation is quite different. They both have an integrated value chain, and sufficient knowledge in their value chain supports their competition in the fruit-drink market, i.e., the two firms have supplementary knowledge and do not need to build an alliance. A fast-moving consumer-goods industry does need to establish an alliance.

When a value chain is incomplete, firms have an incentive to be alert to the complementary strengths between their organizations and their potential partners

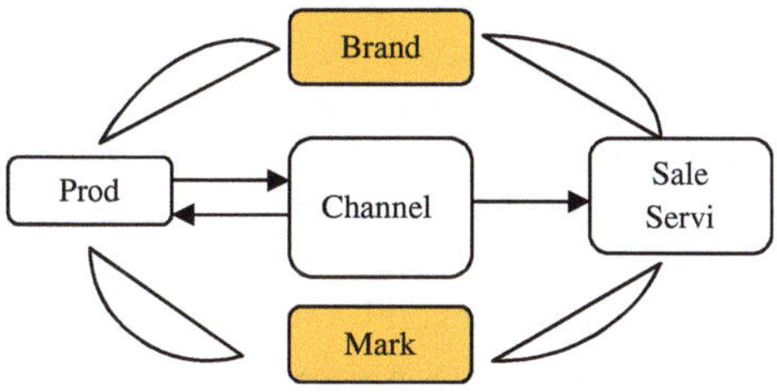

Fig. 1 Incomplete value chain of pepsi carbonated beverage

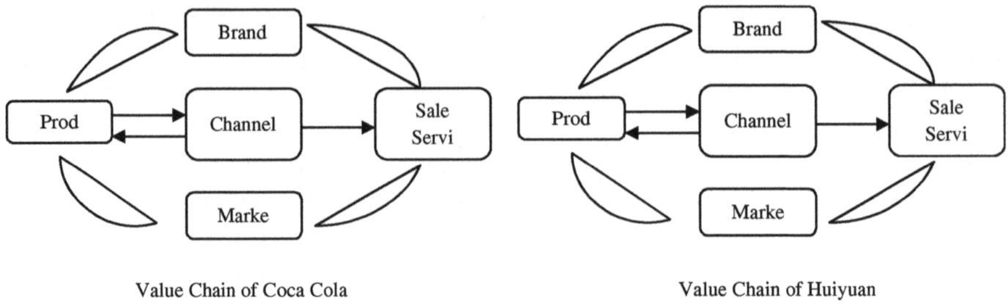

Fig. 2 Integrated value chain of coca cola and huiyuan

because cooperation will pay off when partners can complement each other's weaknesses.

Especially, in a fast-moving consumer-goods industry, integrated value chain resources are idiosyncratic or even indivisible and thus are not readily available in factor markets, but alliance formation can be the primary vehicle for accessing them. Although Teece's argument is set in the context of technology-intensive industries—where emerging firms do not yet have distribution capabilities, and established firms may not have exclusive control over new rent-generating technologies—its basic premises can be extended to fast-moving consumer-goods firms [12]. In the case of Pepsi and Master Kong, neither of the firms have a complete value chain (Pepsi lacks of knowledge of efficient production and sales-channel management, whereas Master Kong lacks knowledge about carbonated soft drinks and brand operation).

3 Method

Sample and data. The sample companies were among the 200 largest fast-moving consumer-goods companies in China's top listed companies from 2000 to 2011 in one of four categories: sales, profits, assets, and market value. We included only those firms with at least 70 % business in the fast-moving consumer-goods sector. All alliance and acquisition activities conducted by these firms from 2000 to 2011 were collected from the China Statistical Yearbook.

Variables. Alliance or acquisitions. We use alternative sources such as mergers and acquisitions. Firm-specific financial data were collected from XXX. For merger and acquisition transactions, we included both complete merger and partial merger.

We adopted six categories as complete value chain of the fast-moving consumer-goods industry: inbound logistic, produce, outbound logistic (channel), sales, advertising, and marketing.

Control variables: We included several control variables in our analyses. The number of subsidiaries in a host country was used to control for the effect of density dependence on disbanding, time, and firm-level attributes.

Time: Dummy variables for each year captured effects of temporal trends.

Total alliances: To assess claims that growth in alliances may be the result of a "bandwagon effect," each dyad-year record this variable is the total number of alliances announced in the industry during the previous year.

Previous acquisition/alliances experiences.

Previous acquisition/alliances experiences: These have often been used as proxy for a firm's acquisition/alliance capability.

Size: Size indicates whether companies that share an alliance are found in the same or in different size categories according to their number of employees and was measured as total sales in the industry in Renminbi.

Performance: Performance indicates the firm's degree of success in the marketplace and was measured as the return on assets normalized to the industry mean.

Liquidity: Liquidity reflects the short-term resources available to a firm and was measured as the "quick ratio," which is defined as current assets minus inventory divided by current liabilities.

Solvency: Solvency addresses similar concerns but with regard to long-term resources and was determined by the total amount of long-term debt divided by the firm's current assets.

4 Conclusion

Firms with an incomplete value chain are more likely to enter into alliances versus acquisitions. Different firms either share common value activities in their value chain or have uncommon value activities. For instance, a firm with an incomplete value chain can choose certain details in the value chain—such as technique development, marketing, or service integration—and take other firms advantages for their own use to create a more competitive advantage. Therefore, every firm can benefit by strategic alliance and eventually reflect expanded value and increased profit. The relation between incomplete value chain and alliance is moderated by market growth such that the positive relation will be stronger at a relatively high level of market growth. With the integration development of a value chain, a strategic alliances will more likely lead to the formation of acquisition, whereas a strategic alliance is transformed from shared to single ownership.

5 Limitations

1. Collaborations between firms can take many forms, whereas we consider only alliances and mergers; if other forms are included, the results will be more promising.
2. There may be additional managerial factors (e.g., demographic characteristics, strategic-process issues) that also play a role in the choice between forming an alliance versus engaging in an acquisition.

References

1. Williamson O (1991) Comparative economic organization: the analysis of discrete structural alternatives. Adm Sci Q 36:269–296
2. Williamson O (1975) Markets and Hierarchies: Analysis and Antitrust Implications. Free Press, New York
3. Hennart J-F (1993) Explaining the swollen middle: why most transactions are a mix of market andhierarchy. Organ Sci 4(4):529–547
4. Kogut B, Singh H (1988) The effect of national culture on the choice of entry mode. J Int Bus Stud 19(3):411–432
5. Hennart J-F (1988) A transaction cost theory of equity joint ventures. Strateg Manag J 9 (4):361–374
6. Balakrishnan S, Koza M (1993) Information asymmetry, adverse selection, and joint ventures. J Econ Behav Organ 20:99–117
7. Porter Michael E (1985) Competitive advantage. Free Press, New York
8. Kogut B, Zander U (1992) Knowledge of the firm, combinative capabilities, and the replication of technology. Organ Sci 3:383–397
9. Bowman EH, Hurry D (1993) Strategy through the options lens: An integrated view of resource investments and the incremental-choice process. Acad Manag Rev 18:760–782
10. Penrose E (1959) The Theory of the Growth of the Firm. Wiley, New York
11. Gulati R (1995) Does familiarity breed trust? The implications of repeated ties for contractual choice in alliances. Acad Manag J 38:85–112
12. Teece DJ (1986) Transactions cost economics and the multinational enterprise. J Econ Behav Organ 7:21–45

Economic Effects of the Modern Information Service Industry in China: Evidence from Beijing

Fenfei Chen, Chenchen Wang and Yuan Ren

Abstract The Modern information service industry plays an important role in the development of China's national economy. In this chapter, from five aspects of the output effect, capital formation effect, technology progress effect, employment effect and industrial effect, an economic effect evaluation index system including 16 indicators is constructed. And taking Beijing as an example, economic effect evaluation is studied from 1995 to 2013 using factor analysis. The results show that the 3 common factors extracted—called technology factor, output factor, and capital factor by rotation—explains 87.775 % of the original variable information, in which the weight of the Technology factor is the largest at 0.789, and the effect on local economy continuously improves. Therefore, technological progress and employment effects should be paid more attention for modern information service industry in China.

Keywords Modern information services industry · Economic effect · Factor analysis

The original version of the reference section in this book was revised: The surname of one of authors was changed from "Anderdassen" to "Andergassen". The erratum to the reference is available at DOI 10.1007/978-981-287-655-3_63

F. Chen (✉) · C. Wang · Y. Ren
School of Economics and Management, Beijing Jiaotong University,
Beijing 100044, China
e-mail: ffchen@bjtu.edu.cn

C. Wang
e-mail: 12120510@bjtu.edu.cn

Y. Ren
e-mail: 13120498@bjtu.edu.cn

F. Chen
Beijing Center for Industrial Security and Development Research,
Beijing Jiaotong University, Beijing 100044, China

M. Li et al. (eds.), *Proceedings of 2015 2nd International Conference on Industrial Economics System and Industrial Security Engineering*,
DOI 10.1007/978-981-287-655-3_28

1 Introduction

With the promotion of infrastructure, expansion of new markets, and diversified improvement of Internet programs, the modern information service industry in China has developed rapidly to a new rational stage. In Beijing, modern information service, with the advantages of location, policy, and talent, gains more opportunity for priority development. In 2013, the growth rate of information transmission, computer services, and software industry was 7.9 % in Beijing, and gross industrial production is 174.96 billion RMB, which is 9.0 % of Beijing's GDP and 11.67 % of the service sector's GDP. Thus, the development degree of modern information service industry has become an important measurement of a country's or region's development level n terms of modernization and comprehensive economic strength.

Some scholars have studied the development of the information service industry from different aspects. Landrum and Prybutok proposed and tested a model of library success that showed how information service quality relates to other variables associated with success [1]. Andergassen and Nardini proposed a simple model where long waves of innovation arise from the endogenous propagation of information among industries [2]. Hung analyzed the information service industry in Taiwan to build the measurement indicators for IC [3]. Zhao and Wang analyzed the location quotient of industry competitiveness of the modern information service industry in Beijing city and showed that the comprehensive competitiveness was in a leading position in the country, but it was also facing the problem of lack of innovation and content standards and norms, such as the lack of international competitiveness [4]. Hu and Liu focused on the system construction of the information service industry and research of institution operation on the basis of industrial institution. They constructed an input-output model to reflect the relationship between input and output, the correlation effect, and the ripple effect on the modern information service industry [5].

Regarding existing research on modern information service industry, most of the research focused on the development of the industry itself, not deeply involved in association with the economic effect, and lacked a comprehensive analysis of economic effect for the modern information service industry. Because different indexes are often used to investigate the ability of the economic effect, a unified economic effect evaluation index system has not yet been developed for the purpose of improving regional industry development and providing strong support for economic development.

2 Construction of the Evaluation Index System

According to the development characteristics of the industry and the impact on the economy, and following scientific systematic optimization and practical design principles, the economic effect evaluation index system of the modern information

Table 1 Economic-effect evaluation index system

Target layer	Criteria layer		Indicator layer	Calculation method
Economic effect	Direct economic effect	Output effect	Profit ratio of sales (X1)	Gross profit/sales revenue (%)
			Proportion of output value of industry (X2)	Output value of industry/output value of region (%)
			Sales revenue growth rate (X3)	Sales revenue growth/sales revenue last year (%)
			Output growth rate (X4)	Output growth/output value last year (%)
		Capital formation effect	Growth rate of fixed-asset investment (X5)	Fixed-asset investment growth/fixed-asset investment last year (%)
			Proportion of fixed-asset investment (X6)	Fixed-asset investment/output value (%)
			Proportion of fixed-asset investment of industry (X7)	Fixed-asset investment of industry/fixed-asset investment of region (%)
	Indirect economic effect	Technical progress effect	Labor productivity (X8)	Industrial-added value/average number of employees (million Yuan/person)
			R&D expenditure (X9)	R&D internal expenditures (billion Yuan)
			Quantity of scientific achievement (X10)	Number of chapter and book (number)
			Quality of scientific achievement (X11)	Number of award-winning achievements (number)
		Employment effect	Quantity of labor force (X12)	Number of urban employees at the end of the year (person)
			Labor cost (X13)	Average salary of urban employees (Yuan)
			Quality of labor force (X14)	Number of R&D personnel (person)
		Industrial effect	Industrial structure effect (X15)	Output value of industry/output of tertiary industry (%)
			Industrial agglomeration effect (X16)	Output value of Zhongguancun Science and Technology Garden (billion Yuan)

service industry is constructed from two aspects of direct effect and indirect effect. The economic effect is further divided into the output, capital formation, technical progress, employment, and industrial effects. Among them, the output and capital formation effects are regarded as the direct economic effect; the technical progress, employment, and industrial effects are regarded as indirect economic effects. Through analysis of the main influence factors on the 5 aspects, 16 indicators were selected as the subordinate indicators as shown in Table 1.

3 Empirical Analysis

In this chapter, the definition of modern information service industry is made according to "Beijing information service industry classification." Considering the availability of relative data, the information transmission, computer services, and software industry as variables is analyzed in China from 1995 to 2013. The data come from China Statistical Yearbook, Beijing Statistical Yearbook, Electronic Information Industry Statistics Yearbook, and China's software and information development report. Because there are strong correlations between the evaluation indicators, the factor analysis method was adopted.

3.1 KMO and Bartlett's Test

Using SPSS software, factor analysis was performed on the original data. The results of KMO and Bartlett tests on correlation between variables showed that the KMO value is 0.71, the Bartlett value is 559.432, and the probability value $P < 0.000$. Therefore, the correlation coefficient matrix is not a unit matrix, and the factor analysis method is suitable.

3.2 Factor Extraction

Table 2 shows the list of the total variance explained in 16 original variables during the process of factor analysis. It can be seen that the characteristic roots of the first factor (F1) is 11.079, which explains 69.244 % of the total variance; the characteristic root of the second factor (F2) is 1.802, which explains 11.259 % of the total variance; and the characteristic root of the third factor (F3) is 1.163, which explains 7.272 % of the total variance; and the cumulative variance contribution rate of the first three factors is 80.503 %. Therefore, F1, F2, and F3, the three common factors, are extracted in the next step, which basically retain the original variables information and has the role to reduce the dimension.

Table 2 Total variance explained

Initial eigenvalues				Extraction sums of squared loadings		
Component	Total	% of variance	Cumulative %	Total	% of variance	Cumulative %
1	11.079	69.244	69.24	11.079	69.24	69.244
2	1.802	11.259	80.5	1.802	11.26	80.503
3	1.163	7.272	87.78	1.163	7.272	87.775
4	0.857	5.356	93.13			
5	0.52	3.248	96.38			
6	0.237	1.479	97.86			

3.3 *Factor Rotation*

The further transformation of rotation is implemented using Varimax rotation with the Kaiser normalization method. Table 3 shows the rotated component matrix.

As shown in Table 3, the factor loadings of the indicators of labor productivity (X8), R&D expenditure (X9), scientific achievement (X10, X11), quantity of labor force (X12), labor cost (X13), and quality of labor force (X14) are higher in the first factor (F1), so it can be called the "technology factor." The higher factor loadings for the second factor (F2) appear in the indicators of the proportion of output value of industry (X2), output growth rate (X4), and industrial structure effect (X15). Therefore, the second factor (F2) can be called the "output factor." The third factor (F3), which has higher factor loadings in the growth rate of fixed-asset investment (X5), proportion of fixed-asset investment (X6), and proportion of fixed-asset investment of industry (X7), is regarded as the "capital factor." Compared with the component matrix, factor rotation has obvious separation effect for factor loadings so that each factor represents the more prominent economic significance.

Table 3 Rotated component matrix

	Component				Component		
	1	2	3		1	2	3
X1	0.904	−0.123	0.155	X9	0.986	−0.028	0.061
X2	0.884	0.345	−0.187	X10	0.971	−0.01	0.126
X3	0.399	−0.341	−0.354	X11	0.962	−0.044	0.137
X4	−0.354	0.828	−0.204	X12	0.983	−0.051	0.053
X5	0.199	−0.163	0.888	X13	0.996	0.028	0.002
X6	−0.75	0.186	0.385	X14	0.954	−0.129	0.104
X7	−0.813	−0.244	0.346	X15	0.583	0.736	0.037
X8	0.957	0.002	0.088	X16	0.984	0.004	0.049

Table 4 Composite score from 1995 to 2013

Year	F1	F2	F3	Total score	Year	F1	F2	F3	Total score
1995	−1.1209	−1.6179	0.0454	−1.07	2005	0.0423	1.3238	0.3082	0.09
1996	−1.0927	−1.6623	0.1618	−1.04	2006	0.1714	0.6920	0.9683	0.12
1997	−0.9924	−1.1981	−0.0472	−0.93	2007	0.4433	0.8453	0.2694	0.47
1998	−0.8716	−0.7303	0.2367	−0.75	2008	0.6839	0.1108	0.4611	0.51
1999	−0.9229	0.1595	0.3424	−0.68	2009	0.9437	0.4909	1.1384	0.8
2000	−0.9497	0.63019	1.0768	−0.57	2010	1.1183	1.2960	1.6766	0.57
2001	−0.7469	1.50241	1.4120	−0.28	2011	1.2550	0.6009	1.5037	0.91
2002	−0.6397	1.0759	−0.5981	−0.44	2012	1.6548	0.4501	1.6064	1.41
2003	−0.5090	1.0497	−0.2813	−0.31	2013	1.8968	0.5180	1.1548	1.55
2004	−0.2789	−0.0269	−1.6074	−0.38					

3.4 *Factor Scores*

Next, according to formula (1), total factor scores from 1995 to 2013 are calculated to reflect the economic effect of the modern information service industry in Beijing; the results are shown in Table 4.

$$Y = \frac{\lambda_1 F_1 + \lambda_2 F_2 + \lambda_3 F_3}{\lambda_1 + \lambda_2 + \lambda_3} = 0.789F_1 + 0.115F_2 + 0.096F_3 \tag{1}$$

From the calculation results, the total scores of economic effect indicate that the impact on the local economy of the modern information service industry continues to improve. From formula (1), the weight of the technology factor is the largest, 0.789, which shows that the impact of technological progress and employment should be paid more attention in the modern information service industry. Compared with the other two factors, the technology factor has maintained steady growth in the past 10 years. In fact, both the amount of labor and labor quality have obviously improved, which makes a great contribution to the comprehensive factor score from 1995 to 2013. Because of the impact of the financial crisis, in recent years the score of the output factor appears to show a decreasing trend. In addition, the score of the capital factor violates significance, and it is not difficult to see that investment in the modern information service industry has improved since 2012.

4 Conclusion

The influence of the modern information service industry on the national economy shows both direct and indirect effects, which are specifically reflected in the five aspects of the output, capital formation, technology progress, employment, and industrial effects. As an example, using factor analysis, the economic effect in Beijing is measured by the evaluation index system including 16 indicators. The 3 common factors extracted by factor rotation—technology, output, and capital—together explain 87.775 % of the original variable information from 1995 to 2013, and the impact on the local economy of the modern information service industry in Beijing continues to improve from the view of comprehensive scores. In fact, the value of the modern information service industry remains at approximately 8 % of the area's GDP in recent years, and its growth rate is significantly higher than that of all industries in most years.

However, in Beijing, most of the information service enterprises are still in relatively small in scale, in which the number of employees is fewer than 100 people, and the enterprises with >500 employees accounts for only 5.3 % of all. At present, the proportion of corporate R&D investment for the modern information service industry in Beijing is less than one third of the international well-known enterprises. Basic software R&D ability is relatively backward: For example, in 2012 China's mobile intelligent terminal shipments amounted to 600 million, ranking first in the world, but all of the native operating system belonged to foreign brands. These are known to be the main factors hindering the economic effect of the industry from the above-mentioned factors. Therefore, it should further be emphasized to expand the variety of financing channels, strengthen technology R&D, pay attention to personnel training, and improve the competitiveness of the industry.

References

1. Landrum H, Prybutok VR (2004) A service quality and success model for the information service industry. Eur J Oper Res 156(3):628–642
2. Andergassen R, Nardini F (2005) Endogenous innovation waves and economic growth. Struct Chang Econ Dyn 3:1–18
3. Hung Y-C (2006) Measurement scale of intellectual capital for the information service industry. Int J Serv Stand 2:154–175
4. Zhao H, Wang J-l (2009) Analysis of Beijing information service industry competitiveness from the comparative perspective. Forum Sci Technol China 7:59–63
5. Hu C-p, Liu K-x (2013) Institution construction and dual-track development of knowledge information service industry in national innovation development. Libr Tribune 33(2):1–5

How Did Lenovo Build Its Global Brand?

Jian Li

Abstract Lenovo is currently one of the world's leading personal technology companies. In the last 30 years, Lenovo executed its global branding strategy, e.g., it acquired IBM's PCD, and the company is now no. 286 on Fortune 500 list. How did Lenovo build its global brand? In this essay, I will explore the process of global branding for Lenovo, analyze the approach Lenovo has taken into corporate brand management, and discuss the corporate identity and image of Lenovo by using the brand identity planning model. The objective of this essay is to find a way to achieve global corporate branding for Chinese enterprises through analyzing Lenovo's case.

Keywords Lenovo · Corporate brand · Corporate identity

1 Introduction

Lenovo is one of the world's leading personal technology companies today producing innovative PCs and mobile internet devices. Now it is no. 286 on the Fortune 500 list, and Lenovo is the world's largest PC vendor and fourth largest smart phone company [1]. How did Lenovo acquire these achievements during last 30 years?

Actually, Lenovo made an interesting business move in acquiring IBM's PCD in May 2005. It is profound milestone in the emergence of Chinese brands, and today it has a great impact on Western consumers. It symbolizes Chinese companies beginning to build top global brands and reflects the current trend in business strategy: changing focus from acquiring revenue toward corporate growth. In this

---

J. Li (✉)
Finance Department, Harbin Finance University, Harbin 150030, China
e-mail: lj_tiger2000@163.com

M. Li et al. (eds.), *Proceedings of 2015 2nd International Conference on Industrial Economics System and Industrial Security Engineering*,
DOI 10.1007/978-981-287-655-3_29

case study, I observe the lure of global branding for Chinese enterprise Lenovo, as well as the processes and challenges of building global brands, by analyzing the case of Lenovo.

2 What Approach Has Lenovo Taken into Corporate Brand Management?

Generally speaking, Lenovo's corporate brand management follows the vision of the company. In such a sense, Lenovo can be considered as having a "vision-driven approach" depending on different political environments.

2.1 The Stages of Corporate Branding

Simply put, Lenovo's corporate brand changes over four stages (Fig. 1). The corporate brand of Lenovo was created in 1984 when the company started business in China and, in fact, focused on doing business only in China, Lenovo used Chinese characters as its corporate brand before 1988.

Along with the expansion of its business, Lenovo established the first subsidiary outside of Mainland China, in Hong Kong, and the company decided to use the word "Legend" as an interpretation of the original Chinese into English. At that time, the Chinese characters brand was continued to be used in Mainland China as the corporate brand and was further applied to some of the main products, whereas "Legend" was applied in the market outside of Mainland China, such as Taiwan and Hong Kong, until 2003.

In 2002, a new significant corporate strategy was decided. The company was determined to go international, even global, settling only in China. The company found that it was not possible to accomplish further development using the former brand name because "Legend" had been registered in most major courtiers.

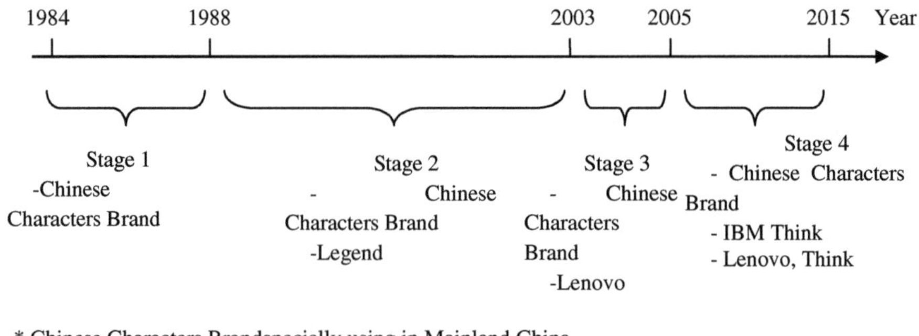

Fig. 1 Corporate brand establishment

The company took 1 year to rename its corporate brand from Legend to Lenovo, taking the "Le" from "Legend," a nod to its heritage, and adding "novo." the Latin word for "new," to reflect the spirit of innovation at the core of the company. The company registered the brand name on April 28, 2003. To reduce the negative effect of changing its name, the company used "Lenovo" both in the Chinese market as well as other markets. In 2005, Lenovo acquired IBM's PCD. Per the agreement, Lenovo has the right to use "IBM" for 5 years and is free to use "ThinkCentre" and "ThinkPad" as long as it wants.

2.2 *Framework of the Corporate Brand*

Figure 2 presents the three elements of corporate brand for Lenovo. Corporate branding can be understood as underpinned by processes linking strategic vision, organisation culture, and corporate image. "Lenovo" was used to build a rather prosperous relationship with image and culture in China. The existing image-culture gap comes from the large number of new employees previously with IBM. Lenovo needed time to educate the incoming "new" employees. The top managers aspire for "Lenovo" to become a well-respected brand known worldwide for innovative technology with adequate quality at a low price.

In the understanding of the new structure of Lenovo, an image-vision gap arose. There were two overall impressions from the outside key stakeholders because (1) the company has not been in the international market for a long time and

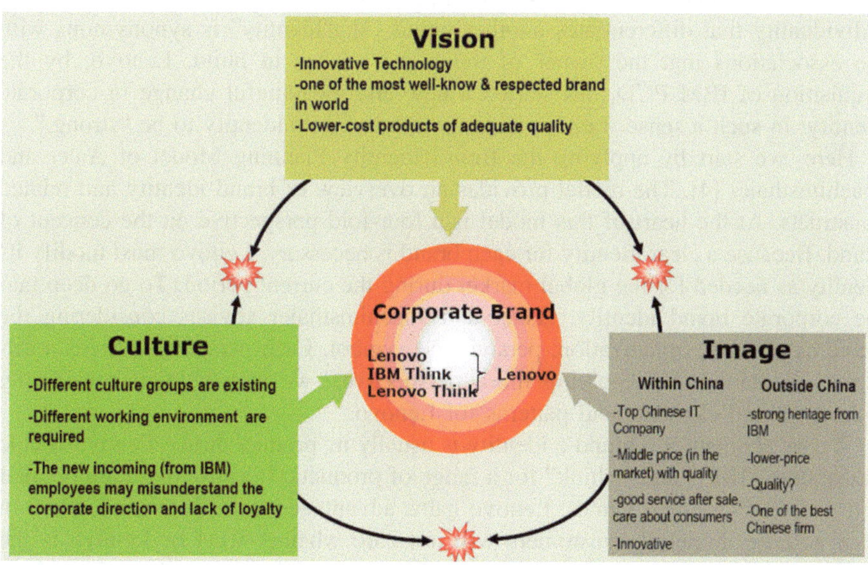

Fig. 2 Lenovo's corporate brand

(2) differences is existing between China and the outside world. However, of these has gap with the company's vision. The outside world requires time to become cognizant of Lenovo, and the Chinese need to adjust their new concept of globalbranding. Because of the huge investment in branding by Lenovo and Lenovo's successful experience in China and surrounding countries, it is possible for Lenovo to close the gaps between image, vision, and culture.

3 Is There a Gap Between Lenovo's Identity and Image?

A corporate brand will start with an image, but it will want to move that image toward a brand identity, which is a set of aspirational associations for the corporate brand to perform its assigned roles [2]. To test whether or not a gap exists between brand name and brand identity, we start discussion and analysis of corporate identity.

3.1 Corporate Identity

Lenovo applies leverages the combination of the brand names "Lenovo" and "Lenovo Think." By extending the original IBM "Think" brand, Lenovo develops and reinforces its own brand identity. An identity is not what a marketer creates but rather what consumers perceive has been created. Corporate identity is interpreted as an organisation's ethos, aims, and values, which together create a sense of individuality that differentiates another brand [3]. "Identity" is synonymous with the associations that the owner of the brand wishes to build. Lenovo, by the acquisition of IBM PCD, has started a new and meaningful change in corporate identity; in such a sense, Lenovo expects to rebuild its identity to be "strong."

Here, we start by applying the Brand Identity Planning Model of Aaler and Joachimsthaler [4]. The model provides an overview of brand identity and related constructs. At the heart of this model is a four-fold perspective on the concept of brand. Because a clear identity for each brand is necessary, Lenovo must modify its identity as needed for the global market during the current period. To go deep into the corporate brand identity, Asker and Joachimsthaler advises considering the brand as product, organization, person, and symbol. Each perspective holds a different orientation. By applying those perspectives, we gain help in considering different brand elements and patterns for Lenovo.

A core element of a brand's identity is usually its product thrust. Lenovo is able to use both "IBM" and "Think" for a range of products. Because these are popular and well-known brand names, Lenovo gains advantages from their application. In such a sense, hopefully customers will not care whether IBM or Lenovo is the producer. Moreover, this also brings Lenovo to a higher-level class. Therefore, Lenovo builds a healthy financial image.

The perspective of brand as organization focuses on the attributes of the organization rather than those of product, e.g., innovation. The brand as organization has more staying power than product attributes, because "it is much easier to copy a product than to duplicate an organization with unique values, people and grogram [5]." Lenovo is striving hard to establish its global vision instead of being a Chinese local firm.

Brand as person is perceived as having a unique personality. From a marketing point of view, a brand must have a distinct personality. There are three ways for a brand personality create a stronger brand. The first is to create a self-expressive benefit for customers: Lenovo expects to bring new technology and innovation to customers. The second is built from the basis of a relationship between customers and the brand. To build a strong brand, Lenovo must begin with a customer and work backward to shape the brand to achieve maximum impact on those customers. The third is helping to communicate a product attribute and thus contribute to a functional benefit. Lenovo is try to apply different and well-designed products to meet consumer requirements. Because the former IBM designs are already recognized by many consumers, Lenovo has inherited those advantages now and expects to carry its designs forward to a new level.

Last but not least is brand as symbol. A good symbol can help a brand to be easier to recognize and recall; therefore, it is a key ingredient of brand development. As mentioned previously, the symbol of "Lenovo" was introduced in 2003. The symbol is basic under Aaker's highlights of symbols: visual imagery, metaphors, and brand heritage.

3.2 *Image*

As discussed previously, identity is the way Lenovo aspires to be perceived, and image is another side of the coin, i.e., how Lenovo is actually perceived. Image is generally conceived of as the outcome of a transaction whereby signals emitted by a marketing unit are received by a receptor and organized into a mental perception of the sending unit. Lenovo is challenged in its way of branding. According to As traditional thinking, Chinese enterprises produce cheap products with poor quality or Chinese enterprises are the suppliers for the world. In the above-presented identity analysis, it is obvious that Lenovo is eager to produce quality goods and be an innovator of design. Lenovo has done much work to build its brand image, and therefore its corporate image; one of them is linking to IBM. By linking to IBM, one of the top three worldwide brands with US$53,791 million in 2004, Lenovo saves time and cost in presenting what is Lenovo and why Lenovo is here in the IT industry and created the possibility and chance for Lenovo to become a top brand overnight.

Because Lenovo uses the IBM label directly in its new products, the original IBM image exists naturally with the product; however, does the image still need to be changed or shifted in some way? Does the producer, Lenovo, convey the strong point of the image, or do "Lenovo" and "IMB" supplement each one another? Of cause, the latter situation is what Lenovo expects to achieve. Corporate image is defined two-dimensionally on the basis of functional attributes and emotional meanings [6]. In terms of functional attributes, Lenovo should change the perception "Chinese product is poor quality" to "Lenovo produces the fastest supercomputers." Lenovo is regarded as the best Chinese enterprise, and there is much positive news and reports on the company, which could give the Lenovo brand a positive emotional meaning to customers.

A strong corporate brand acts as a focal point for the attention, interest, and activity stakeholders bring to a corporation. When corporate branding works, it is because it expresses the values and/or sources of desire that attract key stakeholders to the organization and encourages them to fell a sense of belonging with the brand [7]. Consumers buy brands rather than products, so brand image gives a strong recommendation in purchasing actions. Some marketing researchers have claimed that a strong corporate brand has a significant impact in creating positive consumer perceptions of existing products and as well as in new product extension. As Blomqvist and Kumar's [8] advise, an effective brand-measurement system helps businesses to understand how the brand is performing against customer expectations and determine how the brand is performing against competitors. Therefore, Lenovo should create a brand-measurement system to gain those two advantages. Actually, because Lenovo started as an unknown brand to the public and has grown into a well-known brand following this strategy, Lenovo gains a better brand position and creates a competitive advantage in dealing with other market players worldwide.

4 Conclusion

Lenovo is a trailblazer leading the footsteps of Chinese firms who have moved from being Chinese local suppliers to being worldwide brands. Today Lenovo is one of the world's leading personal technology companies. There are four stages in Lenovo's branding process. Before 1988, Lenovo only had a foothold in the Chinese market, and therefore it used Chinese characters alone as its logo. Now, Lenovo has employed a world-class agency to help build global brands with innovative design. By applying a brand identity planning model, we considered four perspectives of the concept of brand. Lenovo expects to be innovative and design-oriented, which is far from the common perceptions of a Chinese firm. After a study of image, we realize the gap between brand image and brand identity, but we believe the Lenovo is working hard to minimize the gap.

References

1. Lenovo homepage (2015) http://www.lenovo.com/lenovo/uk/en/our-company.shtml
2. Aaker D (1991) Managing brand equity: capitalizing on the value of a brand name. The Free Press, New York
3. Harris F, Chernatony L (2001) Corporate branding and corporate brand performance. Eur J Mark 35(3/4):112–119
4. Aaler D, Joachimsthaler E (2000) Brand leadership. The Free Press, New York
5. Aaker D (1996) Building strong brands. The Free Press, New York
6. Stern B, Zinkhan G, Jaju A (2001) Marketing images construct definition, measurement issues, and theory development. Eur J Mark 36(1/2):201–224
7. Hatch M, Schultz M (2003) Bringing the corporate into corporate branding. Eur J Mark 37 (7/8):1041–1064
8. Kumar S, Blomqvist KH (2004) Mergers and acquisitions: making brand equity a key factor in M&A. Strateg Leadersh 32:142–145

Analysis of Monopoly Power in China's Civil Aviation Industry

Hongchang Li and Yaqi Cai

Abstract Market-oriented reform of aviation industry is always a field that deserves research. The analysis of monopoly power in China's civil aviation industry is the precondition of reforms. This chapter measures the market structure of China's civil aviation industry by means of three indicators: market concentration rate, product differentiation, and market share. It also illuminates the reasons why monopoly power occurs. This research may be of help for Chinese policy with regard to aviation reform.

Keywords Civil aviation · Monopoly power · Market structure · Regulation

1 Introduction

China's airline reform began in the late 1970s as part of China's general economic reform and "open door" policy. It is clear that the government policy of encouraging both industry entry and route entry has facilitated the large expansion of new routes and total traffic. Furthermore, the reform has affected the market structure at both the industry and route levels, which were conducive to airline competition [1]. However, there still exists monopoly pricing and social welfare damage. Therefore, understanding monopoly power in China's aviation industry is of great importance to improve the existing conditions.

H. Li · Y. Cai (✉)
School of Economics and Management, Beijing Jiaotong University,
Beijing 100044, China
e-mail: yayakero@126.com

H. Li
e-mail: 283129726@qq.com

M. Li et al. (eds.), *Proceedings of 2015 2nd International Conference on Industrial Economics System and Industrial Security Engineering*,
DOI 10.1007/978-981-287-655-3_30

2 Method for Measuring Monopoly Power

Monopoly power refers to a major corporation that limits competitors' power to control the market by means of a nonmarket adjustment. Monopolies derive their power from barriers to entry that prevent or greatly impede a potential competitor's ability to compete in a market. The major types of barriers to entry include economies of scale, capital requirements, technological superiority, control of natural resources, and legal barriers. The main methods for measuring monopoly power include market concentration rate, product differences, and market share. Next this chapter will analyze the aviation industry's market structure and monopoly power in China in the aforementioned three ways.

3 Data Analysis and Proof

3.1 Market Concentration Rate

Market concentration rate is the most common and direct measurable indicator of absolute concentration. It is used to measure the difference between the number of enterprises and relative scale. Market concentration rate also reflects the degree of monopoly in a market. In this chapter, we use industry concentration ratio (CRn) to measure it. Set ×, which means quantitative value (may be production, amount of sales, quantity of labor force, total assets, etc.) of the largest enterprises in a market. Industry concentration ratio refers to × and accounts for the share of the whole market or industry. The calculation formula of industry concentration ratio is as follows:

$$CRn = \sum_{i=1}^{n} Xi \Big/ \sum_{i=1}^{N} Xi \tag{1}$$

In this formula, CRn indicates the industry concentration ratio of the largest n enterprises; Xi indicates the NO.i enterprise's turnover volume, freight volume, main business income, number of employees, or total assets; n indicates the number of the main enterprises in this industry; and N indicates the total number of the whole industry.

Bain classified market structure with regard to different degree of monopoly according to the absolute concentration ratio of the top 4 and top 8 enterprises. Table 1 shows Bain's classification of market structure.

Now choose China Southern Airlines (CSA), Air China (AC), China Eastern Airlines (CEA), and Hainan Airlines (HA) for samples and measure industry concentration ratio in terms of each enterprise's business income and the whole airline industry's business income per year. In 2012, the top four enterprises in the airline industry were China Southern Airlines (CSA), China Eastern Airlines

Table 1 Bain's classification of market structure

Market structure concentration rate	Value of CR4 (%)	Value of CR8 (%)
Style of oligopoly I	CR4 ≥ 85	
Style of oligopoly II	75 ≤ CR4 < 85	CR8 ≥ 85
Style of oligopoly III	50 ≤ CR4 < 75	75 ≤ CR8 < 85
Style of oligopoly IV	35 ≤ CR4 < 50	45 ≤ CR8 < 75
Style of oligopoly V	30 ≤ CR4 < 35	40 ≤ CR8 < 45
Style of competition	CR4 < 30	CR8 < 40

Table 2 The business income of the main airline companies in 2012

Airline company	CSA	AC	CEA	HA
Business income (billion Yuan)	101.48	85.57	99.84	46.86

(CEA), Air China (AC), and Hainan Airlines (HA). Table 2 shows the business income of the main airline companies in 2012.

Because the total business income of all airline companies is 388.98 billion Yuan, we can calculate the market concentration rate as follows:

$$CR_3 = (1014.83 + 998.41 + 855.69)/3889.8 = 73.76\,\% \tag{2}$$

$$CR_4 = (1014.83 + 998.41 + 855.69 + 468.57)/3889.8 = 85.80\,\% \tag{3}$$

According to Bain's classification of market structure, the airline industry belong oligopoly style I. The airline market is highly concentrated. The main airline companies comprise almost the whole industry, which means the main companies have enormous influence with regard to the whole airline market; namely, the airline industry is oligopoly industry.

3.2 *Product Differentiation*

Product differentiation is the process of distinguishing a product or service from others to make it more attractive to a particular target market. This involves differentiating it from competitors' products as well as the firm's own products.

The differentiation of airline industry mainly reflects service differentiation. However, the laws from Chinese airline regulators and existing objective conditions lead to Chinese airline companies' limited capacity to provide personalized and different services. In fact, because the three-major airline companies are state owned, the price of airlines does not become the monopoly price due them because of excessive price intervention from the government. In addition, air transportation is not a consumption mode of the masses in China according to Chinese demand. The price elasticity of airline demand is <1; hence, dropping the price is the main approach to attracting consumers.

3.3 Market Share

With years of development, the Chinese airline industry has gone through a unique evolution. The main airline companies can be broadly separated into three grades. The first grade is China Southern Airlines (CSA), Air China (AC), and China Eastern Airlines (CEA). They have quite strong monopoly power. The second grade includes Hainan Airlines, Sichuan Airlines, Xiamen Airlines, Shenzhen Airlines, Shandong Airlines, etc. The third grade includes some private enterprises such as Spring Airlines, Okay Airlines, and Juneyao Airlines. The three grades have their own advantages (Table 3).

In recent years, mergers and acquisitions have become frequent in the airline industry.

As can be seen from Table 4, 47.4 % of the market share belongs to the top three airline enterprises: The three enterprises possess almost half of the market share of the whole industry. Local private airline companies almost become price followers in the market. In other words, small airline companies keep their profit depending on price strategy. In general, the airline industry in China has almost become an oligopoly market structure because the major companies together have quite strong monopoly power.

Table 3 Classification of Chinese airline companies and their major advantages

Grades	Airline company	Major advantages
First	China Southern Airlines (CSA), Air China (AC), China Eastern Airlines (CEA) (nationwide airline company)	1. Have abundant funds and large-scale of fleets
		2. Based on Chinese largest three cities Beijing, Shanghai, Guangzhou
		3. These companies contain the major international airlines
		Possess advanced technology of electronic ticketing
Second	Hainan Airlines, Sichuan Airlines, Xiamen Airlines, Shenzhen Airlines, Shandong Airlines (regional airline company)	1. Possess the most branch routes in China
		2. Passengers are satisfied with their services
Third	Spring Airlines, Okay Airlines, Juneyao Airlines (private enterprises)	1. Flexible operation management
		2. Attract passengers by low prices

Table 4 Market shares of major airline enterprises

Year	2006 %	2007 %	2008 %
AC	23.12	21.54	20.90
CSA	21.10	20.14	19.66
CEA	16.47	15.08	13.49
HA	3.79	3.92	4.09

Source China Civil Aviation Annual Report (2007/2008), Statistical Data on Civil Aviation of China (2008/2009)

4 The Source of Monopoly Power

4.1 Natural Monopoly

The airline industry in China is a natural monopoly. A natural monopoly comes from scale, scope, and net economies. The production function of the airline industry is increasing returns to scale. In addition, joint production of multiple routes services brings airline companies an obvious scale-economy effect. In recent years, as air transportation turnover in China has expended, the Civil Aviation Administration of China has established construction of a network structure based on a Beijing, Shanghai, Guangzhou axis that reflects the effects of a network economy. Hence, the airline industry in China is natural monopoly.

4.2 Administrative Monopoly

The monopoly of the aviation industry in China is dominated and maintained by the government. From the separation of civil aviation transport and air force transport in 1980 to the separate establishment of airline companies and airports in 1987, restructuring the reform of the airline industry in 2002, all of those strategies came from government regulation. In fact, the civil aviation regulator's management process exists in the managerial offside and dereliction of duty, which leads to anticompetitive conduct, which has harmed consumers' interests greatly and damaged the development of the industry. Regulation of the airline industry hinders social welfare maximization. Furthermore, the civil aviation industry in China has high barriers to entry. The major barrier is administrative approval in addition to funds and technology requirements.

5 Conclusion

Since the Chinese civil aviation industry has moved from government to marketable management and from strict regulation to deregulation, its industry scale and service performance have greatly improved. Since 2005, the administrative barriers of airline industry in China has been somewhat loosened. However, the Chinese airline industry is still an oligopoly monopoly market. To improve operating efficiency and expand consumer's welfare, it still need further deregulation, weakening monopoly power, introduction of mechanisms for competition, lowering barriers to entry, and attracting the entry of private enterprise. Meanwhile, laws and regulations with regard to antimonopoly and fair competition must be set.

References

1. Zhang A, Chen H (2003) Evolution of China's air transport development and policy towards international liberalization. Transportation Journal. Spring 42:31–49

Enlightenment on India and Japan Grain Safety Policy for China—A Case Study from a Macro Perspective

Ningshan Xu, Yuduo Lu and Haiteng Wang

Abstract Grain safety, which is closely related to social stability and national security in China, is a great strategy issue with overall importance, and it the following current hot topic is determining how to balance grain import promotion with guaranteeing sufficient supply and grain import limits to guarantee grain safety. This work applied net import data and consumption data of rice, wheat, corn, and soybean from FAO database statistics for India and Japan during 1961 to 2013 to calculate the net import dependence index. It also utilized case analysis method and comparative analysis for discussion of grain-trade features and validity of grain safety protection policies. Research shows that the guarantee of grain safety in China could not only depend on a decrease of grain import but shall disperse countries of origin on the premise that proper import was maintained. This article hereby makes policy recommendation to improve China's grain-trade pattern and guarantee grain safety based on research conclusions.

Keywords Grain import · Grain safety · Net import dependency

1 Introduction

In accordance with international trade theory, strengthened grain trade could result in a comparative advantage and accelerate rational resource allocation. However, grain, as a kind of strategic material, with immoderate reliance of import could

N. Xu (✉) · Y. Lu
Economy School, Dalian University of Technology, Dalian 116024, China
e-mail: xuningshan7@163.com

Y. Lu
e-mail: luyuduo@163.com

H. Wang
China Center for Industrial Security Research, Beijing Jiaotong University,
Beijing 100044, China
e-mail: whtcars@126.com

M. Li et al. (eds.), *Proceedings of 2015 2nd International Conference on Industrial Economics System and Industrial Security Engineering*,
DOI 10.1007/978-981-287-655-3_31

possibly lead to risk of a grain embargo occurring or insufficient supply from the global market, inability to satisfy domestic grain demand, and subsequently threaten grain safety.[1] Hence, it can be seen that grain import should be moderately maintained with regard to grain safety, and to the methods of the management of moderate grain import is currently an issue that requires urgent solutions.

Currently research on domestic and international grain safety has mainly concentrated on the aspects of current grain supply status and trend forecast, index and method for grain safety measurement, factors affecting grain safety, impact and effect of international grain trade etc.; cases studies including international comparison still remained insufficient. This present work selected two countries—Japan and India—both of which geographically abut China and share a similar environmental resources status and have a similar cultural and institutional environment but with entirely different grain trade status, and performed a comparative analysis of the grain safety issues of both countries thru the calculation of net grain import dependency.[2]

Japan and India both are confronted with the same environmental resources status, which is population densely in a limited territory and a shortage of resources. However, it is interesting that the basic lines of grain safety for both countries are widely divergent. Japan is a developed country with agricultural modernization achievements and has maintained high grain net import dependency for a long time; it has also imported the most grain around the world.[3] In contrast, India is a developing country that has suffered from a shortage of agricultural products and hence had a large quantity of import previously. However, since the "Green Revolution"[4] in the late 1960s, not only has the gap between grain supply and gradually relaxed, India has had more grain self-sufficiency in recent years, and the net import dependency of grain has dropped to a negative value.[5]

---

[1]"Grain safety" is a concept of dynamic with continuous evolution. The FAO, based on different history background, proposed a different grain safety concept. In consideration of the feasibility of strategy implementation and convenience of statistical data acquisition, "grain safety" analyzed in this chapter mainly refers to sufficient supply at a national level and specifically measured it through net import dependency index of grain. In addition, "grain" as analyzed in this article, unless otherwise specified, refers to aggregation of wheat, rice, corn and soybean; tuber crop is not included.

[2]Also called as "grain foreign trade dependence" or "grain external dependence." It refers to the ratio of grain net import to grain apparent consumption and is used to measure net dependency of grain consumption from an external supply. It is also the indicator to measure grain safety dependency ratio on global grain market for one country [1].

[3]Conclusion is calculated based on relevant data in FAO statistical database <http://faostat.fao.org/site/368/DesktopDefault.aspx?PageID=368#ancor> (accessed March 30, 2015).

[4]"Green revolution" refers to the technological reformation activity, which could increase the grain production, when some western developed countries popularized high cereal and agricultural techniques into some areas in Asia, Africa and South America.

[5]This conclusion is calculated based on relevant data in the FAO statistical database <http://faostat.fao.org/site/368/DesktopDefault.aspx?PageID=368#ancor> (accessed March30, 2015).

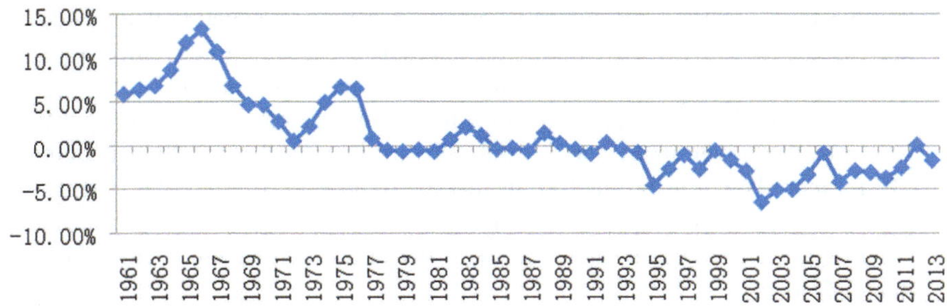

Fig. 1 1961–2013 Net grain import dependency of India (%). Grain data obtained herewith was calculated from rice, wheat, corn, and soybean. Data sources are reorganized and calculated based on relevant data from FAO STAT <http://faostat.fao.org/site/368/deefault.aspx#ancor> (accessed March 2015)

2 Case Study: India—Domestic Self-sufficiency

The FAO compiled the statistics of rice, wheat, corn and soybean for the global top 20 countries[6] in terms of import volume and import value, and India is not included. Thus, India became a country of net grain export and was not included among the world's top import countries in the 1990s. Since the economic revolution of 1991, the net grain import dependency of India has remained a negative value (Fig. 1).

As shown in Fig. 1, the net grain import dependency of India grain basically presented a downtrend comprising three specific periods as follows:

1. In 1961 to 1977, the net grain-import dependency was comparatively high with positive value. Especially during the first 6 years, due to yearly disasters and famines, the net grain import dependency increased year by year and reached 13.86 % in 1966. After 1966, government grain-safety awareness gradually improved, and it established the warning system with a considerable basis through the analysis of sings for planting area, reserves, cultural, procurement, and disaster situation etc. as well as a reflex of the forward market. At the same time, the government developed two mechanisms—buffer stock and economic stock—to ensure stable grain supply [2]. Domestic grain production increased substantially, and grain net import dependency decreased year by year and finally reached 0.85 % in 1977, which is almost close to zero.
2. In 1978 to 1992, grain net import dependency remained close to zero at all times, and grain-safety policies brought obvious effects as well. Domestic grain demand relied primarily on their India's production, with neither surplus stock nor a gap between supply and demand, and grain supply and demand remained a stable relationship.

[6]FAO STAT <http://faostat.fao.org/site/342/default.aspx> (accessed April 3, 2015).

3. In 1993 to 2013, grain net import dependency became a negative value and has decreased year by year thereafter. This is mainly because India decided to carry out economic revolution after the worst economic crisis in 1991, and it substantially cut down agricultural subsidies. But subsequently under political pressure, the government agricultural subsidy increased other than deduction [3]. Farmers began to product grain massively and sell to the government; the grain stock increased rapidly; and an increase export is required for the relaxation of stock.[7] However, negative grain net import dependency declined with fluctuation.

Nevertheless, India, with a half century of efforts, has guaranteed grain supply safety, but it has done so with considerable cost, i.e., the government must undertake huge amounts of grain reserves and consumption subsidies. As estimated by World Bank experts, only approximately one third[8] of the financial subsidies is required to achieve the same wheat-consumption target if it is supplemented with import. Therefore, India's government recently decided to adopt economical methods for a revolution of its grain storage management system followed by a fluctuating marketing environment. Indian grain companies should enter into the global forward market and reduce financial utilization and subsidies. With the transformation of the grain-safety concept, India is currently strengthening the transformation of its early warning forecast, intensifying its early warning mechanism, buffering the stock compression, and guaranteeing future use, etc.

3 Case Study: Japan—Dependency on Import to Guarantee Grain Supply

The situation in Japan is opposite of that in India. Japan lacks grain self-sufficiency and therefore it is largest country of import for agricultural products in the world. Figure 2 shows the net import dependency status for different varieties of grain in Japan since 1961. Analysis shows that corn consumption can basically be satisfied by import, and India's net import independency has remained at 100 % for all of the time since 1967. Soybean is the second largest import, and the net import independency approached 100 %. The soybean net import independency also reached >90 % in recent years, and the standard[9] of grain safety calculated based on

[7]FAO STAT data shows that Indian grain export takes corn as principle, and net import dependency for Indian corn import is −21.79 %, in the annual report.

[8]WTO. Trade Policy Review: India Revision 2007. <http://www.wto.org/english/tratop_e/tpr_e/tpr_e.htm> (accessed April 8, 2015).

[9]Grain-safety degree is reflected by grain net import dependency: <5 % refers to safety status; 5 to 13 % refers to generally safe status; 13 to 15 % refers to unsafely status; and ≥18 % refers to danger status. This standard was deduced based on safety degree of grain self-sufficiency ratio and relationship between self-sufficiency ratio and net import dependency [4].

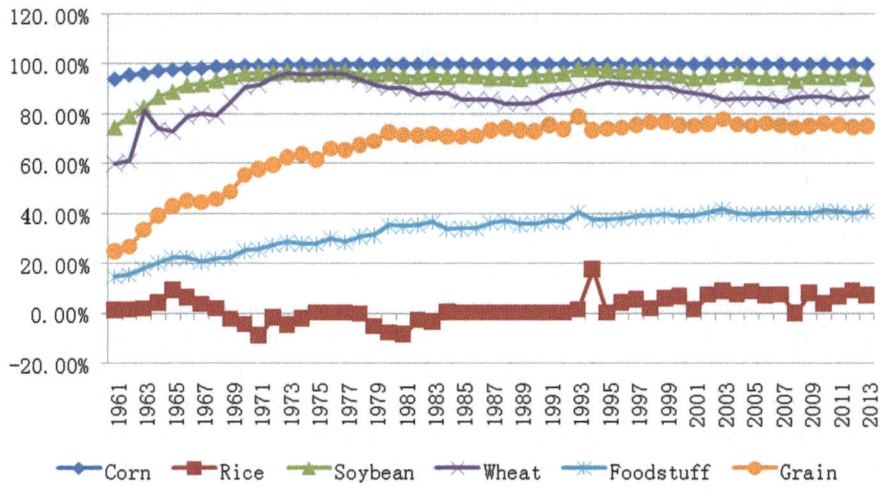

Fig. 2 1961–2013 Grain net import dependency of Japan (%). "Grain" refers to wheat, rice, and coarse grains, which include corn, barley, sorghum etc. Data sources are reorganized and calculated based on relevant data from FAO STAT <http://faostat.fao.org/site/368/deefault.aspx#ancor> (accessed April 14, 2015)

grain net import dependency, corn, soybean and wheat are all under the status of extremely unsafe. However, it should be noted that India's grain net import dependency remained lower during this period, and grain self-sufficiency was generally maintained for a rather long time before 1993; however, it never want beyond the grain-security cordon, i.e., 10 %, though it increased with certain value after 1994. When the above-mentioned four grain crops were taken into overall consideration, net import dependency was basically maintained at approximately 40 %. Therefore, although resources and environmental conditions determined a higher net import dependency, Japan properly controlled the makeup of grain imports, which guaranteed adequate self-sufficiency of the basal ration grain, rice. That is to say, rational makeup of imports is helpful to avoid grain-trade risk.

To ensure adequate rice self-sufficiency, Japan implemented high subsidies. Other than factual subsidies for the reservation cost of government rice, partially storage-expense subsides, these also provided for the grain that is unable to be sold by producers at end of the year. In addition, in 1942 Japan enacted complementary laws and regulations for the Grain Administration Act and its implementing rules with subsequent multiple revisions, which played an important role to regulate grain production and circulation activities. To adapt WTO's relevant rules, Japan enacted the new Grain Administration Act, which changed the previous mode of government-unified management for rice purchase, sale, and price, and predominantly autonomously circulated the rice already in circulation to further fully develop the market mechanism function, reduce government financial subsidies and set the rate formed in the market, relax rice-running restrictions and turn the license system turned into the register system, create diversified circulation channels, and enlarge the degree of marketability and liberalization [5]. In 2003, Japan again enacted an amendment to the Grain Administration Act, which adjusted the rice-production policy and will abolish the current rice production policy in effect since 2008 [6]. The introduction and implementation of these acts played an important role in rice-production stabilization, grain-trade regulation, grain-reserve establishment, grain early warning mechanism operation, and grain-safety system establishment, etc.

Meanwhile, confronted with over-the-top grain net import dependency, and to prevent grain embargo risk or import price traps, Japan took the grain-safety weapon of dispersed import and signed a long-term grain import agreement[10] with many countries that showed favorable results. The path of diversified import sources avoided single import source from America and consolidated global cooperation in the establishment of international grain reserves. Furthermore, Japan also provided technical support for developing countries to develop their agriculture. On one hand, this accelerated agricultural economic development for developing countries; decreased their export requirements; and eliminated competition in the global grain market. On the other hand, it created opportunities for Japan to import other grains from these countries.

4 Conclusion

India's agriculture and development mode of agricultural economy share similarities with those of China. Japan is considered the largest country of grain import, and the threat that confronts it is obvious. Therefore, research on grain-safety problems and the strategies of these two countries to solve them could provide us with an experience reference for immediate solutions of grain safety policy in China. In general, measures that could guarantee long-term national grain safety in China not only consist of the improvement of domestic grain-production capacity, ensuring adequate domestic grain self-sufficiency, and limiting grain import quantity, etc., but also taking into account the objective facts of limited resources and weak growth of grain production. Measures taken by India for guaranteeing grain safety are certainly worth learning from, but storage and management costs incurred from excessive grain production and over-the-top self-sufficiency ratio, on the contrary, would make the problems being confronted much worse. Experience from Japan showed that it is necessary and feasible for to have moderate import. Slightly high net import dependency is not terrible, and grain-trade risk could be avoided effectively with proper makeup of import arrangement as well as diverse import sources. Moreover, these measures are also in favor of the establishment of a

[10]For example, in 2003, the recourse country of import for Japan is as many as 208 countries and regions, which include some islands in the Pacific Ocean and Atlantic Ocean.

well-trade cooperation relationship with more countries and could maintain full and sustainable development of international trade while guaranteeing China's national grain safety.

5 Countermeasure and Suggestion

5.1 Properly Broaden the Target of Grain Net Import Dependency

The reason that China set the target[11] for grain net import dependency (grain and soybean) to no more than 5 and 0 % of grain net import dependency mainly because China is concerned about the risks of grain embargo, natural disasters, decision failures, and grain unavailability, a sharp increase of grain price, in the global market, etc.; however, analyzing the case of India enlightened us to the fact that a super-high grain-safety target will not only incur unnecessary high cost, it will also e adverse to the sustainable development of grain safety. Hence, to properly broaden the target of grain net import dependency until it reaches 10 % will not only reasonably save grain-production cost, so as not cause problem to threaten grain safety, but it will also engage certain economic interests by taking full advantage of the domestic-production comparative advantage.

5.2 Dispersion Import, Reduce the Degree of Concentration for Grain Import[12]

A high degree of import concentration makes it possible for grain-trade risks to threaten the grain safety of China. The research casa of Japan showed that it could optimize the structure for the country of origin and diversify import so that grain safety could be guaranteed under a properly broadened net import dependency. For example, the price is the main but not only consideration. China should also seek a grain-trade partner in multi-azimuth countries, make appropriate annual adjustments for trade-partner countries and put pressure on trade-partner countries; promote and maintain world peace, and maintain normal economic communication with trade-partner countries, etc.

---

[11]Refer to "medium and long term layout plan for national grain safety (2008–2020)" published in November 2008 by the NDRC.

[12]For measuring the degree that grain import concentrates in one or some countries of one country, for example, for China, 98.23 % of its rice import volume is from Thailand, whereas 88.78 % of its soybean import volume is from America and Brazil. This shows that grain import concentration degree is rather high for China. (Data from, Agriculture Department, PRC <http://www.moa.gov.cn/fwllm/sjfw/tjsj_1/gjmy/>).

5.3 Develop Food Trade and Increase Other Commercial Intercourse with Main Grain-Export Countries

Recent research demonstrates that food production has an obvious comparative advantage compared with grain production, and food-trade development could properly avoid grain-trade risk [7]. China should increase other commercial intercourse with main grain-export countries while developing a food trade. This has two following advantages: (1) China could fully play the comparative advantage of its own agricultural production in China, drive and increase agricultural production efficiency, and attract more resources for more efficient production; and (2) China could make the dependency relationship with grain-trade countries closer, and then the relationship of the import country no longer enslaved by the export country; rather the correlative dependence relationship would exist between trade partners without division into principal and subordinate. This is favorable for tempering the powerful position of the export country of export and make grain exporters have no motivation to artificially initiate trade resistance.

References

1. Feng L, Ya X (2008) Grain supply and demand and price trend in China (1980–2007)—discussion on grain price fluctuation, macro stability and grain safety issues [J]. Manag World 3:70–80
2. Dong Y (2009) Grain safety policy of India and its experience reference to China [J]. Econ Manag Res 3
3. Dong Y, Yu J, Liu Z (2008) Liberalization, poverty and grain safety of agricultural trade: experience from India [J]. Acad J SCAU (JCR) 7(4):21–29
4. Limeg Z (2007) Precautionary analysis for agricultural products import and export and agricultural industrial safety [J]. Financ Econ 6:111–116
5. Lei Z, Huichao Y, Jiaen Z, Kefeng Z, Rui C (2005) Enlightenment to China from security mechanism of grain safety in foreign countries [J]. Chin Agric Sci Bull 2(11):417–421
6. Ke S (2004) Legalization management for grain circulation in Japan and enlightenment for China [J]. Circ Econ China 18(7):29–32
7. Feng L (1997) Comparative advantage and food trade structure—third option for food adjustment policy in China [J]. Econ Res 2:3–11

Part IV
Special Session on Cultural Industry

Research on the Current Status of the Cultural Industry Organization in China and Development Countermeasures

Li Xiao

Abstract Based on the industrial organization theory of SCP paradigm, this work conducts a comprehensive analysis of cultural industry in China and reveals the common problems existing in the cultural industry in China, which has important theoretical and practical significance in perfecting the target and the approach of Chinese cultural industry organization. Meanwhile, the solutions toward the problems can regulate the competition behaviors between and enhance the market performance of enterprises in the cultural industrial organization.

Keywords Cultural industry organization · Market structure · Enterprise behavior · Market performance

1 Analysis of the Market Structure of Cultural Industry in China

1.1 The Market Concentration of the Cultural Industry Is Comparatively Low, and Dispersive Competition Dominates the Market Structure

The cultural industry belongs to the industry of nonprominent scale of economies, which decides the low market concentration. During the development of the cultural industry in all corners of the world, in the country with the most developed and advanced cultural industry, the market concentration is still lower than the industry of prominent scale of economies [1]. In recent years, the cultural industry in China

L. Xiao (✉)
Beijing Institute of Graphic Communication, Beijing 100026, China
e-mail: xli2009@163.com; xlkj2002@163.com

M. Li et al. (eds.), *Proceedings of 2015 2nd International Conference on Industrial Economics System and Industrial Security Engineering*,
DOI 10.1007/978-981-287-655-3_32

has seen great development. However, compared with media enterprises of developed countries, the industrial concentration is too low, placing Chinese cultural enterprises in a position of weakness in inn terms of competition. Although several large-scale cultural corporate groups have recently appeared, the problem of small, dispersive and weak cultural enterprises still exists, and large enterprises still hold a small market share. For example, the market share of China Publishing Group Corp. is only approximately 6 % [2].

1.2 Only a Few Excellent Works Emerge in the Large Quantity of Cultural Products and the Homogeneous Competition of Products Is Common

One common characteristic in the market structure of cultural industry in China lies in the few differences among cultural products, the strong replaceability among cultural products, and the industrial administration system, which still needs to be improved. Take festivals as an example. Despite the large quantity of festivals, there are only a few excellent festival projects, while many projects share repeated subjects and positioning and lack of creativity. Some organizers of local economic festival projects have not emphasized out the local cultural connotation enough, thus leading to similarities in short distance and low-level repetition.

1.3 The Strict Control Over Cultural Industry Results in Too High Barrier for Entry and Exit

One important factor contributing to the low market concentration of the cultural industry in China lies in the high barrier for entry to and exit from the cultural industry. The cultural industry in China has both institutional and economic barriers. The institutional barrier mainly consists of the policies, laws, and regulations formulated by the government. Owing to the strong ideological property of the cultural industry, China exerts a strict approval and administration system for units operating in the cultural industry. Take the newspaper as an example. The newspaper in China adopts a minority entry permit with strict restrictions, whereas the newspaper market has clear boundaries among the segmented markets with a very high barrier. The exit barrier of newspaper mainly results from the policies. The cancellations of newspaper in China are all led by government administrations for the implementation of administrative measures instead of the market ones.

2 Analysis of Market Conduct of Cultural Enterprises in China

2.1 Price Competition of Cultural Enterprises Relies Mainly Relying on Predatory Pricing

The cultural product price is the major measure by which to adjust the supply and demand of the cultural industry market. Take the newspaper market of the cultural industry as an example. A predatory pricing strategy is often applied by some newspapers or newspaper groups in the competition of newspaper distribution. The term "predatory pricing strategy" in the newspaper market means that newspaper producers set the comprehensive price of the newspaper significantly lower than the general cost of newspaper to drive out competitors from the newspaper market and put off potential entrants; then they raise the cost when competitors disappear to earn high monopolistic profits. The advantages of this strategy lie in that it can help rapidly open the market, increase sales, win over consumer, and produce social benefits. However, in the long term, it is unfavorable for the promotion of comprehensive competitiveness in the cultural industry.

2.2 Lack of Price Competition with Low Competitiveness Between Brands of Cultural Products

The lack of price competition among cultural enterprises refers to the competition conduct by which cultural enterprises improve their competitiveness through development of new products and technologies as well as better marketing to gain greater market share. Due to the long-term effect by the planned economy system, the cultural industry in China led the way of resource consumption by relying on investment and adopted the low-level same-quality competition strategy. China's brand effect is significantly weaker than that of developed countries. Brand awareness of cultural products and services, not to mention brand advantages, has not been well established in China. Compared with the small number of domestic cultural enterprises with strong brand advantages, the number of cultural enterprises that enjoy certain competition advantages and brand recognition in the international market is even smaller. Even some cultural products with certain brand advantages depend much on long-term financial investment by the government and monopolistic running.

3 Analysis of Market Performance of the Cultural Industry in China

3.1 *Small Scale of Cultural Enterprises and Low Industrial Concentration*

In all sectors of the cultural industry in China, there are only a few large-sized cultural enterprises but quite a large number of small-sized cultural enterprises, and each cultural enterprise only occupies a small share in the total circulation. It is difficult for these enterprises to realize scale of economy. In particular, in some industries with prominent scale of economy, the majority of cultural enterprises still fail to reach the minimum level of scale of economy, thus leading to the waste of cultural resources.

Take book publishing as an example. In western developed countries, the publishing industry has become an important industrial contributor to the national economy. Germany publishes >80,000 types of new books each year with an annual sales reaching as high as >8 billion Euros. Currently, among a total of 16,000 publishing houses registered in Germany, the market is mostly dominated by approximately 100 large- and medium-sized publishing houses, the sales of which account for 92 % of total sales in the publishing industry [3]. The large-sized cultural enterprises in China are small in scale as well as number. Book publishing in China is one of the cultural industries experiencing rapid growth. Until 2008, the annual sales income of the largest book publishing group in China was approximately 3.8 billion RMB compared with the sale of the Bertelsmann Group in Germany for 16 billion Euros [4].

3.2 *The Output Value of the Cultural Industry Is Increasing Each Year, but the Overall Scale Is not Large*

Although the added value of the cultural industry has been increasing each year, the overall scale is not large, thud showing a great gap to developed countries. Currently the output value of the cultural industry of developed countries occupies a considerable proportion among the GNP, and the industry functions as an important drive force for economic and social development. For example, in the United Kingdom, the average development speed of the cultural industry is twice the speed of economic growth, and the industry produces a value of approximately USD$60 billion each year accounting for 11 % of GNP. Exceeding the output value created by any of the traditional manufacturing industries, the cultural creative industry has become the second largest industry in terms of output value in the UK with an employment of 1.95 million people and ranking first in the country [5].

3.3 *Low Efficiency of Resource Allocation of the Cultural Industry and Serious Waste of Cultural Resources*

Due to the long-term influence of the planned economy system and restrictions by the market admission system, the resource allocation for the cultural industry of China is not realized through market adjustment but largely depends on administrative measures. The barriers arisen from the administrative policies have seriously inhibited the rational flow of resources for the cultural industry and led to the block separation of the cultural industry market in China. In China, some sectors in the cultural industry are subject to strict market admission control. Currently the cultural administration authorities in China are the Ministry of Culture, the State Administration of Press, Publication, Radio, Film and Television, the Ministry of Industry and Information Technology, and the National Tourism Administration and each administers programs in its own way. The irrational administration pattern and the long-term integration of government and business have resulted in the monopoly of department industries within the cultural area, serious regional blockage, segmented cultural market, impeded flow channel, and artificial break of the cultural industry chain, thus leading to low efficiency of resource allocation of the cultural industry and serious waste of cultural resources.

3.4 *Only a Few Original Technological Inventions in the Cultural Industry and Slow Progress of the Technologies in the Industry*

Much of the market structure and conduct for the cultural industry show the factors for technological progress of the cultural industry, which are finally reflected by the development of the cultural industry. The technological progress reflects the dynamic economic efficiency and is therefore considered as an important indicator for economic performance. China has only a few original technological inventions and innovations in the cultural industry, whereas the technological progress of the cultural industry is mainly reflected by the use and acceptance of the new technologies in the cultural industry. For example, in the industry of on-line games, foreign firms possess advantages on technologies and operations and have been dominant in the domestic market for a longtime, whereas domestic firms mostly serve as the sales agent by localization and outsourcing businesses, etc., and lack independently developed products.

4 Countermeasures on Development of Cultural Industry Organization in China

4.1 Promote the Integration of Cultural Creation and Technological Innovation and Improve the Technology Content of Cultural Products

The cultural industry is a technology-intensive industry. Only with knowledge and technological support can it can offer better products and service for cultural consumers. It has become common for multinational cultural enterprises and groups to attach importance to the integration of high technologies and cultural industries as well as focus on application of high and new technologies to the creation, production, and dissemination of cultural products.

The cultural industry in China reveals the problem of low technological content and low application of technology in cultural enterprises. The integration of cultural creation and technological innovation is most significant in films. Currently, 3D films are developed very fast in China, and 8 out of global top 10 box-office films are 3D films. However, Chinese 3D films are still weak in terms of core technologies, the excellent films are small in number. China must highly emphasize the application of technologies in the cultural industry, strengthen the technological innovation of cultural enterprises, and perfect the incentive mechanism for technological innovation, thus promoting development of the cultural industry in China.

4.2 Lower the Exit Barrier for the Cultural Industry and Create Conditions for Market Concentration of the Cultural Industry

To lower the exit barrier for state-owned cultural enterprises, it is necessary to sort out the relation between the central and local government and between the government and enterprises, establish a normal bankruptcy mechanism for cultural enterprises, make full use of the law of value to adjust and guide resource flow, remove the institutional obstacles for the trans-regional, trans-industrial, and trans-ownership merger of advantageous cultural enterprises, and create conditions for market concentration of the cultural industry.

In contrast, state-owned cultural enterprises should facilitate the reform of separating government administration and enterprise management and transformation in line with the stock system, establish a modern enterprise system, give full play to the decisive role of the market in resources allocation, promote the coalition and reorganization of state-owned cultural enterprises, cultivate a batch of large-sized cultural enterprises and groups, and lead the strategic structural reorganization and upgrading of the cultural industry.

4.3 Promote the Market Concentration of the Cultural Industry and Break Barriers with Respect to Region, Industry, and Ownership

China should encourage state-owned cultural enterprises to realize strategic reorganization by issuing supportive policies on finance and taxation, broaden financing channels for cultural enterprises, and render financial support to strategic reorganization. In contrast, to fight the block separation of cultural enterprises, China should explore the sharing mode of financial and taxation interests of cultural enterprises after trans-regional reorganization, promote the merger and reorganization of cultural enterprises through the capital market, and support enterprises with financing through stocks and bond issuance.

References

1. Luo J, Li T, Zhu E (2008) Situation of cultural industry organizations and target analysis of their improvement and development. Product Res 1:110–112
2. Beijing Institute of Graphic Communication Institute for Culture Industrial Security Research (2014) China cultural industry security report 2014. Social Sciences Academic Press, Beijing, pp 38–39
3. Xiong C (2012) World cultural industry research. Tsinghua University Press, Beijing, pp 4–6
4. Liu J, Chen H (2010) Introduction to cultural industry. Capital University of Economics & Business Press, Beijing, pp 76–78
5. Xiong C (2012) World cultural industry research. Tsinghua University Press, Beijing, pp 2–5

Study on the Government Regulation of Chinese Cultural Industry in the Times of Globalization

Liwen Han

Abstract In the era of globalization, the cultural industry has become a necessary way in which the culture spreads widely and develops prosperously. Government regulation of the cultural industry should standardize the order of the cultural market, the essence of which is reasonable to deal with the relationship between the government and the market. In China, cultural industry regulatory organizations lack independence and authority and do not have a unified regulatory organizations, which causes the culture industry market to be confused and chaotic. Chinese cultural industry government regulations need legal norms to improve and increase social regulation.

Keywords Cultural industry · Government regulation · Era of globalization

1 Introduce

The cultural industry has obtained considerable development and is currently known as the "sunrise industry" of the twenty-first Century. Government regulation of the cultural industry is becoming more serious to promotes the healthy development of the cultural industry. In periods of transition, Chinese cultural industry government regulation is more prominent and urgent. This chapter will explore Chinese defects in and ways to improve the strategy of cultural industry government regulation.

L. Han (✉)
Cultural Industry Security Institute, Beijing Institute of Graphic Communication, Beijing 102600, China
e-mail: hlw519@sina.com

M. Li et al. (eds.), *Proceedings of 2015 2nd International Conference on Industrial Economics System and Industrial Security Engineering*,
DOI 10.1007/978-981-287-655-3_33

2 The Cultural Industry and Cultural Industry Government Regulation in the Era of Globalization

2.1 The Cultural Industry in the Era of Globalization

The cultural industry is a content industry, the key of which is innovation, which is the source of the vitality of spiritual and cultural products, the main source of market popularity, and a fundamental measure of spiritual product value. In the era of globalization, when the effect of the cultural industry expands day by day, the cultural industry is closely related to the national traditional culture and the national culture security; therefore, the prosperity of culture cannot be separated from the cultural industry of China's particular culture. For example,

"In human history the cultural industry is only a carrier of the culture system that can spread quickly different ethnic cultural achievements to the whole world, which promote rapid communication of different civilization achievement and the overall progress of human civilization. Especially following the application of information technology the cultural industry not only makes the time and space as hindering culture forces lost its meaning but also makes zero space and time become an important strategic forces of modern culture" [1].

The changes determine that the development of cultural industry are not only to meet the needs of the development of economy and culture in general but to determine the dominance of the competition of new knowledge and a new culture, of which the competition results will directly determine the future and destiny of a culture in the era of economic globalization.Therefore, the development of cultural industry has risen to safeguard national cultural security level in the context of globalization.

2.2 Cultural Industry Government Regulation in the Era of Globalization

The cultural industry competition between national countries has become more intense such that government regulation is an important strategy choice for national culture security and the national traditional culture. Government regulation, by which the government or regulatory agencies uses coercive state power of the state to achieve direct or indirect economic regulation, social control, or intervention, can overcome market failure and achieve the maximization of social welfare, namely the goal of "public interest maximization" [2].

Generally government regulation is divided into economic regulation and social regulation. The term "economic regulation" means that the government regulates and restricts enterprise in terms of price, yield, entry and exit, and other acts to prevent low efficiency of resource allocation and to ensure fairness of service supply. "Social regulation" means that the government normalizes and restrictions the social economy main body's behavior to ensure the education, culture, and other social welfare to provide safety, health, and health care to all citizens to prevent pollution [3] Cultural industry government regulation exists not only "to overcome the market failure," but also to maintain the prosperity, development, and security of the national culture. Therefore, the reasonable boundary of cultural industry government regulation is to make up for market failure, of which the primary task is to reasonably standardize the cultural market order. Thus, the core problem of cultural industry government regulation is to handle well the relationship between government and the market. Cultural industry needs government regulation because the cultural market has strong externality, for example, monopoly, nonexclusive and noncompetitive behavior, and information asymmetry [4]. Cultural industry government regulation has played a decisive role in the era of globalization as follows. First, government regulation has a role in promoting cultural industry. Government regulation promotes the development of cultural industry through optimization of the development of cultural industry market environment and stimulation of the internal structure of cultural industry. Government regulation supports the development of cultural industries through the fiscal, tax, finance, investment, and other aspects of the preferential policies as the main means [5]. Second, government regulation has the function of regulating the cultural industry. Under the background of globalization, countries all over the world guide and standardize cultural industry through government regulation to prevent the impact of foreign culture and to ensure the safety of the domestic cultural industry. Through the legislation of cultural industry, the government ensures the core assets of the cultural industry, e.g., the safety and value of intellectual property rights, prevents and reduces the social harm of an unhealthy culture, and influences cultural industry factors through the supervision and examination of the content of cultural products.

3 The Evolution Process of Chinese Cultural Industry Government Regulation

Chinese cultural industry started relatively late, from spontaneous, conscious disorder to industry-management practices and moved effectively forward. Before the reform and opening-up, whether in the political, economic, social, or cultural aspect, China has implemented a strict planned economy system. Government regulation of the cultural industry has been in the planned economic system under the framework of implementation of direct, rigorous, and comprehensive intervention. All activities of enterprise culture are to be decided by the government [6]. Culture management is not meant to solve the "market failure" but to eliminate the market entirely.

After the reform and opening-up, Chinese cultural system reform experienced a gradual deepening process. In 2002, 16 large reports of the party were put forward clearly to develop cultural undertakings and, continue to deepen the reform of the cultural system, to establish and improve the cultural market management system, and to create a good social environment for the prosperity of socialist culture. In 2003, the Cultural System Reform Work Conference held and studied to deploy pilot reform of the cultural system by introducing a number of reforms of cultural system and the relaxation of the cultural market access policy. In this period, the Chinese cultural market entered a rapid development period. Government regulation cultivated and guided the cultural industry to become a leading industry. From the establishment of specialized departments in the Cultural Industry Department of the Ministry of Culture to make a clear distinction between cultural undertakings and cultural industries, and allowing cultural industries to participate in the overall planning of the development of the national economy, China is paying more attention to the construction of the legitimacy of the cultural industry, whereas the basic orientation of government regulation is construction and specification [7].

In 2005, the CPC Central Committee, the State Council Office, issued the document On Further Strengthening and Improving the Export of Cultural Products and Services Work Opinions, which marked the formation of the Chinese cultural industry's "going out" strategy [8]. In August 2006, the country's "eleven five" period, the cultural development plan was clearly the priority development of public cultural services and public cultural undertakings and the active and the steady push forward of the reform of the cultural system. During this period, Chinese government regulation of the cultural industry gradually began to relax, and the policy of governmental support of diverse sources of funding for emerging cultural industries rapidly developed.

4 Defects and Improvement Strategy of Chinese Cultural Industry Government Regulation

4.1 Defects and Shortcomings of Chinese Cultural Industry Government Regulation

Because Chinese cultural industry government regulation is mainly composed of a cultural system, cultural system reform is not thorough enough, so path dependence will still exist. For regulatory agencies that are not directly produced by legislation, for which power is not generated by the Constitution and the laws, the mechanism of cultural industry government rgulation lacks independence, authority, and stability, which easily leads to regulatory agencies being created by a political or business group and therefore a lack of fairness and stability.

The over-regulation of the economy by the government reduces the degree of marketization of the cultural industry and reduces cultural enterprises' economic vitality and competitive advantage. There are two main ways of regulation the culture industry: (1) entry regulation mainly in the form of national monopoly, declaration, examination and approval, license, business licensing, and setting standards; and (2) price regulation mainly for the monopoly of telecommunications monopoly, which includes setting legal price, local government pricing, industry guidance, standards, etc. Access regulation strictly prevents potential competitors, especially for private enterprises that restrict the normal development of a private enterprise culture.

4.2 Improvement Strategies for Chinese Cultural Industry Government Regulation

Cultural industry government regulation effectively optimizes the allocation of resources and guarantees the normal operation of the cultural industry and the healthy development of the society by improving the overall efficiency and cultural welfare level. However, ineffective and inappropriate regulation will not only hinder the prosperity and development of cultural industry, but the speed of economic growth will be slowed by the correlation effect due to the large domestic industry [9]. The key to government regulation is to make the regulation mechanism to maintain neutrality, avoid excessive regulation, and improve regulation efficiency.

China will further promote the reform of the cultural system in depth using legal means to divide the boundary between government and market. Independence comes when regulatory agencies can be maintained with government administrative departments and business contact, but the independence of the power must be maintained [10]. Based on scientific orientation of government cultural function, comprehensive cultural industry regulation is relatively independent, authoritative, transparent, fair, efficient, accountable [11]. In the government regulation of all kinds of cultural industries, we must further transform the function of government and the company system of cultural enterprise. Government legislation strengthens the regulation of the cultural industry gradually builds the formulation, execution, error correction, and supervision of regulation into the operation and management of the legal system.

Strengthening and improving social regulation of the cultural industry. The supervision system for the formation of boundaries by the industry is regarded as the main means of self-regulation [12]. This is conducive to building a complete cultural industry supervision system and to maintaining the independence and neutrality of the government regulation mechanism of the cultural industry, which provides protection for the normative market order and promotes the healthy development of China's cultural industry.

References

1. Gu J (2009) The cultural industry economics. Nanjing University Press, Nanjing, p 240
2. Caoyi Z (1992) Micro regulation economics. Publishing House of Chinese development, Beijing
3. Zhang B (2012) On the present situation and problems of the regulation of culture industry in our country. Books and information, no. 4:85
4. Han P, Yan H (2014) Regulation of cultural industry government. J Harbin Univ Commer China 1:82–83
5. Zhang B, Hou X (2013) The theory of the government regulation on the development of cultural industries Under the background of globalization, no. 5:16
6. Gu J (2009) The cultural industry economics. Nanjing University Press, Nanjing, p 340
7. Cai S, He P (2010) Review and prospect: the evolution of innovation of China's cultural industry policy. J Chengdu Univ (Social Science Edition), no. 2
8. China Academy of Social Sciences Research Center of culture (2006) 2005–2006 China cultural industry development analysis and suggestions of Chinese economic and trade, no. 6
9. Zhang B (2012) On the present situation and problems of the regulation of culture industry in our country. Books and information, no. 4:83
10. Zhang B (2012) On the present situation and problems of the regulation of culture industry in our country. Books and information, no. 4:85
11. Zhang B (2012) On building the moderate mode of cultural industry government regulation. Theory of the issue, no. 9:113
12. Han P, Yan H (2014) On the government regulation of cultural industry. J Harbin Univ Commer China, no. 1:86

Competitiveness Evaluation of Culture Industry in Four Provinces of Central China

Lin Yu, Changjiang Li and Long Wang

Abstract As the "sunrise industry" of the twenty-first century, the culture industry has been an important engine of social progress and economic growth. The four provinces of central China (Hunan, Hubei, Jiangxi, and Anhui) signed a culture strategic cooperation framework agreement to promote culture exchange and interaction. Considering resource, environment, market, and innovative competitiveness as the first-class indexes and evaluation index system of regional culture industry competitiveness, which is made up of some corresponding secondary indexes, this chapter evaluates and analyzes culture industry competitiveness of the four provinces of central China through factor analysis and offers some corresponding countermeasures and suggestions on how to promote culture industry competitiveness.

Keywords Culture industry competitiveness · Factor analysis · Four provinces of central China

This paper is supported by the Education Research Foundation of Hubei Province (No.2012111) and Humanities and Social Sciences Research Projects of Education Ministry of China. (No. 13YJCZH060)

L. Yu · L. Wang (✉)
School of Economics, Wuhan University of Technology, Wuhan 430070, China
e-mail: zmwl@whut.edu.cn

L. Yu
e-mail: 604559353@qq.com

C. Li
Integrated Management, Guotsing Holding Group Co. Ltd, Beijing 100000, China
e-mail: lcj@cnqc.com

M. Li et al. (eds.), *Proceedings of 2015 2nd International Conference on Industrial Economics System and Industrial Security Engineering*,
DOI 10.1007/978-981-287-655-3_34

1 Introduction

As the development of communication technology and Internet increases, culture industry becomes the new growth point of national economic development as a pillar industry in China. The Chinese government has stated clearly that developing culture industry vigorously and enhancing its overall strength and competitiveness is important. The Culture Department of the four provinces of central China signed a culture strategic cooperation framework agreement in 2011 to integrate their cultural resources and promote culture exchange and interaction of the four provinces. In addition, this agreement aims to build a culture circle with the time and regional characteristics that hold a pivotal status in the economy development pattern of China's culture as well as the world culture. However, the culture industry developments of the four provinces are different due to diverse economic development levels, culture industry investment, and people's cultural demand. This work reveals the four provinces' cultural development characteristics and discrepancies by establishing an evaluating index system. It evaluates the culture industry competitiveness of the four provinces using factor analysis to seek effective ways to improve culture industry competitiveness.

2 Literature Review

At present, there are a number of researches by the academic circle about culture industry competitiveness. Qi [1] established a comprehensive evaluation index system with 17 competitive edges and 67 competitiveness indexes on the basis of Michael Porter's four factors and government actions of his Diamond Model. Song [2] attached importance to the influence of "commercial environment" based on the Diamond Model. Hua [3] divided the evaluation index system of culture industry competitiveness into six parts: industrial strength, benefit, correlation, resources, structure, and environment. Based on the VRIO model, Li [4] established a regional culture industry competitiveness system that consists of 4 first-class, 8 s-class, and 46 third-class indexes. The regional culture industry competitiveness system that Zhuang [5] established consists of realistic competitiveness, potential competitiveness, market demand, and investment level index. The article emphasizes the potency, innovation, scarcity, social attributes, and sustainable development of the culture industry. Ma [6] built a comprehensive competitiveness evaluation index system of tourism culture. Generally, there is less study on the culture industry competitiveness of different regions although diverse culture industry competitiveness evaluation has its own focus.

3 Construction of a Regional Culture Industry Competitiveness Evaluation Index System

Based on the achieved research results and the basic principle of culture industry competitiveness evaluation mentioned above, this work establishes a regional culture industry evaluation index system with four first-class indexes—resources, environment, market, and innovation competitiveness—and corresponding 18 s-class indexes (Table 1).

Table 1 Regional culture industry competitiveness evaluation index system

Object level	First-class level of competitiveness	Second-class level
Regional culture industry competitiveness	Resource	Number of public libraries, museums, cultural centers, art performance group X1
		Number of intangible cultural heritage X2
		Proportion that cultural employee's number accounts for country's employees X3
		Completed capital investment in culture industry X4
	Environment	Proportion that culture expenses account for financial expenditure
		Fiscal appropriation of cultural institutions X6
		Average cultural building area per one hundred thousand people X7
		Special funds to protect intangible cultural heritage X8
	Market	Proportion that added value of culture industry account for GDP X9
		Total assets of operating agencies of cultural market X10
		Proportion that employees of culture industry account for local employees X11
		Proportion that per-capita entertainment and education consumption of urban residents account for total spending X12
		International tourist foreign exchange earnings X13
	Innovation	Completed scientific research items of culture and art research organization X14
		Total number of publications established by culture and art research organizations X15
		Quantity of international and provincial prizes won by cultural industry research projects X16
		Proportion of employees of cultural industry research projects account for the employees of high professional rank X17
		Proportion that scientific research expenditure accounts for cultural expenditure X18

4 Factor Analysis of Culture Industry Competitiveness in Four Provinces of Central China

Because of the difference in the four provinces' dimension and importance, their competitiveness must be analyzed by appropriate methods. Using factor analysis, the basic idea of this work is to study the dependencies between variables and represent basic data structure with a few unobservable variables hat become common factors. These common factors reflect some main information that the original variables represented. In this way, the data can be specified so that it is more convenient to perform the research.

In this work, the related data came from the following sources: the China Statistical Yearbook of Cultural Relics from 2010 to 2012, Chinese Statistical Yearbook of 2012; and the Statistical Yearbooks of Hubei, Hunan, Jiangxi, and Anhui provinces in 2012.

To eliminate differences between the data dimensions and reduce the high variance without affecting the existence of the factor loadings, the original data were standardized and correlation coefficients between each index calculated by means of SPSS 19.0. In this way, the correlation coefficient between the variables is > .4. The result shows good linear relation between variables. It is subject to the prerequisites of factor analysis. According to the correlation coefficient matrix of the sample standardization, the eigenvalues of the correlation matrix and the factor contribution rate can be calculated as shown in Table 2.

The main component is in accordance with the variance in descending order, and researchers usually select some principal components in the front. The variance contribution of every factor corresponds. The cumulative variance contribution rate of the first three factors is 100 %; therefore, choosing these three main factors can reflect the total information of all of the 18 original indexes.

Minor difference between factor loading obtained by initial transformation and their meaning is not obvious, so the variance of original component matrix orthogonally was rotated as much as possible. When a component matrix is obtained after rotation, the number of high-loading variables for each factor reaches the maximum. Therefore, it can simplify and enhance the interpretability of the factor. Table 3 shows that the first factor has large loading in X2, X6, X8, X10, X13, X14, and X16, so it is called the "industry resources factor." The second factor has large loading in X1, X3, X5, X7, X11, and X12, so it is called the "market environment factor." Last, the third factor has large loading in X4, X9, X15, X17, and X18, so it is called the "industry innovation factor."

For the industrial resource factor score, Hubei ranked first followed by Anhui, Hunan, and Jiangxi. Hubei province is abundant in cultural resources, of which the number of nonmaterial cultural heritage areas is 3813, so Hubei ranked first. Total assets of the cultural market management institutions and international tourist foreign exchange earnings in Anhui ranks first among the four provinces at a total

Table 2 Factor eigenvalues and contribution rates

Component	Initial eigenvalues			Extract quadratic sum and load			Rotate quadratic sum and load		
	Summation	Variance %	Accumulation %	Summation	Variance %	Accumulation %	Summation	Variance %	Accumulation %
1	9.128	50.711	50.711	9.128	50.711	50.711	6.491	36.064	36.064
2	5.999	33.327	84.038	5.999	33.327	84.038	6.190	34.386	70.450
3	2.873	15.962	100.000	2.873	15.962	100.000	5.319	29.550	100.000

Table 3 Rotating component matrixes

	Component		
	1	2	3
X1	0.381	−0.921	−0.080
X2	0.895	0.222	0.387
X3	−0.133	0.989	−0.068
X4	0.354	−0.006	0.935
X5	−0.081	0.941	−0.328
X6	0.854	0.476	0.210
X7	−0.083	0.978	−0.193
X8	−0.913	0.407	0.031
X9	0.174	−0.241	0.955
X10	0.767	−0.641	0.029
X11	0.628	0.690	0.360
X12	−0.405	−0.860	−0.312
X13	0.845	−0.360	0.395
X14	−0.976	0.046	−0.214
X15	0.028	−0.276	−0.961
X16	−0.857	0.163	−0.489
X17	0.371	−0.397	0.839
X18	−0.333	−0.046	−0.942

of \$9.10816 and \$1.179 billion, respectively. However, for the number of completed scientific research item by culture and art research organizations, as well as the quantity of international and provincial won by cultural industry research projects, Jiangxi has the absolute advantage (Table 4).

For the score of market environment factor, Hubei has the strongest competitiveness followed by Hunan, Jiangxi, and Anhui. For the average number of cultural places/100,000 people, Hubei has a cultural building area of 183.8 m^2 followed by Jiangxi province at 171.5 m^2 per 10,000 people. As for the industrial environment,

Table 4 Factor score and comprehensive evaluation score

Region	Industry resources		Market environment		Industry innovation			
	Comprehensive competitiveness							
Score	Rank		Score	Rank	Score	Rank	Score	Rank
Hunan	−0.04153	3	0.41445	2	1.44101	1	0.55	1
Hubei	0.92859	1	0.90508	1	−0.75402	4	0.42	2
Anhui	0.48989	2	−1.41424	4	−0.09960	2	−0.34	3
Jiangxi	−1.37696	4	0.09471	3	−0.58739	3	−0.64	4

the first three provinces for which their government attached great importance to culture industry, in turn, is Hubei, Jiangxi, Hunan, and Anhui, and the proportion of fiscal expenditure for which cultural expense accounts is 0.47, 0.43, 0.4, and 0.32 %, respectively. Regarding the proportion that employees of culture industry account for local employees, Hubei is in first with a ratio of 4 %, and Jiang Xi is last with a ratio of 2.15 %. There is a very small discrepancy in the potential demand for cultural consumption among the four provinces' residents because the proportion that the entertainment and education consumption of urban residents accounts for total spending is between 11 % and 12 %.

For the score of industrial innovation factor, Hunan has the absolute degree of competitiveness followed by Anhui, Jiangxi, and Hubei. The completed capital investment in the culture industry of Hunan in 2011 amounted 493.45 million Yuan, and the proportion that added value of culture industry account for accounted for 5.2 % of the GDP, thus making Hunan number one. The second is Anhui province, for which the two data are 328.13 million and 4 %, respectively. For scientific research and innovation, the proportion that employees of the cultural industry research project account for employees of high professional rank in Hunan is 31.55 %, which is more than twice that of Jiangxi. For the proportion that scientific research expenditure accounts for cultural expenditure in Jiangxi is 19.43 %, which is three times more than that in Hunan. As for the total number of publications that culture and art research organization established, Jiangxi, Hubei, and Anhui founded three kinds of periodicals, but Hunan only issued only one kind of publication.

5 Conclusion

The result of analysis shows that cultural industry competitiveness difference among the four provinces is obvious due to the difference of their level of cultural industry development. With implementation strategic cooperation between the four provincial cultures, these four central provinces should seize the opportunity and make use of their own advantages to develop China's cultural industry and improve cultural soft power and competitiveness. Therefore, measures should be taken with regard to the following aspects:

(1) Create a policy environment that is conducive to industrial development: The local government should provide a favorable environment for enterprises and formulate a reasonable cultural market management system. Meanwhile, it should identify and punish illegal cultural activities and behaviors to maintain the health of the cultural market operation and the legitimate rights and interests of cultural enterprises.

(2) Increase investment in cultural industries: China must establish a diversified investment system to provide strong capital support for the development of cultural industry. On one hand, the government should increase financial input; in contrast, it must absorb all kinds of social capital through various channels.

(3) Actively develop local cultural resources: Because of difference in the cultural resources of the four provinces, the local governments of the four provinces should make full use of local cultural resources and develop reasonably, effectively strengthen the bonds between cultural and other industries and expand and form a new industrial chain.

(4) Make use of the cultural industry cluster effect: China should accelerate the integration of the cultural industry and build a benchmark of cultural projects to realize a great leap forward in development as well as promote the formation of an industry cluster culture and expand the scale of the cultural industry and enhance its ability to permeate society.

Acknowledgments This paper was supported by the Education Research Foundation of Hubei Province (No.2012111) and Humanities and Social Sciences Research Projects of Education Ministry of China (No. 13YJCZH060).

References

1. Qi S (2004) Report on the international competitiveness of chinese cultural industry. Social Sciences Academic Press, Beijing
2. Song YL (2006) Study on competitiveness of cultural industry in liaoning province. Harbin: Harbin Engineering University Doctoral Dissertation, 52−61
3. Hua J (2005) Cultural industry competitiveness. Guangdong People's Press, Beijing
4. Li XR (2009) Analysis and evaluation of regional cultural industry competitiveness: based on the modification of VRIO Model. Human Geogra (5):76−80
5. Zhuang K, Wang H (2012) Evaluation and promotion countermeasure of regional cultural industry competitiveness in China. Shandong Soc Sci 8:156–157
6. Ma Y, Chen HY (2012) Research on the construction of comprehensive evaluation index system of tourism cultural industry competitiveness. Acad J South Cent For S&T Univ (Social Science Edition) 1:4–7

Electronic Commerce—Great Impetus for Exhibition Boom

Yujuan Jia

Abstract Exhibition industry is a “sunrise industry” [1] with great vitality. With the advent of the new economy era, the application of electronic commerce has not only enriched the development modes of exhibition industry to a certain extent, it has also significantly improved the economic efficiency of the traditional exhibition industry. Through analysis of the advantages that e-commerce can bring to exhibition industry, this chapter aims to demonstrate that e-commerce is a tremendous impetus for the boom of the exhibition industry.

Keywords Electronic commerce · E-commerce · Exhibition industry

1 Introduction

Renowned as the dominant industry of China’s tertiary industry, the exhibition industry is a highly profitable smokeless industry. However, some problems, such as high cost and poor efficiency, have seriously hindered its development. Fortunately, the existence of electronic commerce has changed this situation.

“Electronic commerce” refers to trading services or products using computer networks [2] such as the Internet. Modern electronic commerce has always utilized the World Wide Web for at least one part of the life cycle of transactions. The application of e-commerce, which is characterized by high efficiency and effectiveness, can provide some solutions to these problems.

In this chapter, e-commerce, the exhibition industry, and the reasons why e-commerce can benefit the exhibition industry and its specific functions in the exhibition field will be introduced and discussed in detail.

Y. Jia (✉)
Beijing Research Institute, Dalian University of Technology, Beijing 100000, China
e-mail: ooaiyaya@hotmail.com

M. Li et al. (eds.), *Proceedings of 2015 2nd International Conference on Industrial Economics System and Industrial Security Engineering*,
DOI 10.1007/978-981-287-655-3_35

2 The Exhibition Industry and E-Commerce

As a general term for conference and exhibition organizations, “exhibition industry” is an emerging service industry with far-reaching influence and a degree of correlation [3]. In recent years, the exhibition industry has gradually become a new growth point and has been considered as one of the industries with great potential in the whole world. According to an investigation, Germany, as the representative and the world’s first exhibition power that has held two thirds of the world’s top professional exhibitions, now holds half of the world’s 10 highest-paid exhibitions and was ranked fourth in the world’s 5 largest exhibition centers in 2007 [4].

“E-commerce,” short for “electronic commerce” [5] has developed due to the demand of the government and business to make full use of computing and better apply computer technology to enhance customer interaction, information exchange, and business processes both within an enterprise and among enterprises. It refers generally to all forms of transactions related to commercial activities, including those of organizations and individuals, on the basis of the transmission and processing of text, sound, and visual images with the help of telecommunications and telecommunications-based tools. Electronic commerce, as a product of economic globalization and information networking, has become the world’s most dynamic economic growth point and method to change over the past few years by vigorously promoting the development and prosperity of the world economy. The application of e-commerce in the online interaction platform of exhibition is a significant aspect of informationization in the exhibition industry.

The main purpose of an exhibition is to disseminate information. The anchor exhibitors can collect and locate the material and demand information of target customers, and visitors can find products that meet their needs and obtain product information, which is a process of information flow. Coincidentally, e-commerce has an advantage in terms of information and data record, transmission, exchange, and processing; therefore, e-commerce can make a big difference in the process of the intelligent information service of the exhibition industry. The second purpose of an exhibition is to promote commodity circulation. Trade and commodity fairs, as a kind of business activity and flow, play a main role in the exhibition industry. Similarly, electronic commerce is the electronization of commerce activities, which undoubtedly makes e-commerce have a share in promoting the development of the exhibition industry. Finally, as commerce activities, exhibition activities are held for making economic profit more or less in over a short or long term. Exactly speaking, sponsors provide anchor exhibitors with efficient, fast, convenient and precise e-commerce service by electronic, programmatic, and formal e-commerce means, which, of course, can help anchor exhibitors gain more profit compared with traditional ways.

3 Problems in Exhibition Industry Development

"Traditional exhibition" means that varies exhibitors gather in a designated place to exchange commodity information. All works, such as exhibition planning, advertising, application, booth reservation, etc., depend on manual operation. Therefore, with the rapid development of the exhibition industry, the problems, in particular high cost and inefficiency must be solved urgently.

In regard to the problem of high cost, it is necessary to seek the fundamental reasons for better solutions. As a matter of fact, the exhibition industry is significantly influenced by short-term economic benefits. Therefore, there is a large number of enterprises lacking brand consciousness as well as long-term plans, which leads to enterprises paying much attention to the scale and number of exhibitions rather than accumulation of connotation and quality as well as the establishment of brands in the operation. In addition, exhibition organizers do not follow the law of market economy and comply with the provisions of the industry. Consequently, they set a differentiated charge standard for overseas exhibitors with high fees, thus the enthusiasm of foreign exhibitors. Quite a few organizers neglect appropriate and rational plans to hold exhibitions based on the practical situation, thus making exhibitions lavish and less profitable ones in the end.

As for the problem of poor efficiency, we have a clear view for this. From the macroscopic view, the exhibition industry in our country lacks a pattern of specialization and cooperation, which causes organizers play a dual role as both planning and implementation personnel. Only a certain group of people assume the duties of collecting the exhibits, promoting activities, creating the exhibition layout, as well as providing accommodations and services of all sorts, which tends to have a negative effect on the quality and efficiency of exhibitions. Moreover, from the microscopic view, to be honest, the essence of competition in modern times is no doubt the competition of talents in all fields, and the exhibition industry is no exception. As usual, the exhibition industry requires employees with a good service concept, broad mindset, keen insight, foremost prediction, as well as fluent and skilled foreign language ability. However, the current situations do have a striking contrast to the requirement. The majority of staff in the exhibition industry come from all walks of life, and they lack the relative systematic exhibition knowledge and working skills. Thus, the shortage of specialized personnel greatly limits the development of the exhibition industry and causes the low efficiency and less competitive situation of exhibition industry in a large scale.

Having known clearly of these existing problems facing exhibition industry development, what should we change in terms of timely and drastic measures to solve them for the purpose of creating a better future for China's exhibition industry?

4 Solutions Provided by E-Commerce

The arrival of the Internet era, especially the influence from e-commerce, has brought a new opportunity to the exhibition industry.

4.1 E-Commerce Is Integrated with Exhibition Industry

With the rapid spread of e-commerce, Internet and electronic technologies—such as email, corporate website, e-payment technology and service, Internet identity security authentication, online circulation, automatic processing of information and data, and as online trade—have been especially integrated with the exhibition industry within only a few years.

4.2 E-Commerce Involves in Every Stage of Exhibition Activity

The Internet plays an import role in transferring massive amounts of information and data of exhibition promotion, project selecting, and the business between anchor exhibitors and organizers. In addition, during the exhibition, an "intelligent card" is applied to gather information about visitors and clients. Therefore, it can be said that e-commerce has its place in any stage that involves information gathering and the delivery and processing of exhibition, exhibit, and exhibition participant details. With the help of e-commerce, each stage of an exhibition can become more efficient. In addition, e-commerce can give full play to the exhibition itself.

4.3 E-Commerce Simplifies Exhibition Activities and Makes Them More Efficient

With the unique feature of convenience and high-efficiency, e-commerce has realized the electronization and automation of information gathering, delivery, and processing during exhibition preparation, which has significantly improved the efficiency. At the same time, it greatly simplifies the procedure of information gathering, delivery, and processing. This shows that e-commerce has its unique advantage over traditional methods. As online information release has realized the direct communication between anchor exhibitors and organizers, potential mistakes and time-consuming middle stages can be eliminated.

4.4 E-Commerce Is More Cost-Effective

On one hand, e-commerce converts the traditional and expensive communication methods, such as telephone, fax, and mail among organizer, anchor exhibitors, and visitors into various options. Traditional advertising always costs much money. However, by applying e-commerce, advertising costs can be significantly lowered. Meanwhile, e-commerce advertising can reach a larger target audience. In addition, through e-commerce the exhibition information can reach a wider audience. Organizer, anchor exhibitors, and visitors can obtain more comprehensive information that makes their choices suit their purposes while avoiding unnecessary cost.

4.5 E-Commerce Makes Information Feedback, Gathering, and Processing as Well as Statistics Collection More Automatic

High work efficiency means high profit, which benefits both the organizers and anchor exhibitors. E-commerce highly promotes the stylized standardization of exhibition management, which will subsequently improve the exhibition organizing and management, so that every stage can go smoothly in an orderly way as planned. In addition, the stylized standardization of exhibition management is based on the storability and reusability of e-commerce.

4.6 E-Commerce Leads the Exhibition Industry to a More Scientific Future

Fierce competition within the exhibition industry resulted from the wide spread of e-commerce information. Effective information accumulation and standardized electronic technique will ultimately contribute to the standard management of an exhibition. That is to say, e-commerce can realize standard exhibition management. Meanwhile, the coordination and management in exhibition industry is based on massive amounts of information and data, and its function is performed by selecting the superior ones among various anchor exhibitors. Management standardization and normalization provide a solid scientific foundation for coordination and management in the exhibition industry.

4.7 E-Commerce Globalizes the Exhibition Industry

E-commerce makes the promotion and organization of an project more global and international. People can obtain comprehensive exhibition information conveniently wherever they can access the Internet [6]. With e-commerce, competition in the exhibition industry is becoming more international and fierce.

Along with the advantages that e-commerce has brought to the industry, there also are several issues in need of exhibition organizers' attention. Primarily, in the new economy era the speed of product development and updating is faster than before, and the activities of exhibitors are have become more frequent. In this case, the marketers should pay more attention to industry and client changes. Because the industry is entering prosperous times gradually, it is urgent for exhibition organizers to speed up their process of project development. Last but not least, as there are an increasing number of exhibition companies from which exhibitors can choose, and more innovative ideas are indispensable if a company wants to play a role.

5 Conclusion

To make progress and grow healthy and strong, the traditional exhibition industry must face up to its problems and solve them. It can be concluded from the previous discussion that the exhibition industry can apply e-commerce comprehensively and effectively from the following three aspects: exhibition guild, exhibition organizer, and exhibition exhibitor. By using e-commerce, the exhibition industry can settle business problems effectively to improve management significantly, decrease cost largely, and grasp market opportunity rapidly to establish a consequent long-term and profitable business pattern. E-commerce simplifies the actual exhibition mechanism, decreases expenditure, saves time, improves competitiveness, and boosts the rapid development of exhibition [7].

As for an exhibition company itself, just like Dancygier Bob [8] holds, more effective marketing, more rapid development, and more creative methods must be applied under conditions of fiercer competition.

References

1. Ladkin A (1999) Strategies for career planning and development in the convention and exhibition industry. Wrox, China
2. Gefen D (1979) E-commerce: the role of familiarity and trust. Wiley, UK
3. Wu Q (2001) New century faces new opportunity, convention and exhibition industry needs Standardization. Chinese Economy Times, China
4. Na M (2007) Development and demand analysis of convention and exhibition Industry standardization. China Academic Journal Eletronic PublishingHouse, China

5. Xin X (2000) E-commerce and circulating revolution. China Economy Publishing Company, China
6. Manman Y (2010) Diagnosis on exhibition industry. China Academic Journal Eletronic Publishing House, China
7. Wu J, Xiaoqin Huang (2004) E-commerce and exhibition industry go hand in hand- the ideal combination in new century. University of Chicago Press, Chicago
8. Dancygier B (2010) The Analysis on world exhibition industry. Ohio State University, Diss

Analysis of the Current Status of Development of the Cultural Industry in China

Xiaofei Liu

Abstract Currently the cultural industry in China has ushered in the opportunities for great development and prosperity and is gradually becoming the pillar industry of the national economy. Contrary to various opinions on the concept of cultural industry, this article first gives the definition of cultural industry and then analyzes the current status of development of the cultural industry in China from seven aspects: strategic opportunities, development speed, constitution of the cultural industry, investment on fixed assets, development tendency, financial support, and cultural system reform.

Keywords Culture · Cultural industry · Current status

1 Cultural Industry

The cultural industry is a developing concept, the scope of which changes constantly along with the reform of the national administration system as well as social and economic development. Different countries of the world define it in different ways due to the different characteristics of the countries and different ideas about the cultural industry.

The United Nations Educational, Scientific and Cultural Organization (UNESCO) defines cultural industry as a series of activities on the production, reproduction, storage, and distribution of cultural products and services according to industry standards. As an industry, it has its own features and laws of development as well as specific production factors or resources required by development. Based on the definition given by UNESCO, the cultural industry is studied as a form of industry instead of merely on the basis of its cultural property.

X. Liu (✉)
Beijing Institute of Graphic Communication, Beijing, China
e-mail: lxiaofei0806@163.com

M. Li et al. (eds.), *Proceedings of 2015 2nd International Conference on Industrial Economics System and Industrial Security Engineering*,
DOI 10.1007/978-981-287-655-3_36

In July 2012, the National Bureau of Statistics published the Category of Cultural and Related Industries 2012 to further revise and improve the definitions of cultural and relevant industries in 2004; it also defined the cultural and relevant industries as a set of production activities to offer cultural products and culture-related products to the social public. Based on the definition given above, the cultural and relevant industries in China incorporate (1) culture as the core content, the production activities of cultural products (including goods and services) such as creation, manufacturing, propaganda, and display, etc., to directly satisfy people's spiritual demands; (2) auxiliary production activities required for the production of cultural products; (3) production activities (including manufacturing and sale) of stationery that is the physical carrier or production (use, propaganda, and display) tool of cultural products; and (4) production activities (including manufacturing and sale) for special equipment necessary for the production of cultural products.

2 Current Status of Development of the Cultural Industry

2.1 Great Opportunities for the Development of the Cultural Industry

The Proposal of Central Committee of the Chinese Communist Party on the 10th Five-Year Plan for National Economic and Social Development passed by the Fifth Plenary Session of the Fifteenth Central Committee first proposed the concept of "cultural industry." In recent years, the government has attached high importance to cultural construction and gradually raised it to the level of national development strategy along with the continuous advancement of the cultural system reform and gradual development of the cultural industry, and the development of the cultural industry has ushered in unprecedented opportunities. By reviewing the national policies in recent years, it can be easily seen that the development of the cultural industry has already been raised to the level of national development strategy and that the promotion of the development of the cultural industry has become the national behaviors, thus showing that the development of the cultural industry has ushered in unprecedented opportunities. Premier Li Keqiang pointed out in the government work report at the 2nd meeting of the 12th National People's Congress that in 2013 that China had promoted the healthy development of cultural under-takings and industry, released a batch of fine cultural works, opened large public cultural facilities to the society for free, deepened the reform of the cultural system, strengthened the construction of the cultural market, and realized the growth of added value of the cultural industry by >15 % [1].

2.2 The Development Speed of the Cultural Industry Higher than the Growth Rate of the National Economy

In 2012, the development rate of the national economy of China had seen a decline compared with the constant high-speed growth of the cultural industry. The added value of the legal entities in the cultural industry had reached 1,807.1 billion Yuan with a growth rate of 16.5 %. From 2004 to 2012, the development speed of the cultural industry in China had far surpassed the growth speed of the national economy. In 2004, the added value of the legal entities in the cultural industry was 310.2 billion Yuan accounting for 2.3 % in GDP. According to the statistics announcement released by the National Bureau of Statistics on August 26, 2013, the added value of the legal entities in the cultural industry had attained 1,807.1 billion Yuan and realized a growth by 16.5 % based on the same caliber and current prices, which was 6.8 % higher than the growth rate of the current prices-based GDP. The ratio of the added value of the legal entities in the cultural industry to GDP was 3.48 %, which realized a growth rate by 0.2 % based on the same caliber. In this year, the contribution of the cultural industry to the growth of the total economic output was 5.5 %.

2.3 The Constitution of the Cultural Industry Maintaining Relatively Steady

The constitution of the cultural industry in China was maintained relatively steady in 2012. In this year, the added value of cultural manufacturing, cultural wholesale and retail, and cultural service had, respectively, attained 725.3 billion, 118.7 billion, and 963.1 billion Yuan reaching a growth rate of, respectively, 17.4 %, 9.4 %, and 16.7 % compared with 2011. The added value of the cultural manufacturing, cultural wholesale and retail, and cultural service in 2012, respectively, accounted for 40.1 %, 6.6 %, and 53.3 % in the added value of legal entities in the cultural industry.

2.4 Investment on Fixed Assets Maintaining Rapid Growth and the Scale of Assets of Enterprises Seeing Constant Growth

In recent years, the government has rendered increasingly greater support to the cultural industry. In particular, the release of a series of policies, represented by the Planning on Rejuvenation of the Cultural Industry, has vigorously promoted the rapid development of the cultural industry in China. Under this background, the investment on fixed assets of the cultural industry in China also welcomed rapid development. Since 2012, the investment on fixed assets of the cultural industry in China has maintained rapid growth with increasing investment scale. In 2012, the investment on fixed assets of the cultural, sports, and entertainment industry attained 429.912 billion Yuan with a growth rate of 36.24 % compared with the previous year.

Between 2006 and 2011, the scale of assets of the cultural industry in China increased each year with an annual growth rate of 5.8 %. In 2010, the assets of legal entities in the cultural industry in China totaled 3,055 billion Yuan. By November 2011, the total assets reached 3,122.4 billion Yuan including 3,050.5 billion Yuan from assets of operating units, accounting for 97.7 % of total legal entities' assets, and 71.8 billion Yuan from assets of public units accounting for 2.3 %.

2.5 All Areas of the Cultural Industry Seeing Favorable Development Momentum

In 2012, as supported by policies such as the Decision on Several Major Issues Related to Development and Prosperity of Socialist Culture and the Multiplication Plan on the Cultural Industry in the 12th Five-Year Plan, the cultural industry in China has maintained rapid growth, and the production of films, TV dramas, new media, and publications as reached a favorable development momentum. Although art works declined by 15 %, other departments of the cultural industry in China have realized growth. Except the growth rate of newspaper and audiovisual products, respectively, by 4.09 % and 8.76 %, other sectors have seen a rapid growth rate into double figures and even three figures. The network music sector, with the most rapid development, has seen a growth rate of 379 % followed by art performance at 75.1 %, electronic publications at 48.58 %, digital publications at 40.47 %, periodicals at 35.82 %, films at 30.18 %, online games at 28.3 %, and animation at 22.23 % (Table 1).

In 2012, the national box office of films reached 17.073 billion Yuan with a growth rate of 30.18 % compared with the previous year. Imported films accounted for 51.5 % of the total box office, and the Chinese films accounted for 48.5 %. The driving factor contributing to the high growth of the box office in China lies in the rapid increase of cinema channels in the previous two years. In 2012, the urban cinema screens of China have increased by 3,832 screens, 10.5 screens each day on average, all of which are digital film halls.

In 2012, the Chinese animation industry developed well. According to the State Administration of Radio Film and Television of China, 580 Chinese domestic cartoons were recorded and published in 2012 reaching 470,751 min in total. In terms of proportion, the sequence of the subjects of the recorded and published domestic cartoons should be fairy tales, education, others, history, reality, science fiction, and mythology.

Table 1 Major constitution and market scale of the cultural Industry in China in 2012

No.	Sector name	Segmented market	Market scale (100 million Yuan)	Growth compared with 2011 (%)
1	Books, periodicals, newspapers	Books	723.51	12.28
		Periodicals	220.86	35.82
		Newspapers	852.32	4.09
2	Audiovisual products		28.34	8.76
3	Electronic publications		9.2	48.58
4	Digital publications		1935.5	40.47
5	Films		170.73	30.18
6	Animation		759.94	22.23
7	Network music	Online music	18.2	379
		Wireless music	27.2	13.3
8	Online games		601.2	28.3
8	Art performance		355.9	75.1
9	Art works		1784	−15

Data source (1) The data of books, periodicals, newspapers, audiovisual products, and electronic and digital publications comes from the State Press and Publication Administration. The market scale refers to the operating revenue. (2) The data of films comes from the State Administration of Radio Film and Television of China. The market scale refers to the box office. (3) Other data comes from the Ministry of Culture. The market scale of performance refers to the total income; the scale of the art works refers to the transaction values; and other scales refer to the market scale

In 2012, the number of TV dramas that were completed and attained the Permit for Release of Domestic TV Dramas reached 506 totaling 17,703 episodes. In terms of subject proportion, realistic subjects attained 284 dramas and 9,274 episodes, respectively, accounting for 56.13 % and 52.39 %; historical subjects attained 216 dramas and 8,189 episodes, respectively, accounting for 42.69 % and 46.26 %, respectively; and major subjects attained 6 dramas and 240 episodes, respectively, accounting for 1.19 % and 1.36 % of total TV dramas.

In 2012, the publication and release service realized an operating income of 1,663.53 billion Yuan, having grown by 14.2 % compared with 206.67 billion Yuan in 2011. It also attained an added value of 461.7 billion Yuan, having increased by 59.53 billion Yuan, and attained a growth rate of 14.8 % compared with 2011. Due to the overall requirements for construction of socialist cultural power and promoting the cultural industry to be the pillar industry of the national economy, the press and publication industry has maintained steady and rapid development.

2.6 *Rapid Development of the Cultural Industrial Zone and Gradual Expansion of Financial Support-Based Cultural Industry*

According to the 2013 Statistics Bulletin of the Cultural Development, published by the Ministry of Culture on May 20, 2014, the Ministry of Culture had cancelled 1 state-level cultural industry experimental zone and 1 title of state cultural industry demonstration base through the tour inspection on state-level cultural industry zones and bases in 2013. As of end of 2013, there were 8 state-level cultural industry demonstration zones, 6 state-level cultural industry experimental zones, and 268 state cultural industry demonstration bases in China.

By the end of 2013, the long-term credit balance of domestic and foreign currencies for the cultural industry reached 157.4 billion Yuan, which had increased by 41.9 billion Yuan at a growth rate of 36.3 % compared with the beginning of the year; the registered bond balance of 185 cultural enterprises attained 287.85 billion Yuan; 77 cultural enterprises went listing at the capital markets in Shanghai and Shenzhen; and the number of investment funds in all types of stocks of the cultural industry increased to be 57 and collecting 135 billion Yuan in total. In addition, the Ministry of Culture is also steadily promoting the establishment of projects in the cultural industry. In 2013, 201 project establishments of all types of art had been completed, with total funds support reaching 28.36 million Yuan [2].

2.7 *Reform of the Cultural System Promoting the Development of the Cultural Industry*

Since the 16th National Congress of the Party, reform of the cultural system has promoted the development of the cultural industry in China, separated the public welfare cultural undertakings and operating cultural industry, driven a large batch of state-operated cultural units to become qualified independent market subjects, inspired the domestic vigor of cultural enterprises, and improved the enterprise competitiveness to a certain degree. By April 2013, there were 698,000 cultural or cultural-related legal entities, including 607,000 operating enterprises and 91,000 public welfare legal entities (including associations and funds). The reform of the cultural system has also advanced the integration of the cultural industry with finance and tourism, promoted the emergence of new forms of culture, and accelerated the upgrading of cultural industry structure to a certain degree. Meanwhile, the constant improvement of systems and increasing support render a favorable environment for development of the cultural industry while the cultural industry itself is constantly expanding in scale and strength. The 18th National Congress of Party proposed to continuously promote reform of the cultural system, push forward the overall prosperity of cultural undertakings and rapid development of the cultural industry, and build the socialist cultural power, which also offers new opportunities for development of the cultural industry.

Acknowledgments This article is funded by the school-level key subject of social sciences of Research on Factors Affecting the Security of the Cultural Industry and the Evaluation under the Background of Opening of Beijing Institute of Graphic Communication.

References

1. China Economic Network: Government Work Report: the Added Value of the Cultural Industry Has Increased by over 15 % in 2013. http://www.ce.cn/culture/gd/201403/05/t20140305_2421107.shtml
2. The Ministry of Culture of the People's Republic of China (2013) Statistical Bulletin on Cultural Development. http://www.ccnt.gov.cn/whzx/whyw/201405/t20140520_433223.html

Countermeasures for Maintaining Security of the Cultural Industry

Xiaofei Liu

Abstract Maintaining security of the cultural industry constitutes an important guarantee to improve the soft cultural power and realize development of China's cultural industry. This article researches countermeasures for maintaining the security of the cultural industry and proposes seven dimensions to promote the security of the cultural industry including actively promoting the cultural industry to a global level; transforming the development method of the cultural industry; supporting the merger and reorganization of cultural enterprises; strengthening financial support; forging copyright protection; and intensifying approval and supervision over foreign-invested cultural enterprises.

Keywords Cultural industry · Industry security · Countermeasures

1 Actively Promoting the Cultural Industry to International Market Level

The development of cultural industry is of great significance and meaning for maintaining security of the national cultural industry, enhancing the international competitiveness of the cultural industry, expanding the international influence of national culture, and promoting the comprehensive strength of the country. Since the Reform and Opening-Up, development of the cultural industry in China has ushered in favorable opportunities. However, compared with countries of advanced culture, the development of China's cultural industry is still relatively backward, the international competitiveness of the cultural industry is still weak, and a cultural trade deficit is common. This situation seriously affects safe development of the

X. Liu (✉)
Beijing Institute of Graphic Communication, Beijing, China
e-mail: lxiaofei0806@163.com

M. Li et al. (eds.), *Proceedings of 2015 2nd International Conference on Industrial Economics System and Industrial Security Engineering*,
DOI 10.1007/978-981-287-655-3_37

cultural industry in China. To seize opportunities for the government's support for the development of cultural industry, it is necessary to actively carry out a strategy for China's cultural industry to "go global."

First, update ideas and be brave in innovation. Based on the profoundness of Chinese culture, the cultural industry shall go global, while the cultural products shall be subject to innovation. Second, explore the diversified international market and meet the consumption habits of exporting countries in terms of the marketing mode. Even for the same type of cultural products, different marketing modes may be necessary for different exporting countries due to differences in consumption habits and cultural customs. As the saying goes, know your enemy, know yourself; a hundred battles, a hundred won. Third, talents are key. It is necessary to enhance cultivation and vigorously develop the innovative talents required by the cultural industry. Finally, perfect laws and regulations for the cultural industry. Based on the practical situation of development of the cultural industry in China and relevant experiences of the power cultural industry, offer an effective legal guarantee for the Chinese cultural industry to "go global."

2 Transform the Development Method of the Cultural Industry and Lead the Road to Connotative Development

First, encourage cultural innovation. Encourage cultural workers to go deep into life and create positive and contagious works that reflect the spirit of the times. Encourage state cultural industry demonstration bases to constantly promote the innovation of cultural content and form as well as propaganda measures, improve the ability for research and development, and promote the originality of products. Strengthen technological breakthroughs and improve the supportive ability of science and technologies toward the development of the cultural industry. From a global view, the development of information technology centered high technologies plays an increasingly important supportive and guiding role for raising China's cultural innovation capability and creating new forms of culture. It has also become the new engine to facilitate cultural development. Improve the ability of technologies to upgrade traditional cultural forms and create emerging cultural forms; facilitate the transformation and industrialization of cultural and technological innovation results; and intensify the incubation and cultivation of cultural and technological enterprises.

Accelerate adjustments of the structure of the cultural industry; advance the quality and efficiency of cultural production; facilitate the transformation and upgrading of the cultural industry; improve the scale, intensification, and professionalism of the cultural industry; and speed up moving from a focus on quantity expansion to an emphasis on quality and efficiency.

3 Support the Merger and Reorganization of Enterprises and Encourage Enterprises to Become Larger and Stronger

Merger and reorganization is the shortcut for enterprises to realize scale expansion and resource optimization. It is an important for the intensified and large-scale development of Chinese cultural industry to study how to accelerate and promote the industrial integration and merger of state-owned cultural enterprises. Against the practical situation of small-enterprise scale of Chinese cultural enterprises and low market concentration, the appropriate departments should issue relevant policies and measures to further spur enterprises, especially state-owned cultural enterprises, as the major force of the cultural industry to conduct multi-region, multi-media, and multi-ownership integration and merger.

After completion of reform of the enterprise systems, state-owned cultural enterprises shall facilitate reform of the shareholding system; realize the separation of governmental functions and enterprise managements; and establish a modern enterprise system. Market integration and administrative allocation, as well as other measures, can be adopted to promote united reorganization among state-owned cultural enterprises, cultivate large-scale cultural enterprise groups with prominent businesses, complete industrial links, achieve strong market control, and lead the strategic adjustment, transformation, and upgrading of the structure of the cultural industry.

To enable state-owned cultural enterprises to serve as the core strength for industrial integration and merger, the following tasks shall be emphasized. First, adopt financial and tax support and build an incentive mechanism for strategic reorganization of the cultural industry. Increase financial investment; set up special funds for strategic reorganization of the cultural industry; and make full use of the leverage effect of financial funds through subsidies for loans, risks, technological transformation, intermediary fees, and employee settlement, etc. Meanwhile, formulate policies on tax preferential for merger and reorganization of cultural enterprises and promote the power force for the merger and reorganization of cultural enterprises. Second, enlarge financial support and explore financing channels for strategic reorganization of the cultural industry. When applying financial tools to promote strategic reorganization of the cultural industry, the government should first approve the application of merger loans; encourage commercial banks to release merger loans to cultural enterprises; and support the multi-national merger of cultural enterprises through methods such as loans from domestic and foreign banking groups as well as interior security and exterior loans; support insurance organizations to establish insurance on overseas stock investments of cultural enterprises and establish a risk-release mechanism for loans on the multi-national merger of cultural enterprises; and encourage commercial banks to offer comprehensive credit and supporting financial services for cultural enterprises after completing merger and reorganization and improve the following integration capability for the merger and reorganization of cultural enterprises. Third, actively explore a model for the sharing of finance and tax benefits among regions after multi-region reorganization. According to the assets scale and profitability of cultural enterprises, sign agreements on the allocation of finance and tax benefits after enterprise merger and reorganization; properly settle the ownership of the statistics including the industrial added value after merger and reorganization of cultural enterprises; and realize sharing of the results of the merger and reorganization of cultural enterprises. Finally, give full play to the role of the capital market to promote the merger and reorganization of cultural enterprises and further support cultural enterprises to finance merger and reorganization through issuing stocks, bonds, convertible bonds etc.; and support qualified merged enterprises to conduct refinancing and intensify the capability of the sustainable development of cultural enterprises [1].

4 Strengthen Financial Support to the Cultural Industry

First, strengthen protection over intellectual property rights; push relevant departments forward to formulate and improve administrative rules on appraisal, pledge, registration, custody, transfer, and realization of intangible assets including patent right and copyright; and render professional service for cultural property rights. Establish an appraisal system for copyrights of the cultural industry; perfect types of secondary transaction markets for intangible assets; give full play to the cultural-copyright exchange; facilitate transaction and communication about cultural copyright and establish professional cultural security companies.

Second, establish an effective credit guarantee system for the cultural industry. Encourage all parties of the society to join credit guarantee organizations for cultural enterprises; lower the degree of information asymmetry; explore diversified guarantee businesses; combine the policy guide, proper financial investment, and market operation of credit guarantee; and ensure financing channels for cultural creative enterprises.

Third, encourage commercial banks to carefully study the operation characteristics and profit models of the cultural industry from the view of policies and transform the operation method and risk-management ideas. Move thinking from "property control" to "future cash flow control"; intensify the development of derivatives such as credit risk release tools; strengthen the innovation of credit products on intellectual property right, benefit right, and order pledge, etc.; cultivate culture- and creation-based enterprises; and lead the development of finance of supply chains for upstream and downstream enterprises.

Fourth, vigorously promote investment funds for the cultural-creation industry. Attract and encourage private and social capitals to join the collection of cultural-creation industry funds; effectively raise the credit level of cultural enterprises by investment from industrial funds and improving the status of assets and liabilities of cultural enterprises; and increase the independent financing ability of enterprises.

Fifth, explore direct financing channels for cultural enterprises; guide powerful cultural enterprises to list for financing; give full play to the guiding role of the Growth Enterprise Market and the exit mechanism; and encourage venture capitals to enter the financing market for medium- and small-sized cultural enterprises. Issue different collective bonds based on the features of different regions and production modes of different cultural products and explore new ideas for financing [2].

5 Strengthen Copyright Protection and Forge Supervision and Law Enforcement on Copyright

Give full play to the guiding role of the government, from central to local, to intensify copyright protection and cultivate and render an environment and platform favorable for development of the cultural industry. First, make unremitting efforts to carry out education on the awareness of copyright of the part of all people and launch broad propaganda and popularization of copyright to the public. Second, strengthen corrections and supervision and solidly push forward the application of original software. Because software is the core and spirit of the information industry, which is closely related to the cultural industry, it is necessary to strengthen the protection of software copyright, promote the application of original software, facilitate the construction of innovative countries, and maintain a favorable international image. Third, intensify the supervision and law enforcement on copyright; reinforce the strike against copyright piracy; make supervision and law enforcement the focus of copyright work; emphasize activities to investigate and treat large cases related to copyright and strike against online copyright infringement; control the momentum of piracy; and maintain the order of copyright. Further strengthen the judicial protection of intellectual property rights and clarify additional details and specific procedures on standards for case acceptance, evidence protection, and enforcement to bring convenience to safeguarding rights and obtaining evidences. Against the too-low costs on infringement, it is necessary to increase the compensation amount for copyright victims, where appropriate, for infringers with evident evil intentions, as well as those committing a long period of infringement, considering the obligees' cost to safeguard rights, the final actual value of the software involved in the piracy, and the evil intention and infringement time of the perpetrators. In addition, it is necessary to further clarify criminal responsibilities for the infringement of computer software. In particular, criminal responsibilities shall be investigated for units and individuals that seriously affect the market order and economic development, who are involved in a huge amount and serious nature of infringement, and who exert great losses on the software owner in order to effectively punish and control infringement against copyright and realize the warning and threatening function of the laws. Finally, implement copyright registration and strengthen social services for copyright [3].

6 Strengthen the Approval and Supervision of Foreign-Invested Cultural Enterprises and Safeguard the Copyright of National Cultural Resources

The entry of foreign-invested culture breaks the current market monopoly of the cultural industry, stimulates the vigor of the Chinese cultural market, and favors the maturity and standardization of the Chinese cultural market. Meanwhile, we shall also attach great importance to the illegal actions and negative influences that may be created by foreign-invested cultural enterprises and establish a scientific and perfect system for the entry of foreign-invested cultural enterprises.

The enterprise-system transformation of operating cultural units mainly defines property rights in a narrow sense. At the same time, we shall establish broad awareness of property rights and pay attention to the construction of a broad property-rights system for the cultural industry. The production factors of the cultural industry are all subject to certain property rights, and, on the whole, the resources on which cultural production relies (including cultural heritage, talent, technology. and capital resources, etc.) belong to the people of China. The property right of cultural resources consists of four aspects: right of possession, right of operation, right of disposal, and right to benefit.

Currently, China mainly focuses on supervision and inspection on the right to dispose of cultural resources, but it relatively ignores the rights of possession, operation, and benefit. After entry to the WTO, China properly and in a timely fashion opened the area of media that (1) has relatively low relation to national ideology and government public opinion guidance or (2) has a strong cultural and professional nature, low property in ideology, and strong risk control—such as advertisement operation, wholesale and retail of newspapers and magazines, and some other operating businesses— while exerting strict control over areas related to national ideology and government guidance on public opinions (such as newspapers, advertisements, and TV). Currently, the cultural market has become the major market for the capital competition of all countries, whereas the cultural region has become the main battlefield for the international political struggle and ideology competition. Countries, whether powerful or weak in culture, adhere to the principle that the right of possession of cultural resources is exclusive to one country and that the right of operating cultural resources can be exchanged to increase cultural trade and investment among countries.

Acknowledgments This article is funded by the school-level key subject of social sciences of Research on Factors Affecting the Security of the Cultural Industry and the Evaluation under the Background of Opening of Beijing Institute of Graphic Communication.

References

1. Pei L, Rendering Vigorous Support to Strategic Re-organization of the Cultural Industry, The People's Daily, 2010-12-08
2. Shiping L and Xiaohua C, How to Benefit from Financial Support to the Cultural Industry, Xinhua Daily Telegraph, 2011-12-05
3. Mingzhao X, Experts Advocate Strengthening Law Enforcement on Software Intellectual Property Rights, China Press and Publishing Journal, 2013-07-18

Part V
Special Session on National Economy

A Study on Strategic Evaluation Index System of Chinese-Enterprise Transformation and Upgrading Based on ISO26000 Embedding in the GVC

Chen Lv, Hanchuan Lin and Fei Wang

Abstract Based on the ISO26000 new social responsibility standard and embedded global value chain, this work designs and constructs 5 essential factors and 25 secondary indexes of the strategic evaluation index system of Chinese-Enterprise transformation and upgrading including economic benefits, green driving, technological innovation, accountable governance, intelligent rate, etc., T also calculates the corresponding index weights with the help of expert evaluation and AHP method. Finally, we put forward some suggestions for Chinese-Enterprise transformation and upgrading strategy.

Keywords Enterprises strategic evaluation index system · Global value chain · AHP method

1 Literature Review

Upgrading is the central concept of the global value chain theory [1]. Porter first regarded enterprise upgrading as a process of obtaining obtain higher returns by manufacturing better products, producing more efficiently, or turning toward higher-tech activities. From the global value chain perspective, Gereffi [2] considered enterprise upgrading as a process whereby an enterprise or economic body

C. Lv (✉) · H. Lin
School of Tourism, Taishan University, Taian 271021, China
e-mail: lvchencd@163.com

H. Lin
e-mail: lhcaj@263.net

F. Wang
International Business School, University of International Business and Economics, Beijing 100029, China
e-mail: mylv21@163.com

M. Li et al. (eds.), *Proceedings of 2015 2nd International Conference on Industrial Economics System and Industrial Security Engineering*,
DOI 10.1007/978-981-287-655-3_38

improves or enters into a capital-intensive or technology economic field with more profitability. Other scholars have also agreed on this view [3].

For enterprise transformation and upgrading paths in the GVC, Amsden [4] and Lee and Chen [5], etc., argued that newly industrializing countries embedding into the global value chain to generally realize enterprise transformation and upgrading was from OEM to ODM and finally ended with OBM.

Kaplinsky and Morris [6] concluded that it was a automatic implementation process including technology upgrade → product upgrade → function upgrade → chain upgrade. This upgrading path was more specific than the one proposed by Gereffi [2]. The practice in India and Brazil proved that there would be more space for improvement of enterprise upgrading and performance, with a concentration on the domestic market's NVC [7], and finally implemented function upgrading in the GVC division system with competition and cooperation between the NVC and the GVC (Fangming Qian 2013). Entering into the market segments was also a feasible path for enterprises upgrading in developing countries.

Scholars hold different views about whether enterprises should implement upgrading by the global value chain. In a Brazil footwear cluster study, Humphrey and Schmitz [8], Schmitz [7], etc., discovered that enterprise upgrading in developing countries did not go well. Although joining the global value chain and achieving product and process upgrading was possible, it was difficult to achieve function or transdepartment upgrading. In a comparative cluster study between the Brazil furniture industry and its footwear industry, Navas-Alemán (2011) proved that the domestic and local value chain would supply more upgrading opportunities, higher value added space, and better reward.

Many domestic scholars have studied the GVC'S upgrading strategy. Liwen Tan [9] figured that most of our domestic enterprises had not really embedded into the GVC mainstream and thus possessed little market competitiveness. Against the background of integrating into the GVC, intraproduct vertical specialization by forward, backward, and horizontal channels, which gave rise to technology spillover effects, improved Chinese enterprises' total factor productivity, technical efficiency, and technical progress [10]. Climbing the value chain cannot be based only on function configuration, namely, the saying "once embed, function upgrading spread"; however, there is also the saying "continually embed, upgrade in embeddedness". Laggard enterprises in developing countries should seek to achieve value power in the continuously growing value chain growing [11].

Based on the ISO26000 New Standard of Social Responsibility, this work attempts (1) to build a strategy evaluation index system for Chinese-Enterprise transformation and upgrading as well as embedding into the GVC and (2) to apply the AHP method to calculate index weights. The final aim is to offer help for effectively evaluating Chinese enterprise-transformation and upgrading strategy.

2 Establishment of an Evaluation Index System

On the background of the international labor division, the global dispersive value nodes (such as R and D nodes, product nodes, marketing nodes, etc.) has changed into an integrative global value chain (“line”) according to the newest development of the GVC. In addition, every value node concentrates horizontally and integrates into several value modules (“plane”). Dynamic differentiation and integration of value modules constitutes the GVN. Against the background of the GVN, the Chinese-Enterprise transformation and upgrading strategy has extended into a dynamic process of node → line → plane → net, which has achieved value-node upgrading (node profit), value-chain upgrading (chain profit), value-modularity upgrading (module profit), and integrated-GVC upgrading (net profit). According to the ISO26000 New Standard of Social Responsibility and embeddedness of Global Value Chain—and following the principles of scientific, comprehensive, rational, and feasible while giving full play to the characteristics of highlighting the main line guiding role and sustainable development—we design and construct 5 essential factors and 25 secondary indexes including economic benefits, green driving, technological innovation, accountable governance, and intelligent rate.

2.1 Economic Benefits

The main evaluation indexes are defined as return on total assets rate, main business value-added growth, main business value-added rate, cost profit rate, labor productivity growth, etc.

2.2 Green Driving

The indexes mainly include decreased rate of O_2, water, and SO_2 emissions, increased rate of comprehensive use of solid waste; rate of sustainable use of resources, rate of environmental protection, and rate of environmentally friendly urban and rural development, etc.

2.3 Technological Innovation

The main indexes include new product output rate, world’s leading technologies, patent achievement transformation rate, rate of income-creation from new products in consumer health and safety, rate of sustainable consumer development, rate of technology development and acquisition, rate of creation of community finance and income, etc.

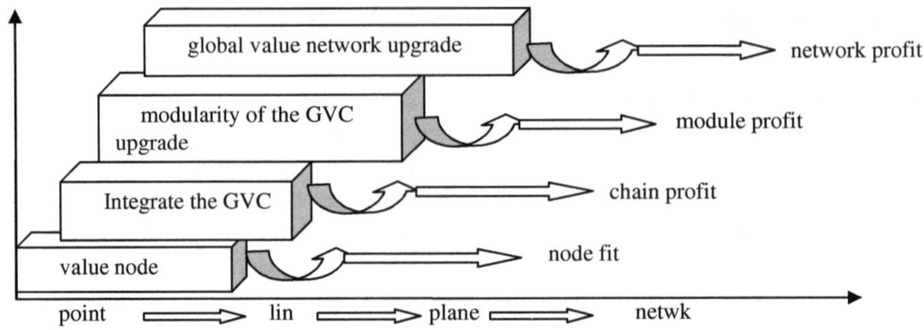

Fig. 1 Dynamic-evolution mechanism research content on Chinese-Enterprise transformation and upgrading strategy in the GVC

2.4 Accountable Governance

The main indexes include rate of conscientious investigation, rate of handling appeals, rate of participation in industry collusion, rate of discrimination against vulnerable groups, rate of social donation, etc.

2.5 Intelligent Rate

The main indexes include rate of key-process numerical control, rate of resource-planning penetration, rate of developing Internet-marketing frontiers, rate of improving health and security at work, etc. (Fig 1).

3 Quantitative Research on the Strategic Evaluation Index System

Based on the strategic evaluation index system mentioned above, this work constructs a linked hierarchy with every index. The strategic evaluation index system of Chinese-Enterprise transformation and upgrading is divided into two stratums: the first factor set is T = [A, B, C, D, E] = [intelligent rate, accountable governance, technological innovation, green driving, and economic benefits]. The second factor set, which is attached to the first stratum, is $A = [A_1, A_2, A_3, A_4, A_5]$, $B = [B_1, B_2, B_3, B_4]$, $C = [C_1, C_2, C_3, C_4, C_5, C_6, C_7]$, $D = [D_1, D_2, D_3, D_4, D_5]$, and E = [E1, E2, E3, E4] (see Fig. 2).

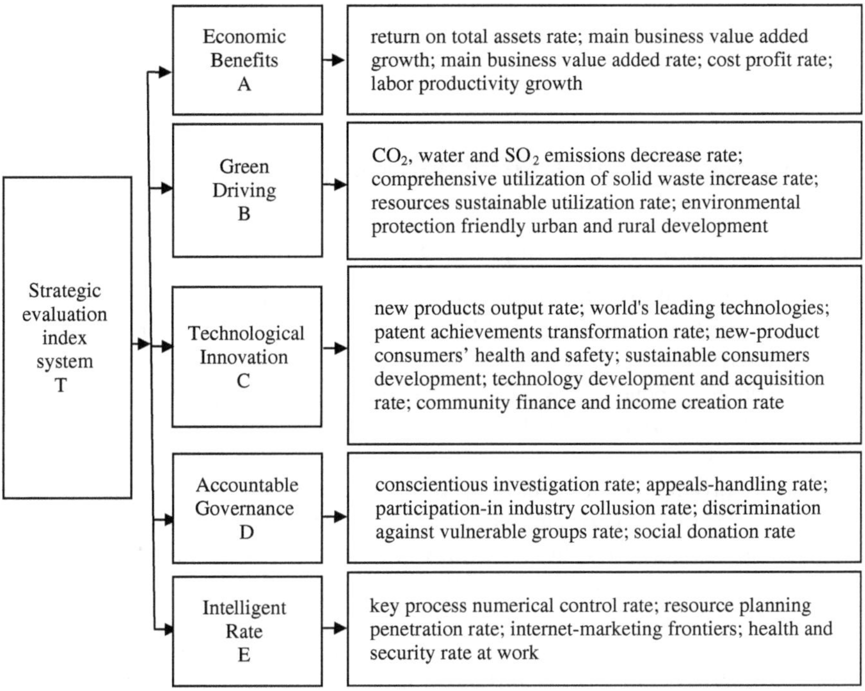

Fig. 2 Strategic evaluation index system of Chinese-Enterprise transformation and upgrading based on ISO26000 embedding in the GVC

3.1 Construct Judgment Matrix of Pairwise Comparison

The relative importance of every factor in each stratum is represented by the numeric value and constructs matrix form. Assume the assessment objective as T, the assessment index set F = [f_1, f_2, …, f_n], and construct judgment matrix P.

$$P = \begin{Bmatrix} f_{11} & f_{12} & f_{1n} \\ f_{21} & f_{22} & f_{2n} \\ f_{n1} & f_{n2} & f_{nn} \end{Bmatrix} \tag{1}$$

where f_{ij} represents the numeric value of every factor's relative importance (i = 1, 2, …, n; j = 1, 2, …, n).

3.2 Check the Consistency

CR = CI/RI, namely, a consistency ratio of judgment matrices serves as a common consistency index for judgment matrices.

$$CI = \frac{\lambda max - n}{n - 1} \tag{2}$$

3.3 Determine the Weights of Assessment Indexes

Use AHP, the concrete operation processes are as follows:

- For the total goal, weight coefficient of the judgment matrix T-A: W = [0.1462, 0.1882, 0.1965, 0.1969, 0.2752], λmax = 5.0461, CI = 0.0115, CR = 0.0103 < 0.1.
- For economic benefits, weight coefficient of the judgment matrix A-Ai: W = [0.1715, 0.1911, 0.2474, 0.2046, 0.1857], λmax = 5.0294, CI = 0.0074, CR = 0.0066 < 0.1.
- For green driving, weight coefficient of the judgment matrix B-Bi: W = [0.2668, 0.1833, 0.2870, 0.2629], λmax = 4.081, CI = 0.0270, CR = 0.0303 < 0.1.
- For technological innovation, weight coefficient of the judgment matrix C-Ci: W = [0.1232, 0.1256, 0.1331, 0.1748.0.1665, 0.1412, 0.1356], λmax = 7.1035, CI = 0.01725, CR = 0.0127 < 0.1.
- For accountable governance, weight coefficient of the judgment matrix D-Di: W = [0.2691, 0.2431, 0.1470, 0.2024, 0.1385], λmax = 5.0766, CI = 0.0192, CR = 0.0171 < 0.1.
- For intelligent rate, weight coefficient of the judgment matrix E-Ei: W = [0.2433, 0.1702, 0.2845, 0.3020], λmax = 4.0876, CI = 0.0292, CR = 0.0328 < 0.1.

All satisfy the consistency checking.
The formula is:

$$CR = \frac{\sum_{i=1}^{5} wiCIi}{\sum_{i=1}^{5} wiRIi} = \frac{0.1432 \times 0.00735 + 0.1882 \times 0.0270 + 0.1965 \times 0.0125 + 0.1969 \times 0.01915 + 0.2752 \times 0.0292}{0.1432 \times 1.12 + 0.1882 \times 0.90 + 0.1965 \times 1.32 + 0.1969 \times 1.12 + 0.2752 \times 0.90} = 0.0166 < 0.1.$$

Satisfy the consistency checking.

3.4 Determination of Final Weights

(See Table 1.)

Table 1 Total sequences of index weights

Criterion level		A	B	C	D	E	Final weight
		0.1432	0.1882	0.1965	0.1969	0.2752	
Index level	A_1	0.1712					0.0245
	A_2	0.1911					0.0274
	A_3	0.2474					0.0354
	A_4	0.2046					0.0293
	A_5	0.1857					0.0266
	B_1		0.2668				0.0502
	B_2		0.1833				0.0345
	B_3		0.2870				0.0540
	B_4		0.2629				0.0495
	C_1			0.1232			0.0242
	C_2			0.1256			0.0247
	C_3			0.1331			0.0261
	C_4			0.1748			0.0344
	C_5			0.1665			0.0327
	C_6			0.1412			0.0277
	C_7			0.1356			0.0266
	D_1			0.2691			0.0530
	D_2			0.2431			0.0479
	D_3			0.1470			0.0289
	D_4			0.2024			0.0398
	D_5			0.1385			0.0273
	E1				0.2433		0.0669
	E2				0.1702		0.0468
	E3				0.2845		0.0783
	E4				0.3020		0.0831

4 Conclusion

The index derives two implications as follows:

- Facing more opportunities than challenges in ISO26000 New Standard of Social Responsibility, Chinese enterprises should seize this opportunity to embed into the global value chain and achieve transformation and upgrading.
- In the strategy-evaluation index system of Chinese-Enterprise transformation and upgrading, the intelligent rate and accountable governance have become basic standpoints.

Therefore, Chinese enterprises should actively participate in and pay attention to the activities about the international social responsibility standard's discourse and power establishment power as well as promote Chinese enterprises transformation and upgrading against the background of global value chain upgrading.

References

1. Gong S (2009) Analysis on the influence on the enterprises' upgrading in the global value chain by the industrial cluster. Econ Manage 44(3):76–78
2. Gereffi G (1999) International trade and industrial upgrading in the apparel commodity chain. J Int Econ 48(1):37–70
3. Kaplinsky R, Readman J (2005) Globalization and upgrading: what can (and cannot) be learnt from international trade statistics in the wood furniture sector? Ind Corp Change 14 (4):679–703
4. Amsden AH (1989) Asia's next giant: South Korea and late industrialization. Oxford University Press
5. Lee JR, Chen JS (2000) Dynamic synergy creation with multiple business activities: toward a competence-based business model for contract manufacturers. Appl Bus Strategy 6:209–228
6. Morris M, Kaplinsky R (2001) In: Kimani PM (ed) A handbook for value chain research. International Development Research Centre. http://www.ids.ac.uk/ids/global/pdfs/vchnov01.pdf
7. Schmitz H (2004) Local upgrading in global chains: recent findings. Paper to be presented at the DRUID summer conference 2004 on industrial dynamics, Innovation and Development. Elsinore, Denmark, 14–16 June 2004
8. Humphrey J, Schmitz H. (2000) Governance and upgrading: linking industrial cluster and global value chain research, IDS Working Paper 120. http://www.ids.ac.uk/download.cfm?objectid=DE302D61-5056-8171-7BD4C8E76F9FBBED
9. Liwen T, Haiyan M, Linqing L (2009) Myths and realities of China's apparel industry-from a global value chain perspective. China Economist, 4(2):103–115
10. Caigang Z, Quan L (2013). Review of Monte Carlo modeling of light transport in tissues. J Biomed Optics 2013:319–326
11. Li H, Zhang Y (2007) The role of managers` political networking and functional experience in new venture performance: Evidence from China's transition economy. Strateg Manag J 28(8):791–804
12. Michael B, Eisenstat Russell A, Bert S (1990) Why change programs don't produce change? Harv Bus Rev 68(6):158–166
13. Davidson WH (1999) Beyond re-engineering: The three phases of business transformation. IBM Syst J 32(1):485–499
14. Keil T, Eloranta E, Holmstrom J, Jarvenpaa E (2001) Information and communication technology driven business transformation-a call for research. Comput Ind 44(3):263–282
15. Porter ME (1980) Competitive Strategy: Techniques for Analyzing Industries and Competitors. Free Press, New York

A Pilot Study on Chinese Flower Industry Cluster

Zi Feng and Biju Xu

Abstract The work adopted location quotient method to identify Chinese flower industry cluster phenomena from different perspectives. The research mainly collected three economic indicators: number of companies, number of employees, and total value of output. Making use of the data, the authors conducted empirical analyses of the phenomena. Results showed that a flower industry cluster was formed in Fujian, Henan, and Hunan, that it is rapidly gathering in Zhejiang and Shandong, and that industrial clusters are gradually emerging.

Keywords Location quotient · Flower industry cluster · Identify

1 Introduction

The flower industry is the general format composed of various related industries that supply flowers and related products for business objects. The scope of the businesses is interconnected by their interests and have different work divisions involving scientific research, production, processing, sales, trade, consumption, etc. [1]. At present, the flower industry has become a labor-, capital-, and technology-intensive emerging green "sunrise" industry and has grown by 6 % to approximately 9 % per year [2]. The world flower trade increased to >$300 billion by 2011 [3]. Flower industry occupies an important position in the global economy.

Now, "industrial cluster" is the main form of development. The notion of industrial cluster was first mentioned by Porter in 1998. He defined the concept of

Z. Feng (✉)
Postdoctoral Programme of China Industrial Economic Security Research Center,
Beijing Jiaotong University, Beijing 100044, China
e-mail: lelezifeng@163.com

B. Xu
University of Sunderland, Sunderland Srl 3sd, UK
e-mail: leftedgar@foxmail.com

M. Li et al. (eds.), *Proceedings of 2015 2nd International Conference on Industrial Economics System and Industrial Security Engineering*,
DOI 10.1007/978-981-287-655-3_39

industrial cluster as "among a certain particular field (usually there is a leading industry as the core), a large number of enterprises with close industrial contact and the relevant supporting organizations cluster and give rise to the phenomenon of powerful and sustainable competitive advantage." Many countries strive to cultivate and update the local industrial cluster for the purpose of enhancing the local competitiveness [4]. The flower industry is looking for cluster development mode in our country. Some reports have mentioned China's flower industrial cluster, but these were not empirical studies. In addition, the future development area of the flower industry in China is not clear. Therefore, the flower industry was studied to better develop it.

2 Identification of a Flower Industry Cluster in China

Since the 1980s, domestic and foreign scholars have developed a large number of theories about industrial cluster. They generally believe that the development of an industrial cluster is an effective way to enhance competitiveness. Therefore, a definition of industry cluster has become especially important. The main method of definition includes qualitative and quantitative analysis. This work empirically analyzed China's flower industry cluster phenomenon by location quotient.

2.1 Location Quotient (LQ)

The idea of LQ, also called "specialization," was first put forward and applied to location analysis by P. Haggett. LQ can measure the spatial distribution of the elements in the area and reflect the degree of specialization of a certain industrial sectors as well as the position and role of a certain area in high-level area. It also can evaluate the size of a certain industry cluster in a region as well as its degree of agglomeration-scale advantages in a high-level area, etc. [5]. LQ is calculated by formula (1).

$$LQ_{ij} = (X_{ij} \div \sum_{i=1}^{n} X_{ij}) \div \left(\sum_{j}^{n} X_{ij} \div \sum_{j=1}^{n} \sum_{i=1}^{n} X_{ij} \right). \quad (1)$$

In the expressions, "i" means an industry; "j" means a region; "x" expresses the total value of output or the number of enterprises and workers, etc. $LQ \geq 1$ indicates the existence of an industry cluster in the region, and $LQ < 1$ shows that a regional industry cluster has not been formed. The larger the LQ number, the higher the area of specialization level and the higher the degree of the industry cluster.

2.2 Identification of a Flower Industry Cluster in China by LQ

According to the current related Statistical Yearbook, this work adopted three location quotient coefficients to comprehensively identify a flower industry cluster in various regions in China. Among them, LQ_1 was based on the number of companies, LQ_2 on the number of workers, and LQ_3 on the total value of output.

Table 1 shows that there were three regions in China for which the LQ coefficients on the flower industry were <1: Fujian, Henan, and Hunan. The results indicated that the specialization level of flower industry in these areas was better than the national average and that an industry cluster had been formed in these areas.

According to application rules of the LQ coefficient, the larger the coefficient, the more obvious the degree of industry cluster in a certain area. Figure 1 illustrates the cluster degree of the flower industry in the three areas. It shows that the three regional industry clusters were just over the horizon. At the same time, it identified that the whole flower industry in our country was still in its infancy.

2.3 An Empirical Analysis of the Chinese Flower Industry Cluster

Previous research of industrial cluster showed that there were the three important characteristics of industrial clusters: enterprises agglomeration in the area, configuration of human resources, and economics of scale in production. Therefore, the authors chose these characteristics to test the above-mentioned results.

Fujian province is the important production base of bonsai plants in China. In 2011, there were 2096 flower enterprises. Among them, large- and medium-sized enterprises accounted for 12.83 %, up 2.85 % from 2010. The number of farmer households reached 55,800 and grew by 53.72 % compared with the previous year. The number of employees increased by 32.16 % from 2010 to 2011 and numbered a total of 183,700. Among them, professional and technical personnel increased by 13.75 % compared with 2010. In addition, gross domestic product arrived at ¥4.224 billion, and increase of 24.8 % compared with 2010. It accounted for approximately 4.59 % of the gross value of national flower production. Potted exports reached $57.665 million accounting for 55.66 % of total exports potted plants.

Henan and Hunan are both important bases of ornamental seedling production. As the fourth-largest flower planting province, Henan had 91,300 ha for flower planting in 2011, and the acreage accounted for 10.59 % of the total area of the national flower cultivation. There were 1699 flower enterprises including 347 large and medium-sized enterprises. The number of farmers reached 94,100 households. The workforce increased to 49,3200 people, up 17.43 % from 2010. The gross domestic product also increased by 22.33 % reaching ¥9.004 billion. In Yanling,

Table 1 *LQ* coefficient of various regions in China's flower industry

Region	Three *LQ* coefficients of flower industry in various regions (2009 to 2011)								
	LQ_1			LQ_2			LQ_3		
	2009	2010	2011	2009	2010	2011	2009	2010	2011
Beijing	–	0.13	0.12	0.22	0.18	0.20	0.53	0.64	0.53
Tianjin	–	0.10	0.09	0.06	0.12	0.10	0.38	0.15	0.12
Hebei	–	0.40	0.42	0.47	0.46	0.49	0.67	0.79	0.60
Shanxi	–	0.34	0.30	0.09	0.08	0.14	0.14	0.07	0.13
Neimenggu	–	0.12	0.09	0.06	0.10	0.06	0.05	0.07	0.06
Liaoning	–	0.26	0.25	0.68	1.72	0.79	1.55	1.82	1.84
Jilin	–	0.25	0.37	0.18	0.18	0.07	0.53	0.43	0.66
Helongjiang	–	3.06	0.05	0.89	0.84	0.04	0.13	0.16	0.16
Shanghai	–	0.13	0.11	0.19	0.16	0.14	0.15	0.20	0.16
Jiangsu	–	0.93	0.92	3.49	2.83	2.86	1.91	1.45	1.80
Zhejiang	–	1.88	2.14	2.19	2.35	2.41	0.49	0.55	0.65
Anhui	–	0.61	0.85	1.25	0.56	0.60	0.76	1.06	0.98
Fujian	–	1.47	1.30	1.16	1.25	1.32	1.35	1.44	1.34
Jiangxi	–	1.33	1.80	0.44	0.43	0.45	3.24	3.73	3.20
Shandong	–	0.55	0.63	0.72	0.97	1.21	0.44	0.61	0.91
Henan	–	1.54	1.09	1.97	1.37	1.41	1.77	1.99	1.86
Hubei	–	0.90	0.82	0.71	0.71	0.55	1.09	0.94	0.87
Hunan	–	1.06	1.39	1.44	1.53	1.43	1.14	1.29	1.35
Guangdong	–	2.14	1.95	0.39	0.40	0.41	0.81	0.92	0.79
Guangxi	–	0.41	0.59	1.89	1.81	2.19	0.96	0.93	0.67
Hainan	–	3.38	2.90	0.57	1.17	1.12	2.00	2.25	2.88
Chongqing	–	1.36	1.20	0.83	0.76	1.16	1.13	1.68	1.39
Sichuan	–	2.85	3.05	0.87	0.87	0.86	1.27	1.47	1.41
Guizhou	–	0.70	0.61	0.08	0.07	0.07	0.60	0.57	0.86
Yunnan	–	1.01	0.93	0.75	0.92	0.74	1.42	2.30	1.28
Xizang	–	–	0.11	–	–	0.02	–	–	0.02
Shanxi	–	0.24	0.17	0.09	0.16	0.13	0.19	0.24	0.20
Gansu	–	0.33	0.33	0.27	0.25	0.20	0.25	0.24	0.16
Qinghai	–	0.03	0.01	0.03	0.03	0.01	0.07	0.06	0.18
Ningxia	–	0.17	0.10	0.13	0.17	0.16	0.65	0.66	0.38
Xinjiang	–	0.07	0.06	0.03	0.03	0.02	0.06	0.09	0.05

Note According to 《the China statistical Yearbook (2010–2012)》, 《the Chinese forestry statistical yearbook (2010–2012)》

which had become the country's largest center of flower and tree production and sales distribution, annual flower sales had reached ¥6 billion [6]. Chenzhai flowers market is the largest comprehensive trading market of flowers, in which trade reached ¥1 billion per year [7]. Hunan was the fifth largest flower-planting province. In 2011, the planting area reached 30,600 ha, and annual sales reached

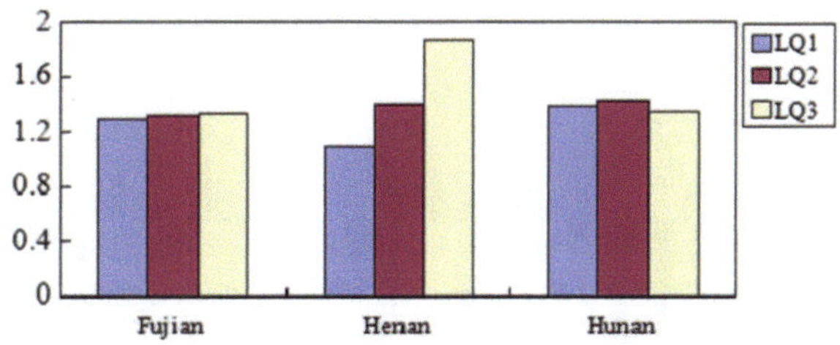

Fig. 1 Cluster degree of the flower industry in the three areas

¥4.912 billion [8]. There were 1557 flower enterprises, including 243 large- and medium-sized enterprises, increasing by 38.65 % compared with 2010. The number farmers increased to 76,800 households, and the number of employees in flower enterprises reached 323,300 people, an increase of 3.92 % from the previous year. Moreover, the value of flower output also increased by 44.84 % compared with 2010 and reached ¥4.776 billion.

The above-mentioned statistical indicators empirically tested the phenomenon of flower industry cluster in Fujian, Henan, and Hunan. The results indicated that the cluster phenomenon was obvious. Furthermore, it proved that the method of LQ is feasible to identify the phenomenon of flower industry cluster in China.

3 The Identification of Flower Industry Cluster in Emerging Region

According to the principle of LQ, the LQ coefficient could not recognize an emerging industry cluster. The main reason was that the concentration of the emerging industry cluster is relatively lower than the national average. However, the new cluster has a strong growth potential; therefore, it could be discerned by the growth rate of the LQ coefficient [9, 10]. This work adopted this rate to identify the emerging area in the future.

3.1 Recognition of the Emerging Flower Industry Cluster

According to the statistical data of 2009 to 2011, the annual growth rate for the LQ coefficient is shown Fig. 2. In Fig. 2, there are only two areas—Shandong, Zhejiang—whose average annual growth rates of the three LQ factors were all greater than zero. This shows that the flower industry is rapidly growing in Zhejiang and Shandong and that industrial clusters are gradually emerging.

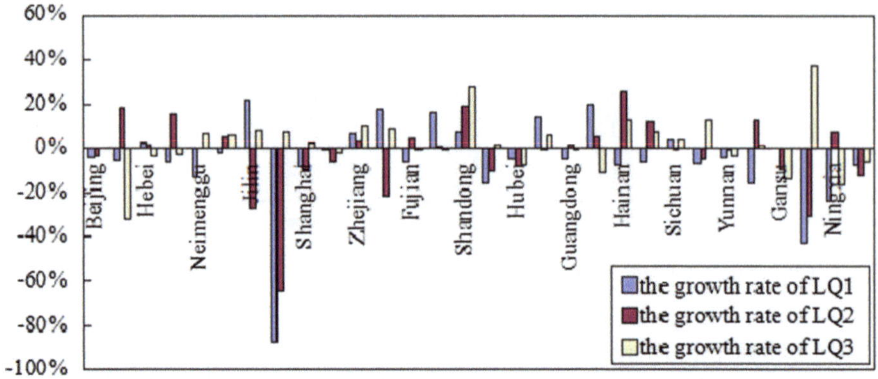

Fig. 2 Average annual growth rate of LQ coefficient about each region in 2009 to 2011

3.2 Empirical Analysis of the Emerging Flower Industry Cluster

Zhejiang has various landscapes, abundant germplasm resources, and convenient traffic conditions. These are advantageous to develop a flower industry. Zhejiang mainly cultivates ornamental seedlings and potted plants. Moreover, some of the major production areas, such as Xiaoshan and Ningbo, have created unique brands. Zhejiang is the sixth largest flower-planting area. In 2011, the region provided 58,200 ha to plant flowers. Flower enterprises increased to 8135, up 26.20 % from 2010. Employees were >501,100 people. The GDP also increased by 54.04 %. The flower business and workforce both exceeded the national average. Thus, the flower industry cluster in Zhejiang is forming.

Shandong also has certain advantages, resource, and location advantages. These have positive significance to development of the local flower industry. Shandong has gradually formed a new pattern of flower industry—highlighting regional distribution and promoting development based on leading enterprise. Shandong has established many obvious regional features of a cut-flower production base such as peony in Heze, Chinese rose in Laizhou, rose in Pingyin, cyclamen in Qingzhou, honeysuckle in Pingyi, cuckoo in Weifang, and lotus in Jining. In addition, there are many garden flowering-shrub production bases including Heze, Yantai, Linyi, Taian, and Weifang. Moreover, ornamental seedling production bases were set up in Jinan, Weifang, Jining, and Taian, etc. In 2011, as the second largest flower-planting area, Shandong had 98,100 ha to plant flowers, and the area has increased by 39.97 % compared 2010. The number of flower enterprises increased by 21.19 % from 2010 reaching 2379. The number of employees increased by 59.41 % compared with the previous year reaching approximately 444,600 people. In addition, the GDP growth rate was as high as 96.07 %. All of these changes show that Shandong is becoming an emerging industrial cluster for flowers.

This work used several indicators—number of companies, number of employees, and output—to test the phenomenon of flower industry agglomeration in Zhejiang and Shandong. The results indicate that the method of measuring average annual growth rate by location quotient can identify an emerging flower industry cluster in China.

4 Conclusion

This work used LQ coefficient method to define the flower industry cluster in China and performed empirical analysis of industrial cluster. The conclusion is that it is reasonable and feasible that LQ coefficient can recognize a flower industry cluster phenomenon in China. In addition, the annual growth rate of LQ coefficient could identify an emerging flower industry cluster phenomenon.

References

1. The state forestry administration (2013) National flower industry development planning (2011–2020). http://www.forestry.gov.cn/uploadfile/main/2013-2/file/2013-2-8-ad8eb1dbcb-5c4a7f97cf43389693b9ce.pdf
2. van Hemert N (2005) E-business and the Dutch Flower Industry—A survey for strategic opportunities. In: Proceedings of the IAMA conference, Chicago. USA, pp 2–4
3. Zhong W (2013) Guangdong flower industry development report in 2012. http://121.8.170.203:8080/subjectView_1795.html
4. Porter ME (1990) The Competitive Advantage of Nations. The Free Press, New York
5. Hong Z (2006). The research of Tianjin industrial layout based on the theory of industrial agglomeration. Tianjin university press, Tianjin
6. Jun D (2012) Flower industry blossoming in Henan province with the second largest planting area. Henan Daily, the fourth section
7. Jun D (2011) Flower industry output value of Henan reaching 9 billion yuan which was the second in nation. http://www.gov.cn/gzdt/2011-09/22/. content_1953591.htm
8. Xumei L (2012) Status and development of flower and seedling industry in Hunan province. Hunan Forestry Science & Technology, 39(6):45–47, 58
9. Jin W, Lan H, Ying Z (2004) The identification method of industrial cluster and empirical research. Science of Science and Management of S.&T., 11:117–120
10. Yanggen L (2008) Analysis of the industry cluster based on a location quotient method—an empirical study on Zhejiang industrial clusters. http://www.paper.edu.cn/html/releasepaper/2008/04/1014/

Discussion of Investment Analysis Method in the New Round of the China Stock Bull Market

Zhaoyu Qiu

Abstract In the area of securities analysis method, the three general methods are basic analysis, technical analysis, and quantitative analysis. This chapter discusses the difficulty and problems that arise when these methods are applied when investing in China's stock bull market, which is currently progressing. In this chapter, I try to find a better way to help investors make the right investment decision, which must be closer to the real market. The analysis and decisions are focused on the driving force of a market based on emotion and behavior. The tool for correct decision-making should be the combination of those three methods; it should also include other information that can affect or determine behaviors and emotions in the market.

Keywords Investment analysis · Securities market · Behavioral finance

1 Market Volatility and Basic Analysis

The point of basic ANALYSIS is that the market price of a security will deviate from the intrinsic value, which is considered to be fixed datum to measure investment opportunity. When this deviation occurs, investors can amen their position and thereby earn profit [1]. Two main hypotheses exist in basic analysis. The first hypothesis is that there is an intrinsic value actually exists and that and this value could be determined. The second hypothesis is that the market will make mistakes and generate an unreasonable price; however if this market could find the mistake and correct it, the market price will regress to the intrinsic value [2]. But everyone doubts how and when this correction occurs.

Z. Qiu (✉)
Faculty of Management and Economics, Dalian University of Technology,
Dalian City 116024, People's Republic of China.
e-mail: 290254702@qq.com

M. Li et al. (eds.), *Proceedings of 2015 2nd International Conference on Industrial Economics System and Industrial Security Engineering*,
DOI 10.1007/978-981-287-655-3_40

Is there a deviation between the price and value of Chinese macroeconomics and the market index point? First we should look at the current economic situation in China. The PMI in March, as reported by China's government in April 2015, rebounded slightly. Researchers from BOCOM International Securities believe that the rebound is a result of workers returning to enterprises after the Spring Festival, but it is still weak compared with past years [3]. Other aspects show that there is little chance for industrial growth in March to rebound. For example, the coal consumption of power plants in March had a negative two-digit decline than the same period of last year, and heavy industry in North China was compelled to limited production due to environmental problems that occurred in late March. The hopeless rebound of industrial growth in March removes the last doubts about the slowdown of GDP growth in the first quarter. It is expected by the central government that the GDP will come down to 7 % of the target [3]. This is the situation the market China this year.

Then how was the GDP of China in last year? As can be seen from Fig. 1, the quarter year-on-year growth rate is on the downside. In addition, CPI, PPI, and PMI indexes also worsened pr worsened. It can be concluded that the economy in all quarters of last year is declining significantly. However, in the fourth quarter of last year, the Shanghai Composite Index rose from 2368 points o 3234 points, showing a growth of 36.84 % in just 3 months, which is quite rare even in the past 10 years. Of course, investors or institutions that chose to "go short" during the increase suffered huge losses. A leading well-known private equity fund in China established a large number of short positions in stock index futures when the index came to > 2700 points and they finally suffered a rather great loss when "one nine quotes" occurred (which means large-cap stock grew far more than other stocks). These institutions might have had complicated reasons for or have done a lot of research before choosing to establish short positions, but their core thoughts must have been that the large-cap point (price) has exceeded the current economic status (intrinsic value) and that the final price would return to its value; however, they were wrong.

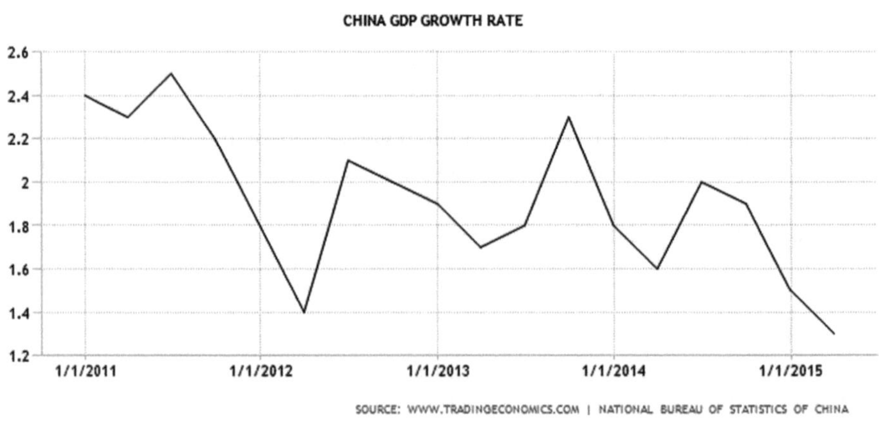

Fig. 1 China's quarterly GDP growth rate

Microanalysis of the company shows that since the substantial increase in the Shanghai securities composite index in November last year, many stockholders of listed companies chose to dump and cash in at the end of last year. This behavior especially occurred on securities company stocks as well as blue-chip stocks. The typical example is CITIC SECURITIES. As the largest security company in Asia, CITIC Securities has the leading-edge team for securities research and investment, and its share price has considerably increased since November 2014. In January 2015, CITIC Securities released a notice of changes to stockholders' equity, which showed that the company received the notice on reducing holding-shares from China CITIC Co., Ltd., the largest stockholder on January 16. From January 13 to 16, 2015, the largest stockholder reduced its holding-shares totaling 348 million through the centralized quotation trading system of Shanghai Stock Exchange, which accounted for 3.16 % of the general capital of the company. The shares held by China CITIC Co., Ltd. were reduced from 20.20 to 17.14 %, but the company still served as the largest stockholder of CITIC Securities. China CITIC Co., Ltd. (which belongs to CITIC Group) posses absolute control over CITIC Securities. However, China CITIC Co., Ltd. did not clearly announce the purpose of the capitals. Actually, the real purpose was to obtain cash earnings against the backdrop of the excessively high share price of CITIC securities. According to the average price of CITIC Securities in the past week, the largest stockholder held 348 million shares, which will cash in at >11 billion Yuan on a price of 32 yuan/share [4]. The reduced holding-shares amounted to >10 billion RMB; such a large-scale reduction of China CITIC Co., Ltd. must have undergone deep consideration. As the largest stockholder of CITIC Securities, China CITIC Co., Ltd. is certain to gain a deep understanding on all aspects of CITIC Securities. The company has the best basic analysis team in China. China CITIC Co., Ltd. reduced the holding-shares because they thought that the share price was higher than the intrinsic value of CITIC Securities. Despite the temporary decline in the share price of CITIC Securities after the reduction holding-shares by the largest stockholder, the share price was raised to 38 Yuan/share on April 9, 2015, after just 3 months. The large stockholders of the listed companies of the Shanghai A-share market and the growth enterprise market substantially reduced their holding-shares in this round of uptrend showing that the group most familiar with the company also thought that the share price was excessively high and had veered from its intrinsic price. Worried about the regression value of share price, they made such financial decisions. Despite the frequent reduction of holding-shares by large stockholders, the market index and particular share price have set a new high.

If the intrinsic values of securities could be measured, the institutions with powerful information channels, research capacity, and good fundamental analysis groups—such as CITIC Securities and the top private-equity mentioned herein—the group with have deepest understanding of the internal conditions of the company would be the largest stockholder of the listed company. After they establish short positions or sell off stock, the share price is still on the rise. If they could not measure the intrinsic values of securities, it is impossible for general investors determine the value.

Even If the internal values could be measured, the degree of deviation, as well as the duration of conditions of the price substantially deviating from the internal value, would still cause a large degree of risk. Like the private-equity fund mentioned in this chapter, there will be great loss on short positions of stock index futures if the degree is too high. If the deviation exists for a long time, the investor who held the real estate stocks and sold them at the end of 2014 will miss opportunities for a long period of time.

If the basic analysis method fails to accurately define and verify the degree and duration of deviation, it will not be able to help investors gain profits. In contrast, it may mislead investors in establishing their position against the trend under market rise or decline and cause great loss.

2 Technical Analysis and Quantitative Analysis

Technical analysis has three basic hypotheses as follows:

- Market action discounts everything
- Prices move in trend.
- History tends to repeat itself.

There always appear continually rising or falling behavior the real market, and even if the reverse trend comes provisionally, the price will still move back to the original direction [5]. This behavior seems effective. If we apply the theory of technical analysis on the real market, we will face many problems. For example, a Double Bottom or Smite Form appears as the technical analysis of the market, and then the inversion of the market does not happen, the technical analysis method still can use many other theories to explain and identify it. Actually there are many contradictions among different theories in technical analysis. The successful application of technical analysis must be based on the user's own experience or understanding. In the continuous market, the method of technical analysis is able to show a good effect; however, when the market is stuck in a trading range, technical analysis methods or strategies—such as breaking strategy, wave theory, average system strategy and other technical indicators—do not work.

From June 2003 to May 2006, the Shanghai copper futures contracts increased overwhelmingly. The price of the contract went up from approximately 17000 to 84649 Yuan, and the growth rate was almost 400 %. From June 2003 to May 2006, the trend of copper future maintained a standard long-order form, and the K lines kept increasing without any large fluctuation. The average lines system also shows that long orders and other technical indicators were bullish as well. The technical analysis investor earned a great profit from this increase of copper futures. When the same investor faced the same contract after 2006, and the market trend became irregular, the investors found that the method that had brought them lots of profit did not work any longer. Finally, most of the investor lost their money and even declared bankruptcy.

Quantitative analysis is based on historical price. The most important method is statistical arbitrage. In the general case, the difference in price of two or more securities shows a particular regularity. Statistical arbitrage can catch this opportunity when the difference in price is abnormal, and the deviation of the difference in price will finally be corrected. However, if the particular regularity or the logic of the difference in price between those securities changes, statistical arbitrage will face massive risk. Long-term capital management is the most famous example of this risk.

Therefore, the technical and quantitative methods both have very obvious limitations. This means that investors will fail if they use these methods mechanistically.

3 The Choice of Investor in Investment Decision

Although those methods include basic, technical, and quantitative analysis, they all have obvious limitations and shortcomings. There are some successful investors who use those methods. Those successful investors may use one or more of the above-mentioned methods, but the core reason for their success is not the method but rather how they use the method. They have a very deep understanding of the market, so they can use many investment methods in a flexible way based on their own understanding. The key point is the understanding the market, and there are some common features between their understanding. The common features are that they can efficiently characterize and realize the market sentiment as well as investor behaviors.

Behavioral finance has three points that can partly explain the phenomenon:

- Investors are imperfectly rational; they make mistakes.
- Most of the time, the rational investor and the imperfectly rational investor will both influence the market, which is contrary to the traditional theory that only the rational investor finally determines the price.
- Because of the deviation, emotion, preference, and other psychological reasons embedded in the cognitive process, investors cannot make unbiased estimations of the market in a rational way [6].

In the real market, capital can fully reflect market behavior. According to the data of the China Securities Registration and Settlement Corporation on April 8, 2015, in the previous 3 weeks the number of new accounts in China's A-Shares market is >million each week. This number exceeded all records in these 3 weeks. In the meantime, the Shanghai securities composite index increased by 18 %, and the market trading volume increased daily. At the end of March, the China Securities Regulatory Commission set free the limitation of mutual funds to invest Hong Kong stocks by the Shanghai Hong Kong Stock Connect program, and then the price of Hong Kong stock of those companies, which hold both A and H shares, increased significantly. This evidences reveals that money is the driving force of the increased stock market; thus, the focus of investment analysis method should be the phenomena and essence of money behaviors to understand how those money behaviors occurred.

I created a test in the real market to determine the use of market behaviors in investment analysis and decision-making. First we should define the relationship between market emotion or behavior and the market price. I assume that the price is underrated only when the trading behavior is rare. The other indicator is market emotion. I can use the information of other market participants, including particular mutual fund and individual investor, when the mutual fund has bad performance and is difficult to sell. When many individual investors' amounts do not trade well for a couple of years, it means that the market emotion is low. The final indicator of this is that the market index has been stuck in a reasonably bounded range for a long period of time. When market appeared to be in this situation, I will establish the position. Because the chance is based on the market index, my position should be highly related to the market index. I choose the portfolio in the financial sector that is most related closely to the market index. Therefore, I build a position in the beginning of 2014 which the market shows is in such a situation. The bull market came at the end of 2014, and those position have a profit >150 %.

As we know, the stock market has periodic fluctuations. In the process of the market moving from depression to craze, much money from other place keeps rushes into the stock market. Those investors do not conduct any deep research, and they are not clear in their actions. The only reason they buy stock is that the market has a rising trend. The only evidence of their investment decision is the stock price. The choice of irrational investors is seriously affected by price information, so price information is an important part in my future research.

4 Conclusion

The technical analysis and quantitative methods are all based on price information, but they could not be combined efficiently with other information. Basic analysis must also be combined with other information to acquire a better performance. The market is an organic whole, so we cannot use a method to separate any part of it. In future work, the combination of information and method should be focused. The combination of information can provide an assumption; the technical method can determine market behavior that can be used to verify the assumption; and finally the portfolio will be established by adequate consideration of risk and profit. The investment system should become an organic whole the same as the market.

References

1. Zhao L, Tianlang Z (2012) Securities Investment Analysis. China Financial Publishing House, BeiJing

2. Xiaoqiu W (2004) Securities Investment Theories. China Renmin University Press Co. LTD, BeiJing
3. BOCOM International Holdings Company Limited. (2015) China's Macro-economic: Forward looking Statistics in March, http://finance.qq.com/a/20150409/052951.htm Accessed 15 April 2015
4. Xiaobo A (2015) Major Shareholder of Citic Securities Reducing Holding-shares. Beijing Times , p 38
5. John M (2010) Technical Analysis of the Financial Markets. Seismological Press, BeiJing
6. YuLei R (2013) Behavioral Finance. Commerce Institute of Changsha Central South University, Chao Sha

Market Power and Scale Economies of the Electricity Industry: Comparison Before and After the Reform

Jing Wang and Yaodong Zhou

Abstract This work uses nonlisted company samples from 1999 to 2007 to study the market power and scale economies before and after electricity market reform. The results of empirical research show increasing market power and slightly greater scale economies, but the whole industry could not share the benefit from the scale economies. From the perspective of the business category, the power-generation sector obtained more benefit from the reform: Its market power increased, and the scale economies were slightly greater; however, the market power of the grid had been constrained but at the cost of decreased scale economies. It is concluded that the market-oriented reform in 2002 was only an adjustment of the "rules of the game" between state-owned enterprises, not involving other stakeholders, so it was difficult to form an energetic, continuous, and predictable market structure.

Keywords Electricity industry · Market power · Scale economies · Electricity reform

1 Introduction

Although the marketization of the electricity industry was explored in the 1970s all over the world, it is hard to describe the institutional arrangements of electricity market reform. The cooperation, technical features, and degree of monopoly of different departments lead to complexity of the institutional arrangements.

The initial institutional arrangements of China's electricity market were similar to the pattern of overall monopoly. From 1985, the government carried out a business model called "province as entity". In 2002, the National Development and

J. Wang · Y. Zhou (✉)
School of Economics and Management, Beijing Jiaotong University, Beijing 100044, China
e-mail: ydzhou@bjtu.edu.cn2

J. Wang
e-mail: 13120488@bjtu.edu.cn1

M. Li et al. (eds.), *Proceedings of 2015 2nd International Conference on Industrial Economics System and Industrial Security Engineering*,
DOI 10.1007/978-981-287-655-3_41

Reform Commission impelled market-oriented reform, and then the electricity market finished the separation of the power plants and the grid. However, the process of reform slowed down during later years. In 2009, the Inner Mongolia multilateral electricity trading market was called off; thus, the reform route of competition was hampered.

Competition and scale economies are not only contradictory, they are also two significant factors. The ultimate goal of reform is the coexistence of efficient competition and economies. Therefore, measuring the market power and scale economies before and after electricity market reform is meaningful for us to learn the target and contents in the next period.

From the perspective of existing articles, most researchers considered the market concentration rate to have declined after reform. Gong and Wang [1] showed that CR_4 dropped from 52 to 45 % during 1998 to 2010 and that the number of firms increased from 96 to 123. Liu and Shen [2] used hierarchical liner models based on power-generating companies, and the results showed that deregulation has a notable effect on the number of firms. However, some researchers thought that the market power was still high. Yu and Fu [3] pointed out that the market power of grid was 0.34 by calculating the approximation of Lerner index in 2010.

In terms of scale economies, most studies employed DEA or SFA. The research results were controversial. Li [4] used SFA to study the scale economies of 15 companies during 2001 to 2007, and the results indicated that output elasticity was 1.021. Liang [5] pointed that the sum of both capital and labor output elasticity was 0.87. Other researchers focused on the relationship between the scale of a firm and its efficiency. Zhang and Xia [6] showed a U-shaped relationship between scale of a firm and its efficiency, and samples appeared to show diseconomies of scale. However, Yang and Yu [7] found that firm size and scale efficiency had positive relationship.

Considering existing researches, most researches studied market power and scale economies by dividing them, and the methods chosen were different. Because the methods have different assumptions, which may lead to negative effects on the credibility of results, this work was based on the model of Klette [8], i.e., calculating market power and scale economies before and after reform in the same equation; then the effect of reform is discussed.

2 The Model and Samples

2.1 Model

The work is based on the model built by Klette [8]. In the model, firms within an industry are assumed to be constrained by a production function $Q_{it} = A_{it}F_t(X_{it})$, where Q_{it} and X_{it} represent the output and a vector of inputs for firm i in year t. A_{it} is a firm-specific productivity factor, whereas $F_t(\bullet)$ is the production function common to all firms. The subscript "t" is time series.

Supposing there is a "representative firm" within an industry whose output is chosen as the median value of output within the industry each year. As well as output, the input of the "representative firm" is chosen as the median value of every kind of input. According to the multivariate generalized mean value theorem, the production function relationship can be expressed in terms of logarithmic deviations between a sample and the representative firm.

$$\hat{q}_{it} = \hat{a}_{it} + \sum_{j \in M} \bar{\alpha}_{it}^{j} \hat{x}_{it}^{j} \tag{1}$$

In Eq. (1), $\hat{q}_{it} = \ln(Q_{it}) - \ln(Q_t)$, where a lower-case letter with a hat is the logarithmic deviation from the "representative firm" of the corresponding upper-case variable. Q_t and X_t represent the output and inputs, respectively, of the firm in the year t. $\bar{\alpha}_{it}^{j}$ is the output elasticity for factor j evaluated at an internal point ($\bar{X}_{it}$) between X_{it} and the reference point X_t.

Further assumptions of the model: (1) The representative firm acts as a price taker in the input markets, whereas it has some market power in the output markets. (2) The representative firm determines the labor employed according to the first-order conditions of profit maximization. (3) Price elasticity of demand ε is the reciprocal of the ratio of price to marginal cost. First-order conditions with these assumptions imply that

$$A_{it} \frac{\partial F_t(X_{it})}{\partial (X_{it}^{j})} = \frac{W_{it}^{j}}{\left(1 - {}^{1}\!/_{\varepsilon_{it}}\right) P_{it}} \tag{2}$$

W_{it}^{j} is the price for input j, P_{it} is the price of output, and ε_{it} is the price elasticity of demand. The factor $(1 - 1/\varepsilon_{it})^{-1}$ represents the ratio of price to marginal cost. Denoting this ratio between price and marginal cost by μ_{it} and using the set of first order conditions in Eq. (2), we have that

$$\bar{\alpha}_{it}^{j} = \mu_{it} \frac{\bar{W}_{it} \bar{X}_{it}}{P_{it} Q_{it}} = \mu_{it} \bar{s}_{it}^{j} \tag{3}$$

S_{it}^{-j} is the cost share of input j relative to total revenue. It can be measured by the cost share of the mean value point between a sample and representative firm. Supposing that $\bar{\eta}_{it} = \sum_{j \in M} \bar{\alpha}_{it}^{j}$ is output elasticity, and then we separate capital from other inputs. Equation (1) can be rewritten as

$$\hat{q}_{it} = \hat{a}_{it} + \mu_{it} \sum_{j \neq K} \bar{s}_{it}^{j} (\hat{x}_{it}^{j} - \hat{x}_{it}^{K}) + \eta_{it} \hat{x}_{it}^{K} \tag{4}$$

Supposing there is a universal market power μ and scale economies η within an industry, using Eq. (4), we obtain that

$$\hat{q}_{it} = \hat{a}_{it} + \mu\hat{x}_{it}^{V} + \eta\hat{x}_{it}^{K} + (\mu_{it} - \mu)\hat{x}_{it}^{V} + (\eta_{it} - \eta)\hat{x}_{it}^{K} \tag{5}$$

Denoting $\hat{x}_{it}^{V} = \sum_{j \neq K} \bar{s}_{it}^{j}(\hat{x}_{it}^{j} - \hat{x}_{it}^{K}), v_{it} = (\mu_{it} - \mu)\hat{x}_{it}^{v} + (\eta_{it} - \eta)\hat{x}_{it}^{v}$, we obtain Eq. (6) where $\hat{a}_{it} + v_{it}$ is the intercept of equation.

$$\hat{q}_{it} = \hat{a}_{it} + \mu\hat{x}_{it}^{V} + \eta\hat{x}_{it}^{K} + v_{it} \tag{6}$$

2.2 *Data*

According to the suggestions by Klette [8], we divided inputs into three categories: labor, capital and energy. The data in this paper are from the GTA nonlisted companies database. We divided the samples into two parts according to time: samples from 1999 to 2002 and samples from 2003 to 2007. Furthermore, we divided the samples into two parts according to the business: power-generation firms and grid firms.

The theoretical model presented includes the factor s_{it}^{-j}, which is evaluated at one internal point in the domain between the point of the representative firm and the observed level of operation for the establishment. Because the location of this particular point and the corresponding shares are unknown, according to the suggestion of Klette [8], we have approximated the shares by taking the average value of the share for the observed establishment and the time industry median share.

As in a panel model, the close correlation of the error, inputs, and productivity is likely to cause and endogenous problem, and OLS may lead to biased estimation. Therefore, the work used the generalized method of moments (2SLS and GMM) to estimate the equation. The advantage of the generalized method of moments is that in an endogenous condition, it can still produce unbiased estimators. 2SLS is the best choice if there exists spherical perturbation, while GMM is optimal if errors have heteroskedasticity.

3 Tests and Empirical Analysis

This work used FE(2SLS) and FE(GMM) to estimate the equation, and the results of market power and scale economies—including power-generation firms, grid firms, and total firms—are listed in Table 1. The estimate 1, 3, 5 are calculated by 2SLS, and estimate 2, 4, 6 are calculated by GMM. The results of overidentification test are also listed in Table 1.

Table 1 The Estimate of Market Power and Scale Economies

	Power-generation sector		Grid sector		Total firms	
	Estimate 1	Estimate 2	Estimate 3	Estimate 4	Estimate 5	Estimate 6
μ (1999–2002)	1.1304*** (23.7081)	1.1122*** (26.3584)	1.2406*** (17.2003)	1.2487*** (17.7768)	1.1164*** (21.1682)	1.1149*** (21.1549)
μ (2003–2007)	1.2273*** (22.0936)	1.2310*** (22.1891)	1.0911*** (22.7355)	1.0549*** (27.5375)	1.1867*** (25.2752)	1.1867*** (25.3517)
η (1999–2002)	0.9460*** (25.4888)	0.9300 *** (29.4601)	0.9961*** (15.9278)	1.0061*** (16.9882)	0.9087*** (21.3993)	0.9087*** (21.4005)
η (2003–2007)	0.9626*** (17.9804)	0.9665*** (18.0763)	0.8964*** (18.2347)	0.8560*** (23.0846)	0.9459*** (21.8290)	0.9459*** (21.8990)
R2 (1999–2002)	0.6288	0.6353	0.3979	0.3905	0.5947	0.5947
R2 (2003–2007)	0.7643	0.7627	0.6627	0.6775	0.7797	0.7797
Overidentification test (1999–2002)	0.671 (0.4127)	0.671 (0.4127)	0.244 (0.6210)	0.244 (0.6210)	0.567 (0.4514)	0.567 (0.4514)
Overidentification test (2003–2007)	1.762 (0.1844)	1.762 (0.1844)	1.565 (0.2110)	1.565 (0.2110)	0.000 (0.9990)	0.000 (0.9990)

***, **, and * reflects levels of 1 %, 5 %, 10 %, respectively. They are statistically significant. In parentheses are the corresponding z-values. Robust standard errors are adopted in 2SLS regression. The results of Hansen test are used for overidentification test

From the perspective of market power, there was a certain market power within electricity industry. As estimates 5 and 6 show in Table 1, the market power of electricity had increased and reached 1.1867. Considering business category, the market power of power-generation firms was 1.23 and increased by 10 to 12 % after reform. The market power of grid was 1.05 to 1.09, while compared with the market power before reform it declined by 15 to 19 %.

The whole industry showed diseconomies of scale both before and after reform. Both generation and grid firms had diseconomies of scale. Compared with the degree of scale economies before reform within the whole industry, they increased by 3.72 %. From the perspective of business category, the increasing degree of scale economies for generation firms was 1.66 to 3.65 %, and in two periods generation firms showed diseconomies of scale. The decline of scale economies for grid firms was obvious at approximately 9.97–15 %. Therefore, we concluded that the degree of scale economies increased, but only by a little.

In fact, the empirical results reflect some features of this reform.

- The reform introduced limited competition into the power-generation business. Private and other types of enterprises had no way to share the benefits of competition, and they had no inspiration to participate in the competition. Therefore, this reform had very little effect on market power for the power-generation sector.
- The change of pricing rules was another main aspect of the reform. Because grid firms had little power over pricing, the market power of grid firms declined obviously. In contrast, the market power of the power-generation sector increased. It may have close relation with the strategic merger behaviors of main firms, e.g., the repeated merger events after 2002.
- The reform offered limited inspiration to scale economies. Reform merely split power-generation business and supply business, then made asymmetric regulation for two kinds business. But reform ignored the "longitudinal economic characteristics" of the whole industry, and there was no effective institution maintaining the scale economies. Thus, the industry was still in the stage of diseconomies of scale.

4 Conclusions

This work used nonlisted electricity company samples from 1999 to 2007 to study the market power and scale economies before and after electricity market reform. The results of empirical research showed increasing market power and slightly larger scale economies, but the whole industry could not share benefit from the scale economies. From the perspective of business category, the power-generation sector obtained more benefit: Its market power had been increasing, and the scale economies were slightly larger, whereas the market power of the power-transmission sector had been constrained but at the cost of decreasing scale economies. Therefore, only a few stakeholders could obtain benefits from the reorganization of assets in the reform, and the reform ultimately had a limited effect on competition and scale economies.

Based on the results of market power and scale economies, the reform in 2002 must be completed, and the reform still has risk.

- With the rapid investment of new energy and adjustment of market rules, traditional power-generation sector will be under pressure of structural competition, and its investment risk will increase.
- The market power of grid firms is limited by price control and corresponding subsidies. Because grid firms lack of a clear standard of price control and subsidy, the market power is likely change much according to the tendency of regulatory pricing.

Furthermore, because the institution lacks of system checks and balances, the relationship between stakeholders can change at any time, and this will cause a change in market structure. After all, we believe that the market structure of electricity is unstable, and it will change according to the tendency of policy and the market environment.

Acknowledgments I appreciate the support of Joint Project of scientific research and graduate student training by Beijing Board of Education-City traffic environment externality in Beijing: empirical study and governance mechanism (No. B13H100010).

References

1. Gong RK, Wang ZG (2013) Market competition, regulatory and reform of property rights and growth of China's power industry. Inq Econ Issues. 7:31–40
2. Liu XM, Shen L, Zhang FX (2011) Empirical study on the impact of deregulation in Chinese infrastructure industry on corporate efficiency. Manag Rev. 2:11–18
3. Yu LC, Fu Q (2012) Analysis and measure on the market strength of China's power grid. China Indu Econ (11): 44–57
4. Li T (2009) Productive factors, power price regulation reform and the efficiency of thermal power enterprises: stochastic frontier evidence from a-share listed companies. J Finance Econ 4:107–118
5. Liang SG (2012) An empirical study of effects of regulatory reform and performance of China's power generation industry listed companies-based on panel data and stochastic frontier analysis. Rev Indu Econ 3:35–56
6. Zhang GX, Xia DW (2011) Ownership structure, environment regulation and efficiency of Chinese power generation industry- stochastic frontier production function analysis based on provincial panel data during 2003–2009. China Indu Econ 6:130–140
7. Yang SY, Yu LC (2008) The efficiency and productivity change in China's electricity supply system. Collect Essays Finance Econ 5:15–22
8. Klette TJ (1999) Market power, scale economies and productivity: estimates from a panel of establishment data. J Indu Econ 47(4):451–476

Problems and Countermeasures of PPP Mode in China

Xuejing Hou and Yanzhang Wang

Abstract China's current financial situation is grim; PPP mode has become an important channel for local financing; and PPP projects will grow rapidly this year in China. The authors believe that PPP should improve top-level design, strengthen the financial-sector PPP project veto, and establish and improve relevant legislation in order to play the role of intermediary organizations, supervise service, and protect the interests of both government and business interests to prevent the BCC from triggering a local debt systemic risk.

Keywords PPP mode · Debt risk · Local debt

1 Introduction

Currently, China is accelerating the launch of PPP (public private partnership) mode in urban infrastructure, public rental housing, and local financing-platform project restructuring, which effectively releases social investment potential [1]. However, due to the lack of top-level design, many problems exist around the PPP mode. The outbreak of PPP projects will lead to a number of hidden risks for the local financial situation.

X. Hou (✉)
Faculty of Management and Economics, Dalian University of Technology Dalian, Dalian 116024, China
e-mail: houxuejing@xinhua.org

Y. Wang
Information and Decision Technology Institute, Dalian University of Technology Dalian, Dalian 116024, China
e-mail: yzwang@dlut.edu.cn

M. Li et al. (eds.), *Proceedings of 2015 2nd International Conference on Industrial Economics System and Industrial Security Engineering*,
DOI 10.1007/978-981-287-655-3_42

1.1 The Status of the Current PPP Mode

Based on the news learned that this year that China's fiscal revenue and expenditure pressure is very large, there is an urgent need for the local government to complete the investment plan to find new sources of funding. As the construction, operation, and management of public infrastructure and a channel to provide a public service project–financing mode, PPP helps to resolve debt and improve public services, e.g., this year, some of the local government budget was used in "promoting the use of the PPP mode."

1.2 PPP Mode Creating a Cooperative "Win-Win-Win" Situation

At present, all of the provinces in China have been announced that they plan to invest nearly 3.4 trillion Yuan in PPP projects in 2015. PPP mode is popular in China and has its own advantages. For example, PPP mode broke the traditional investment pattern by financial funds in the public infrastructure [2]. Local governments, society, and social capital can benefit from PPP mode to achieve a "win-win-win" situation.

In addition, PPP will help promote China government functions by acting as both a "player" and "referee." As a referee, PPP allows the market to play a decisive role in the allocation of resources, improve supply efficiency, and improve the level of public-service government public goods [3].

PPP mode improves the efficiency of financial funds by leveraging more social capital because financial capital has played a game of "skillfull deflection of the question" regarding leverage and guidance. From a social point of view, the public can benefit from the increased supply of public services.

1.3 PPP Project Elusive Small Social Capital

Current PPP projects exist for "investment, low income, long period, not quit, professional high threshold" and other investment characteristics, which require relatively decentralized social capital funds with strong risk tolerance.

From an investment perspective, all provinces have recently launched large-scale PPP projects, and the largest single one cost nearly 30 billion Yuan; therefore, 500 million Yuan should be considered minimal.

Regarding the investment cycle, projects involving transportation, water conservancy, municipal construction, protection of housing projects, health care, and other fields require pension funds to exit profitably from the general fund within 3 or years or so.

In addition, PPP projects require a higher professional threshold. Just entering the early stages requires a higher investment, e.g., project design generates many costs. Although the two sides are jointly funded under the contract, in fact, when the project is finished, the investors' money, for these design costs will often be for naught.

2 Problems of PPP Mode in China

With the investment and financing system gradually being shown to depend mainly on financial land malpractice, the establishment of new financing mechanisms, in particular promoting PPP mode, has become the main means of financing for local governments. However, some local governments have encountered many problems from the level of institutional mechanisms during the vigorous promotion of PPP projects.

2.1 *Lack Cooperation Between Government Departments*

Based on the research found regarding the process of promoting PPP projects, lack of top-level design, reform of financial-sector development, and lack of coordination can sometimes become obstacles.

For example, on December 4, 2014, when the Development and Reform Commission and the Ministry of Finance issued two documents—"Guidance on the government to carry out cooperation and social capital" and "The government and the Social Capital Mode of Cooperation Operations Guide (Trial)"—status quo thinking became evident when the two departments each promoted their own PPP project. The author subsequently learned that the policies, selection criteria, and procedures of the entities are also different.

Policy conflict with social capital increases the difficulty of docking, the number of different sector-led PPP projects and programs, and the difficulties that arise in participating in social capital, thus creating entry barriers for businesses PPP project-setting.

2.2 *The Existing Laws Lag Behind the PPP Mode*

At present, China has no specific PPP laws and regulations. PPP regulation currently is based on contract law and regulation-franchising law as well as informing the relevant ministries. The existing laws do not meet the practical needs of PPP development, and some laws even conflict, thus producing a number of obstacles in procuring social capital.

Wastewater treatment is currently under PPP mode based on most typical conflicts of the law. In terms of consideration of public safety, such projects are usually registered under a corporate-identity partnership, so PPP problems can be traced back to an unlimited partner. However, the actual operation of the existing Partnership Enterprise Law of People's Republic of China poses conflicts. Partnership enterprise law stipulates that an investor can set up a partnership but that general listed companies cannot be a partner. This creates disorder in some companies involved in PPP projects if participation is necessary to bypass the law.

PPP is one of the reasons for avoiding the legal department because in PPP projects related regulatory thinking is not uniform. The treasury is currently working on legislation to promote the PPP Government and Business Cooperation Law. At the beginning of the Development and Reform Commission on the "infrastructure and public utility franchise management approach," the community was consulted.

In addition, due to PPP projects, investment, low income, long cycle, quit easy, high professional threshold, and influence by policy factors, particularly regarding the need for investment, are mechanisms for the protection of human rights.

3 The Local Governments Debt Risk of PPP Mode

Due to the lack of top-level design around PPP mode, there the phenomenon of "take PPP as a basket, and everything entered" appeared, which did not leverage private capital and diversify investment risks. In addition, local government debt risks will transfer to the social capital bank resulting in a domino effect and triggering a series of risks.

3.1 Social Capital Is Facing Policy Risks

Risk is faced by the subjective factors of alternative market rules leading to social capital. PPP projects have no national standardized guidelines for cooperation. For example, PPP projects in different parts of the same building are not necessarily conducted in the same way, To safeguard enterprises' interests, there must be competent leadership by the government while the project is carried out.

Regarding PPP project–cycle length, and after the general local government sometimes encountered the phenomenon of "new official to ignore the old account," especially because local financial pressures have increased, the number of original agreements will be discounted.

On investigation it was learned that smaller-social capital does not have the fund strength to participate in PPP projects. Along with financial strength of the business, PPP projects often require rigid payment guarantee, and the policy uncertainty and returns affected cooperation between community capital and the government.

3.2 The Financial Risks Faced by Local Government

PPP modes are currently trying to promote their extraordinary development throughout China [4]. The authors believe that the local debt is already high at the moment and that government regulation of PPP-mode development will generate a heavier debt burden.

Regulators have perfected the indicator system because there is no scientific basis for the approval of the subjective factor of greater regulation and lack of scientific validity. Once the local government funding strand breaks and slows down other items, a domino effect will occur.

Regarding vaguely defined benefit-prone BCC, the current portion of PPP projects by local governments have a rather vague definition: Specific rules are not very mature, and joint social capital by PPP mode has disguised financing. For example, some people rely on personal relationships to engage in PPP investment.

In addition, the current category for "social capital" is also not clearly defined. At present, the ratio of the Ministry of Finance–funded state-owned enterprises can not be >50 %, whereas the Development and Reform Commission has not made specific provisions.

If private capital does not indicate effective participation in PPP projects, it will be difficult to achieve the original purpose of risk diversification.

4 The Problem for the Current PPP-Mode Strategies

The authors believe that China should gain international experience in promoting PPP mode by building public participation as well as a risk-allocation mechanism. At present, there is a need for early introduction of top-level design and promotion of PPP mode to achieve scientific and orderly development.

4.1 Giving Definition of PPP

Because PPP still has no clear boundaries, there exist some local government policy loopholes regarding the transfer of benefits and seeking rental space. It is proposed to clarify the boundaries of the PPP premise and projects and further transform the government function to guard against damage to the public interest as well as the appearance of "gray trade" and other issues. The definition of PPP mode is accurate, but it has not been vetted. PPP projects in developing countries are more mature because they have been clearly defined [2].

Taking these perspectives, the authors believe that China should define PPP under two basic conditions:

1. Stipulate the government's public investment program and its responsibility throughout PPP projects. Furthermore, products, services, and expenses linked to government performance should be and provided.
2. Complete the acquisition of social capital.

4.2 Establish a Sound Legal System Related to PPP

Currently, the provisions of relevant ministries on PPP projects within their scope of management only apply to part of the industry; often the cannot link to each other, and there is a global and systemic management. PPP projects that related to the public interest and involve the government of this special legal body must promote specific legislation at the national level.

In particular, PPP projects must make decisions about project company, bidding, tax incentives, and other issues to create special laws. PPP projects are recommended to ensure that the interests of all parties involved in the project will protected, particularly that of the public, while clearly growing the PPP field and making the way through legislative PPP applications.

4.3 Giving the Financial Sector Veto of PPP Project

The IMF stressed that the financial sector of PPP-project approval has veto power, but China's financial sector and the current binding is relatively weak in this regard. The author suggests that China strengthen the financial sector to veto PPP-project approval so as to prevent financial risks that the project might incur.

In addition, PPP mode has the unique need of long-term payment mechanisms supporting its budget, but the annual budgetary mechanisms implemented make it difficult to guarantee long-term payments. If China were to establish long-term budget planning, part of the government expenditure responsibilities could more easily be regulated to ensure that the project has the ability to pay its costs.

4.4 Prominent Position of Industry Associations

The author recommends the establishment of a national PPP industry association that would provide project preparation, mid-term guidance, and a series of post-supervision professional services. In addition, PPP dishonesty regarding statistical reporting and its public-exposure system, government, and corporate defaults are available through a "black list" that will be publicized.

Acknowledgments I am grateful to the experts, as well as my mentor Wang Yan Zhang, who gave so much of themselves in assisting me in writing this chapter.

References

1. China's government work report (2015) National People's Congress, Beijing
2. Jiakang S (2009) The concept of public-private partnership (PPP), the origin, characteristics and functions. Public Finance Research, Beijing
3. Jiakang (2013), China needs to build and develop the core of the new reform supply-side economics. Public Finance Research, Beijing
4. Gushengzu (2014) The new "Budget Law" and the local government debt risk control. Social Science Front, Beijing

Study of the Effect of Private Enterprises to Absorb the Employment of Guangdong

Mengchen Li

Abstract Since the 1990s, with the development of the private enterprise, it has become a new growth point of the national and local economy and the important carrier of technology innovation in China. Private enterprise provides important insurance of increasing employment and the stabilization of society, and it has made great contribution to alleviate pressure on employment and maintain social stability. Since the Reform and Opening-Up, the growth of the private enterprise in Guangdong has been rapid; in 2015, the number and development of private enterprises in Guangdong ranks first place in China. The rapid growth of private enterprises has become an important factor stimulating employment. For quantitative analysis of the employment effect of private enterprise development, this work used two indexes—private enterprise fixed assets investment and social employment. We found that the development of private enterprises has a significant role in promoting the economy' employment.

Keywords Private enterprise · Employment absorption effect · Fixed capital investment

1 Introduction

The pressure of the economic downturn increases as the economic development of China enters a period of new normal. In the government work report of 2015, China government put forward a target of offering >10 million new jobs in cities and towns and maintaining the registered urban unemployment rate within 4.5 %. Since the 1990s, with the development of private enterprise, private enterprises has become a new growth point of the national and local economy as well as the important carrier of technology innovation in China, and it has also deepened the

M. Li (✉)
School of Economics and Management, Beijing Jiaotong University, Beijing 100044, China
e-mail: 13120560@bjtu.edu.cn

M. Li et al. (eds.), *Proceedings of 2015 2nd International Conference on Industrial Economics System and Industrial Security Engineering*,
DOI 10.1007/978-981-287-655-3_43

reform of state-owned enterprises. Private enterprise is important to ensure increasing employment and the stabilization of society, and it has made great contributions to alleviate pressure on employment and maintain social stability.

Since the Reform and Opening-Up, the growth of private enterprise in Guangdong has been rapid: In 2015, the development of private enterprise in Guangdong ranks first place in China. By the end of February 2015, the total number of Guangdong enterprises generated $2.3294 million, up 25.31 % from the previous year. In 2015, the number and development of private enterprise in Guangdong ranks first place in China, and the number of enterprises is >2 million.

According to data from the industrial and commercial bureau of Guangdong province, the actual number of private enterprises in Guangdong province has reached 2.0348 million, more than the same period in 2014 and 2013, respectively, 457,700 and 758,600, and an average annual growth of 29.72 %. Private enterprises accounted for 87.35 % of the total number of enterprises in Guangdong province and accounts for, respectively, 4.98 % and 2.51 % growth during the same period in 2013 and 2014, which also plays positive role in promoting the growth of market players, optimizing industrial structure, promoting employment and entrepreneurship, and the revitalization of the eastern northwest and other aspects. Thus, we can see that private enterprise has become an important driving factor in promoting entrepreneurship and employment. Therefore, study of the development of private enterprises in Guangdong Province on employment has important practical significance.

Since the Reform and Opening-Up, China's economic system has always insisted on market reform, and a diversified ownership system is the inevitable direction of economic development. Therefore, private enterprises grew fast like bamboo shoots and became an important support for employment against the social background of the Reform and Opening-Up, which is like a spring rain. Zhang Mao, director of the State Administration for Industry and Commerce, mentioned that under the pressure of the economic downturn, individual industrial and private enterprises still absorbed 250 million jobs, which increased employment. We can determine that private enterprise plays an important role in attracting employment and stabilizing China's economy. In Yu Chun [1] and Ying [2] both believe that private economy is the fastest-developing and dynamic and economic group to contribute to the Chinese economy, and they also mentioned that private enterprise has already become, and will continue in the future, the main way to increase employment. Jing [3] mentioned that the increasing amount of capital owned by private enterprise will have assignable effect on absorbing employment. In a word, all of these researchers agree that private enterprise plays an important role in the rapid develop of economic and social stability and absorbs employment [4].

There is no doubt that studying the empirical relationship between fixed capital investment and private enterprises to absorb the employment in Guangdong province, which leads in the development of private enterprises, and to give a corresponding theoretical explanation and policy suggestions will have important theoretical significance and application value [5]. Therefore, cointegration test, Granger causality test, and generalized impulse response function were employed to test the internal relationship between the fixed capital investment of private enterprises and their ability to absorb employment in Guangdong Province from 2004 to 2013 [6]. Meanwhile, the mechanism and path of fixed capital investment of private enterprises to absorb the employment of private enterprises were analyzed to provide a theoretical basis and empirical analysis for the formulation of relevant policies [7].

2 The Empirical Analysis and Explanation

The purpose of this work was testing the relationship between the fixed capital investment of private enterprises and the employment that is being absorbed by private enterprises, not the employment-decision factors, so this work only considers the influence of important variables. According to the changing trend of variable data, we built a modified C-D Cobb Douglas function model:

$$\mathrm{LZ} = \mathrm{f}(\mathrm{FIXZ}) = \mathrm{C} * \mathrm{FIXZ}^{\mathrm{C}(1)} \tag{1}$$

$$\text{In logs}: \ln(\mathrm{LZ}) = \mathrm{C} + \mathrm{C}(1) * \ln(\mathrm{FIXZ}) + \mathrm{U_t} \tag{2}$$

According to this framework, we construct the ln(LZ), ln(FIXZ) model of the variable system. LZ represents the number of the employment absorbed by private enterprises, and ln(LZ) is the representative of the natural logarithm. FIXZ is the representative of the fixed investment of private enterprises in Guangdong Province, and ln (FIXZ) is the representative of the natural logarithm of FIXZ.

In this work, the data are from the Statistical Yearbook "China" (1996–2013). To eliminate the influence of original variables' heteroskedasticity, we logged the data.

3 Empirical Analysis

(1) Unit Root Test
The time series analysis is the premise of smooth sequence, and a nonstationary time series in the regression analysis can lead to a spurious regression problem. Therefore, before analysis, unit root test of the variables was used to determine the sequence of stability (Table 1).
(2) Cointegration Test
When the time series analysis of the variables are nonstationary series, in addition to differential treatment being performed on the original sequence of variables, cointegration analysis is used to determine if these linear combinations of nonstationary variables are a stationary sequence. If the linear combinations of nonstationary time series enter into stationary variables, then there is a long-term

Table 1 Results of unit root test table

Variable	The type of examination	ADF	P	Result
lnLZ	(c,t,3)	−3.263484	0.1127	Not smooth
lnFIXZ	(c,t,3)	−3.645780	0.0628	Not smooth

equilibrium relationship between these variables. In the analysis of the relationships between multiple variables association, the most commonly used method is the Johansen cointegration test method (Table 2).

As judged by the test trace statistic, the null hypothesis "None" means that there is no cointegration relationship. The probability of P trace statistic is 0.0017, so the null hypothesis can be rejected, and thus there is at least one cointegration relationship. The next, "At most 1," means that there is at most one cointegration relationship, and the trace statistic probability P calculation of the assumption of a value of 0.1497, which means we cannot reject the null hypothesis (Table 3).

From the test, we can determine that one of the adjustments is a negative number, which means that the cointegration relationship is effective, and the cointegration equation is:

$$\mathrm{lnLZ} = 0.887\mathrm{lnFIXZ} \tag{3}$$

(3) Vector error—correction model

Because the expression from the Johansen cointegration test and the term of error correction is nearly the same, we can list the formula:

$$\mathrm{CointEQ1} = \mathrm{lnLZ} + 0.8682\mathrm{lnFIXZ} - 1.5397 \tag{4}$$

Table 2 Part results of Johansen cointegration

Unrestricted cointegration rank test (Trace)				
Hypothesized		Trace	0.05	
No. of CE(s)	Eigenvalue	Statistic	Critical value	Prob.**
None*	0.799216	24.55222	5.49471	0.0017
At most 1	0.137744	2.074848	3.841466	0.1497

Table 3 Part results of Johansen cointegration

Unrestricted adjustment coefficients (alpha)			
D(lnLZ)	0.043886	0.000719	
D(lnFIXZ)	−0.0052	0.019012	
1 cointegrating equation(s)		Log likelihood	55.24509
Normalized cointegrating coefficients (SE in parentheses)			
lnLZ	lnFIXZ		
1.000000	−0.887		
	(0.01303)		

VEC model:

$$\Delta Y_{t-1} = \begin{bmatrix} -0.29 \\ -0.41 \end{bmatrix} \text{CointEQ}_{t-1} + \Delta Y_{t-1} + \cdots + \Delta Y_{t-3} + \theta + \varepsilon_t; \quad \Delta Y = \begin{bmatrix} \text{DinLZ} \\ \text{DinFIXZ} \end{bmatrix} \tag{5}$$

According to VECM, we can know there is a significant positive correlation between the fixed capital investment in private enterprises of Guangdong Province and the employment absorbed by private enterprises.

(4) Granger Test
From the test results, in the short term, fixed asset investment in private enterprises is the Granger reason for the employment absorbed by private enterprises; the P value is 0.1125; and the Granger Test was significant (15 % confidence level).

4 Summary

Research shows that, whether in the long- or short-term, the fixed capital investment of private enterprise growth will significantly promote the growth of the employment absorbed by private enterprises. From stimulating effect of the development of private economy on employment, the development of private enterprises has a positive effect of great impact on employment. Guangdong Province commercial system reform relaxes the establishing conditions of private enterprises and promotes private enterprises to increase their investment and number, which increases the number of employment. So to increase employment, we should promote the development of private enterprises, especially emerging Internet companies as well as small and micro businesses.

References

1. Yu Chun T (2005) Private enterprise development and the employment enterprise economy. Enterprise, vol 1, p 78
2. Ying L (2005) Trend analysis of characteristics of employment in private enterprise [J]. J Hebei Univ Econ Bus 02:25–30
3. Jing S (2013) Analysis of the impact of capital accumulation and economic problems of employment of different ownership structure, 04:39–42
4. Ying Z, Shaoyong Z (2012) The employment of Jiangxi private enterprises [J]. Jiangxi Soc Sci 6:75–78
5. Mingyue F, Huihua N, Ting J, Songtao T (2010) Estimation of the China industry employment elasticity [J]. World Econ 8:3–16

6. Qicheng S, Chunli D (2011) The relationship between the development of private enterprises and employment—based on the perspective of time series [J]. J Chongqing Univ Technol (Soc Sci) 7:26–31
7. Yanyi L, Rong L, Weidai D (2012) EViews analysis and application of statistics [M].Publishing House of Electronics Industry, Beijing, pp 168–177

Models, Risks, and Regulations of P2P Lending in China

Baolin Ma and Zheyi Wen

Abstract Yixin Inc. was founded in 2006 as the first P2P lending company in China, which started a new Internet-financing era in China. The P2P lending market has show explosive growth since 2012, and Internet-financing has become the most popular phrase on TV and in newspapers, government documents, etc. P2P lending energizes the traditional finance market and creates market opportunities in China, but it also brings approximately many problems and government regulatory challenges, such as money theft, illegal fund-raising, and fraudulent financing. This work describes different P2P lending models, of which the main risks are analyzed, and suggests regulatory policies to protect against risk.

Keywords P2P lending · Models · Internet financing · Regulation · China

1 Introduction

Since the first P2P lending company, Zopa, was founded in Britain in 2005, the P2P lending market has developed rapidly all over the world. There are both profit and nonprofit P2P lending platforms in the United States. The two main for-profit platforms are Prosper and LendingClub. As March of 2011, these two platforms combined had facilitated approximately 63000 unsecured, fixed-term, and fixed-rate loans totaling approximately $469 million, most of which involved consumer loans. The main nonprofit platform, Kiva, facilitated approximately 273000 interest-free loans totaling $200 million [1]. As December 2014, the loans of LendingClub were >$62 billion, while those of Prosper were >$20 billion [2].

B. Ma (✉)
China Center for Industrial Security Research of Beijing Jiaotong University, Beijing, China
e-mail: mabaolin203@126.com

Z. Wen
Rongyou Wealth Inc., Beijing, China

M. Li et al. (eds.), *Proceedings of 2015 2nd International Conference on Industrial Economics System and Industrial Security Engineering*,
DOI 10.1007/978-981-287-655-3_44

On December 12, 2014, LendingClub was listed on the NYSE, which was a great milestone of the Internet-lending market. At the same time, it was also a shot in the arm for P2P lending in China. In 2006, the first P2P lending company, Yixin Inc., which imitated Zopa, was founded in China. According to Internet statistics, the number of P2P platforms was China is >1,575 by the end of 2014. The total loans reached 252 billion Yuan [3]. The rapid development of P2P lending energized the market, but it brings causes many problems and challenges.

2 The Connotation of P2P Lending

P2P lending is a lending practice through an Internet platform or uses network technology, and it has typical features of the Internet thinking rather than a simple process of lending through Internet technology. P2P lending is a decentralized and multilinear interaction of information and cash rather than single or linear point-to-point interaction.

3 Models and Risks of P2P Lending in China

The model of P2P lending in Europe and the United States is a purely online platform, which is fully consistent with the popularity characteristic that is the spiritual core of Internet finance and the long-tail theory, which is the theoretical basis of Internet finance. However, in China, due to the imperfect social-credit environment, P2P lending has developed many models. Generally speaking, there are three types: pure online-platform mode, online and offline integrated mode, and pure offline mode. Each mode has its own advantages, disadvantages, and main risks.

3.1 Pure Online-Platform Mode

This model is a typical pattern of P2P lending, of which the main process is shown in Fig. 1. First, the borrower registers an account on the platform and submits an ID card, house card, income proof, and other credit information. The borrower requests the desired loan information—such as amount, term, and purpose—and the platform system presents a credit level and reaches a final rate based on the borrower's information. Second, a lender enters a bid according to the platform data, and the lenders' money is put into an escrow account because the transaction is locked by the system. The transaction is successful until the borrower's financing needs are met. The last step is that the lender's funds in the escrow account are then

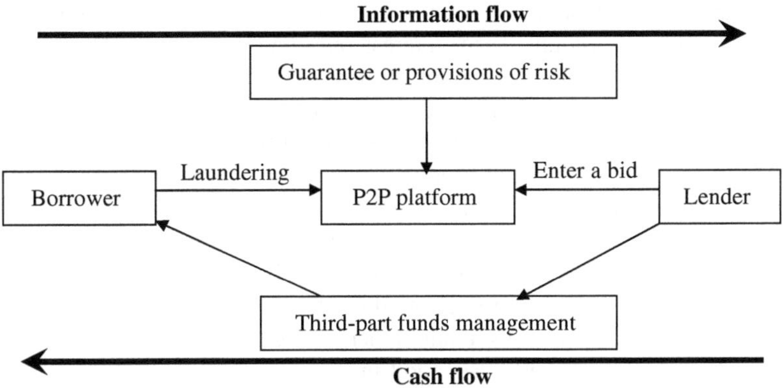

Fig. 1 Pure online-platform mode

transferred into the borrower's account from the escrow account. The platform then charges fees to both the borrower and the lender.

The main risks of this mode are credit default and system technology risks. Credit risk refers to the borrower failing to repay resulting in bad debt and investor losses. Technical risk is data leakage or loss due to network vulnerabilities and hacker's attack. To avoid credit risk, increase loan credit, and attract investors, platforms have increasingly begun to offer credit guarantees or risk reserve, which are closely related with the platform.

3.2 *Pure Offline Mode (Bonds Offline-Transfer Mode)*

Features of this mode include offline completion of the three steps of borrowing, trading, and lending, strict risk control, little or to no bad debt, bonds transferring by way of a personal account, and high cost of due diligence offline. Yixin Inc. is a representative of the mode. The main process is shown in Fig. 2. The company's

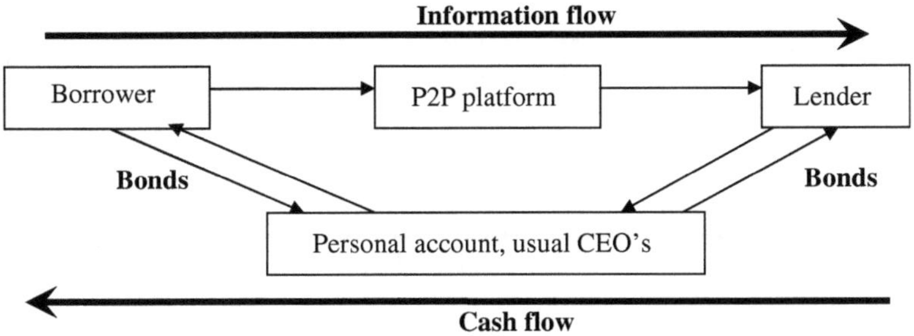

Fig. 2 Pure offline mode

credit department sales staff looks for suitable borrowers, who negotiate with the company about loan amount, term, interest rate, and repayment. Then borrowers obtain funds from a personal account closely related with the platform, usually belonging to the CEO, and then the loan contract is signed.

At last, bonds are formed. Then the person, usually the CEO, splits or combines the bonds to form fixed financial products, which are sold to investors. This mode usually facilitates loans of millions of dollars rather than thousands of dollars. Adequate borrower collateral is required in this mode, so no funds are managed by a third party. Finally, investors or lenders obtain a benefit certificate of the bonds. The main risk of this mode lies in debt-trading. Because that action is completed in a personal account, some risks could emerge such as money theft, selling by self-assessment, and increased false credit increasing. There will also be a liquidity risk created by a mismatch of loan amount and timing due to the split and combination of bonds. If a large amount of loan is default, cash flow would be broken, and then platform would become bankrupt.

3.3 Online and Offline Mode (O2O Mode)

The process of O2O mode is shown in Fig. 3. The P2P platform launches the borrower's information on its website after a review. Then investors purchase notes on the platform. In this process, the P2P platform introduces guarantee companies to attract investors. Many scholars believe that this mode makes it easy to disrupt financial markets resulting in difficult supervising and counting. This study suggests that innovative financing model of small loan companies combined with P2P platforms is consistent with China's social environment.

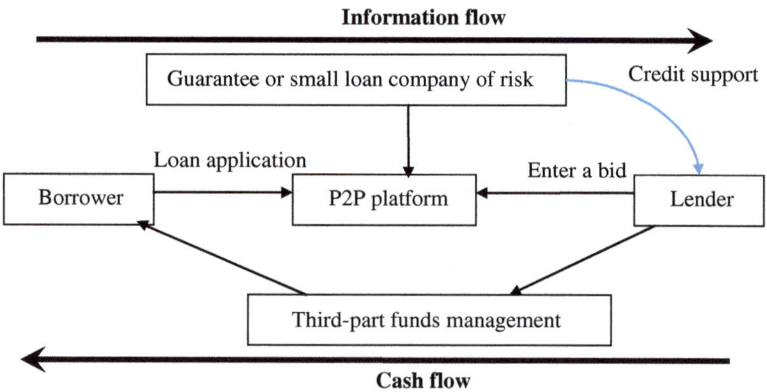

Fig. 3 Online and offline mode

4 Analysis of Risks of P2P Lending in China

Referring to the views of Ye Xiangrong and Lijun [4, 5], this study presents nine major problems attributed to four risks as listed in Table 1.

4.1 Qualitative Analysis on Risks

All P2P platform models pose credit risk. Credit risk is the main risk of pure online mode, because lending is based entirely on data, and no one has yet completely mastered credit data and money use of borrowers. The risk is small in pure offline mode because borrowers generally provide a mortgage to the P2P platform. In O2O mode, the credit risk is still a larger proportion than in the other two modes.

Financial data disclosure and no supervision of intermediate accounts results funds precipitation and liquidity risk. In pure online mode, the funds are generally hosted by a third party, so there are few problems. The risk is very high in pure offline mode because the transaction is completed in personal account leading to the cash flow being entirely controlled by one individual. In O2O mode, the transaction is still automatically matched on the platform, so the risk is relatively small.

Fraud, money-laundering, information leaks, illegal fund-raising, asset securitization, and guarantees: All of these issues are attributed to legal risk in this study. In pure online mode, the risk is little. In pure offline mode, the controller participates in the transaction, so the artificial risk is high because it is easy to result in leaks, illegal fundraising, and asset securitization. In O2O mode, credit-guarantees risks, asset securitization, information leaks, and other issues can easily occur.

Product distortion, such as second and circle transactions, poses no problem in terms of customer trial and advertisement. If there are many second and circle transactions, the platform would convey false prosperity, false credit, and increased leverage. In pure online mode, the risk is high because there are many second or circle transactions. In pure offline mode, transactions are small due to there being

Table 1 Risk score rating system

Problems	Risks	Online	Offline	O2O
Debt default	Credit risk	5	1	3
Account regulation, disclosure of finance data	Funds precipitation and liquidity risk	1	5	1
Fraud, money-laundering, information leaks, etc.	Legal risk	1	4	3
Second and circle transactions	False prosperity, increased leverage	3	2	2
Total score of risks		10	12	9

sufficient sales staff in the credit and financing department. In O2O mode, the financing department is still on the P2P platform, so risk still exists.

4.2 *Quantitative Analysis of Risks*

Through analyzing the properties of various risks, this study proposes a risk-rating system for P2P lending as shown in Table 1, in which the individual risks increase gradually from 1 to 5. For a certain mode, the gross risks can be calculated through adding the scores up to down. The result of calculation is that the risks are highest in a pure offline mode, higher in pure online mode, and lowest in O2O mode.

5 Regulation Suggestions

5.1 *Inclusive Attitude*

P2P lending promotes information flow in the process of cash flow. It involves not only a reformation of financing reformation but also a reformation of thinking. Because of banks' own criteria, obtaining SMEs and personal loans from banks is difficult. Now P2P lending breaks the traditional banking monopoly. Questions of "to lend or not lend," "lend to who, "how much benefits" are decided by the lender, not by the bank, which confers the dignity of wealth itself and restores the people's right to master wealth and bargaining rights for benefits, neither of which can be achieved with the traditional bank.

As a new thing, Internet financing requires the improvement of regulations and service and payment systems, all of which are definitely delayed at the beginning of the lending transaction. However, it meets the needs of the commerce and is in line with the trend of social development. Many market mechanisms, business models, and trading courses exist in the gray area of the law, but why not create an innovative financial model? Why not improve the relevant laws and regulations of reform? Therefore, this study proposed that the regulation of P2P lending should be in an inclusive principle.

5.2 *Clarity of Main Supervisor and Regulatory Framework*

In consideration of current confusion in the financing industry, this study suggests that the whole market should be led by the China Banking Regulatory Commission and specific business be regulated at present by industry associations.

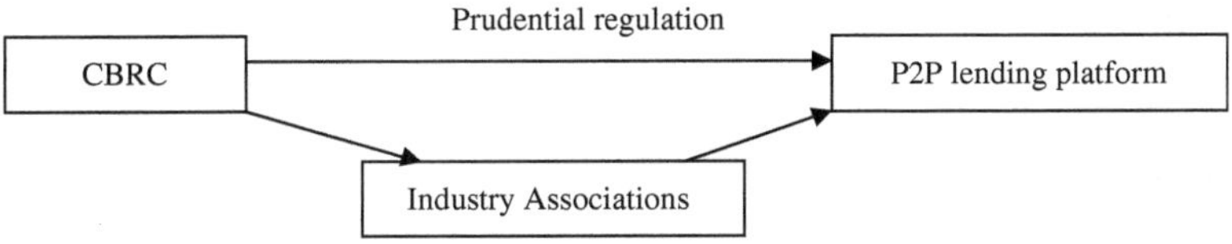

Fig. 4 Regulation framework

5.3 Regulations

(1) Inclusive regulatory measures

These mainly relate to entrance barriers, associated business-cutting, third-party fund management, disclosure of financial data, industry self-regulation, industry standards, etc. If high standard barriers are set up, grass-roots entrepreneurs will be not be able to start businesses, which results in decreased innovation and vitality, which in turn is harms the market system. The CBRC should enforce funds management and financial data disclosure for large platforms. For small platforms, due to limited industry influence, enforcement is not realistic and will increase operating costs. As shown in Fig. 4, The CBRC should establish P2P lending industry associations, become involved in their management, and treat them as agencies of establishing industry standards, industry self-regulation, and inclusive supervision to P2P platforms.

(2) Prudential regulations

These mainly relate to financial flow, bad debt, financial data of the associated guarantee company, and immediate account monitoring in offline mode. Guarantee and small loan companies are often associated with the P2P platform in O2O mode, and these must be separated and independent managed. Mandatory data reporting is necessary for every business in the associated companies. In offline mode, due to cash flow in a personal account, the potential for moral hazard is high, so the platform should be strictly regulated in terms of financial flow, default rate, personal account, and the data regarding bond transfer. All of the data should be mandatorily provided to CBRC or its branches.

The opportunities and challenges from P2P require higher regulator higher criteria, which also contributes to the improvement of regulation. For some models with small risks, inclusive regulations are suitable, whereas prudent regulations should be taken for those with huge risks. The rules should be constantly adapted to the development of social economy. In the course of continual adjustment, an optimal state will be reached.

References

1. GAO (2011) Person-to-person lending—New regulatory challenges could emerge as the industry grows
2. 2015. http://www.wangdaizhijia.com/news/guowai/15844.html
3. 2015. http://shuju.wangdaizhijia.com/industry-type-2-3-2015.html
4. Ye X (2014) The research on risks and regulations of P2P lending. Financ Regul Res 3:71–82
5. Li J (2013) P2P lending: characteristics, risks and regulations. Financ Dev Rev 3:35–50

Part VI
Special Session on Finance Group

Studies on Capital-Buffer Mechanism of China's Commercial Banks from the Perspective of Liquidity

Shaofeng Yuan, Yun Zeng and Wang Jing

Abstract This work uses the data of 23 commercial banks from 2003 to 2011 concluding that the capital-buffer level of commercial banks is pro-cyclical, and the enhancing liquidity-creation function of large commercial banks help to perfect the capital-buffer mechanism. However, the enhancing liquidity-creation function of joint-equity commercial banks will decrease the capital-buffer level, and the liquidity-creation function of the entire financial system is negatively correlated with the capital-buffer level. On this basis, we propose the recommendation of consummating counter-cyclical capital-buffer mechanisms.

Keywords Commercial banks · Liquidity creation · Capital buffer

1 Introduction

"Capital buffer" refers to the difference between the actual capital adequacy ratio of commercial banks and the minimum regulatory standards set by Basel New Capital Accord. Establishing a counter-cyclical capital-buffer mechanism is an important part of strengthening counter-cyclical capital and liquidity regulations. From the perspective of bank profitability, asset quality, the economic cycle, and credit

---

S. Yuan (✉)
China Center for Industrial Security Research, Beijing Jiaotong University, Beijing 100044, China
e-mail: yuanshaofeng1982@126.com

Y. Zeng
School of Finance & Public Administration, Shanghai Finance University, Shanghai 201209, China
e-mail: zengyun1202@126.com

W. Jing
School of Finance, Dongbei University of Finance and Economics, Liaoning 116025, China

M. Li et al. (eds.), *Proceedings of 2015 2nd International Conference on Industrial Economics System and Industrial Security Engineering*,
DOI 10.1007/978-981-287-655-3_45

activities, currently domestic scholars mainly use return on assets, the economic cycle, the NPL ratio, the size of banks, and other key indicators to study the effect on the capital-buffer of operating activities and the external environment of commercial banks, and then they propose the policy framework of building counter-cyclical capital-buffer mechanisms. Considering the complex interrelationships between bank capital and liquidity level intertwining with the external economic environment, liquidity risk will be one of the sources of bank crisis. This work studies the mechanism how capital buffer affects liquidity levels and thus provides reference to the perfection of commercial banks' counter-cyclical regulations.

2 Literature Review

With the application of innovative financial instruments, commercial banks' abilities of using financial tools to create liquidity are gradually enhanced. The relationship between liquidity-creation and capital buffer has been brought to the attention in many fields, and two competing opinions have been formed during the process:

- The hypothesis of financial-fragility crowding-out proposed by Diamond and Rajan [3, 4] suggested that commercial banks' enhancing abilities of liquidity creation could increase the degree of liquidity mismatch and the weight of risk assets (such as credit assets) and that this contributed to the negative correlations between bank capital and liquidity level.
- The risk-absorption hypothesis proposed by Allen and Gale [1] suggested that liquidity creation could improve the quality of credit assets and that this contributed to the positive correlations between bank capital and liquidity level.

Recently, based on the data of all commercial banks from 1993 to 2003, Berger and Bouwman [2] concluded that capital ratios of large commercial banks were positively correlated with liquidity creation, whereas capital ratios of small commercial banks were negatively correlated with liquidity creation. Fungacova et al. [5] found that high liquidity creation significantly increases the probability of bank failure.

Based on the view of studies of Berger and Bouwman [2], this work divides the liquidity level of commercial banks' balance sheets items into three categories—liquid, semiliquid, and illiquid—and give different weights to each category. We set the weight of the liquid assets to be 0.5, the weight of semi liquid assets to be 0, and the weight of illiquid assets to be −0.5. We can conclude that when liabilities are applied to financing for the same amount of illiquid assets, the liquidity creation can be calculated as $0.5*1 + 0.5*1 = 1$, and vice versa as −1. In addition, when liquid liabilities are applied to financing for liquid assets or illiquid liabilities are applied to financing for illiquid assets, no extra liquidity will be created. Considering the immature development of the securitization market, the incomplete data of off balance sheet business and many other factors, this work classifies housing mortgage loans, consumer loans, and some long-term assets as illiquid assets and ignores off balance sheet business. This work focuses on the relationship between capital buffers and liquidity levels of commercial banks; therefore, the indicator of the liquidity level (LC) should also serve as a controlling factor of the amount of capital buffers. In addition, as shown in previous studies, economic expectations, risk appetite, banks' debt structure, and loan demands also affect capital buffers. The final model is concluded as followed:

$$\begin{aligned} BUF_{it} = {} & \beta_0 BUF_{it-1} + \beta_1 ROE_{it} + \beta_2 NPL_{it} + \beta_3 ROA_{it} + \beta_4 LOAN_{it} \\ & + \beta_5 MSKD_{it} + \beta_6 GGDP_{it} + \beta_7 ASSET_{it} + \beta_8 LC_{it} + \mu_{it} \end{aligned} \tag{1}$$

We define BUF_{it} to be the capital buffer at year t, which is the part exceed the minimum capital requirement set by Basel New Capital Accord, and it is calculated by subtracting 8 % (the minimum capital-adequacy ratio set by Basel New Capital Accord) from the risk-weighted assets ratio (calculated by dividing risk assets by the total amount of capital held by banks); LC_{it} is set to be the liquidity level for bank i in year t; ROA_{it} is set to be the weighted average return on assets in year t; NPL_{it} is set to be the NPL ratio in year t; NPL = NPL/Gross Loan; ROE_{it} is set to be weighted average return on net assets in year t; $LOAN_{it}$ is set to be loan-to-assets ratio, which reflects the demand for loans, the operating model, and the main source of profits of banks; $MSKD_{it}$ is set to be market debt structure in year t, which reflects the funding structure of banks, MSKTD = (total liability-total deposit and short funding)/total liability; $ASSET_{it}$ is set to be the amount of total assets in year t of bank i, which reflects the scale of commercial banks, and ASSET = 1n (Total Asset); $GGDP_{it}$ is set to be the growth rate of domestic GDP indicating the situations of macroeconomic operations; and GGDPt = GDPt − GDPt − 1/GDPt; μ_{it} is set to be the constant amount of capital buffers and random error term.

3 Fundamental Models and Data

3.1 Research Methods

Considering the dynamic continuity of changes of capital buffers, explanatory variables on the right of the formula (1) include lags of dependent variables, thus leading to the interpretation of the explanatory variables and random disturbance, which result in the endogeneity of explanatory variables and (2) consider that estimated coefficients are biased and nonuniform. Therefore, to obtain consistent estimators and overcome endogeneity and heteroscedasticity of the model, we use appropriate instrumental variables (first-order lag endogenous variables) and GMM methods for estimation. In this work, dynamic panel models are established on STATA 10.0 statistical software. Because of the small size of cross-section sample data, short length of time series, and downward bias caused by two-step systematic GMM (generalized method of moments), we choose one-step GMM method for estimation. In the regression model (1), the liquidity indicator LC, the size variable ASSET, and the economic cycle variable GGDP are considered as exogenous variables and by themselves can be used as tools, whereas explanatory variables BUFt−1, ROAt, ROEt, Loan, NPLt, and MSKDt are considered as endogenous variables, and their first-order lag variables can be used as tools.

3.2 *Data and Descriptive Statistics*

Taking the availability of samples into account, this work calculates the capital-buffer level of 23 commercial banks from 2003 to 2013 including 4 large joint-equity commercial banks (ICBC CCB, and BCM); 9 small- and medium sized joint-equity commercial banks (CITIC, CEB, HXB, CMBC, CMB, SPDB, PAB, CIB, and GDB); and 10 city commercial banks (Bank of Beijing, Bank of Nanjing, Bank of Ningbo, Bank of Shanghai, Bank of Hangzhou, Bank of Dalian, Bank of Chongqing, Bank of Tianjin, Huishang Bank, and Hankou Bank).

As shown in Fig. 1, before 2008 excess capital-buffer levels of commercial banks continued to increase. However, after the global financial crisis in 2008, capital-buffer levels of commercial banks decreased significantly. Since then, with the slow recovery of the economy, capital-buffer levels have appeared to restore growth, but they showed signs of adjustment in 2013. However, in horizontal comparisons capital buffers of large commercial banks are the most stable, whereas capital buffers of city commercial banks are the most violate. Before 2007, liquidity-creation functions of large commercial banks affected that of joint-equity commercial and city commercial banks. Since 2008, liquidity-creation functions of

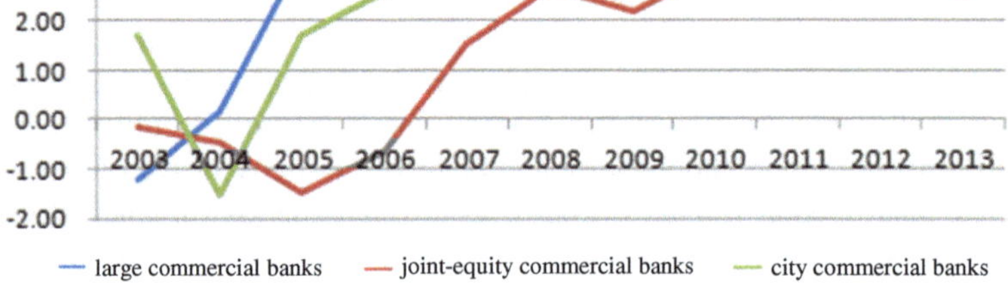

Fig. 1 Capital buffers of listed commercial banks from 2003 to 2013

city commercial banks have affected that of large commercial and joint-equity commercial banks.

4 Empirical Tests and Analysis

4.1 *The Impact of Liquidity-Creation Functions of Commercial Banks on Capital Buffers*

As shown in Table 1, bank capital buffers are significantly negatively correlated to the liquidity level of commercial banks because decreasing liquidity level helps to increase capital buffers, and that corresponds to the financial-fragility crowding-out hypothesis. Among them, the coefficients of liquidity creation is −0.0249, indicating that a 1 % increase in liquidity-creation level causes 0.025 % decrease in capital buffers; thus, the impact of liquidity expectations should be taken into considerations while calculating counter-cyclical capital buffers. Second, capital buffers have a certain "path dependence" property. In this work, we include a first-order lag capital buffer (buf (−1)) into the independent variables. The coefficient of this variable is 0.2072, the value of P is 0, and the coefficient of lag 1 capital buffer is significantly positive, thus indicating that capital-buffer level at an early stage has a significant impact on the current period. Third, the decreasing NPL ratio of commercial banks increases capital-buffer level. The coefficient of this is −0.234, thus indicating that 1 % increase in NPL ratio causes a 0.234 % decrease in capital buffers. Fourth, the improving profitability of commercial banks improves capital-buffer level, and 1 % increase in the weighted average return on assets (ROA) causes a 3.970 % increase in capital buffers. Fifth, the return of equity has less impact on capital buffers, whereas it has a higher significance level and a correlation coefficient of 0.0274. Sixth, due to economic capitals taken up by loans of commercial banks, the increase in loan assets decreases capital-buffer level. The correlation coefficient is −0.092, and that equals to a provision standard of commercial banks' core capital. In addition, the impact of the size of the commercial banks, debt structure, and economic growth are not significant on capital buffers.

Table 1 Estimation results of liquidity buffers of commercial banks

Explanatory variables	Coef.	P > |z|	Explanatory variables	Coef.	P > |z|
BUF(-1)	0.2072^{***}	0.00	MSKD	−0.0218	0.875
ROA	3.9702^{**}	0.026	GGDP	0.2064	0.116
ROE	0.0274^{***}	0.00	ASSET	−0.1819	0.568
LOAN	−0.0921**	0.038	LC	-0.0249^{***}	0.010
NPL	-0.2341^{*}	0.098	_CONS	5.1683	0.152

Note ***, **, and * represent that t values in the table are at 1, 5 and 10 % significance levels, respectively

4.2 The Impact of Commercial Banks' Qualities on Capital Buffers

To study whether the relationships between bank capital buffers and liquidity level differs with different natures of commercial banks, this work studies the asymmetry properties of commercial banks' capital buffers. As shown in Table 2, coefficients of the liquidity creation of large commercial banks are positive and are significant at the 1 % significance level, which corresponds to the risk-absorption theory. Coefficients of liquidity creation of joint-equity commercial and regional commercial banks are negative and are significant at the 5 and 10 % significance levels, which corresponds to the financial-fragility crowding-out hypothesis.

One possible reason for the above phenomenon is that since 2003 China has stepped into a "golden age" when the business scale has rapidly expanded, the quality of assets has been obviously improved, and the ability of earning profit has heightened substantially. During the process, large commercial banks based on large scales of branches have more stable deposits, and the quality of bank assets and the capital-sufficient ratio have improved after stripping nonperforming assets, injecting capital, and reforming joint-stock systems, all of which make the absorption effect override the crowding-out effect in the process of liquidity creation. As for shareholding and regional banks, their relatively limited business area, fewer branches, less stability of deposits, and lower degree of reliance on the refunding resource strengthen the liquidity level of the liabilities end. Meanwhile, more reliance on less liquid credit assets, such as loans at the assets end, will occupy the economic capitals of commercial banks, increase the liquidity risk of commercial banks, and makes the crowding out effect override the risk-absorption effect in the process of liquidity creation.

Table 2 Asymmetric regression results of relationships between capital buffers and liquidity level

	Large commercial banks	Joint-equity commercial banks	City commercial banks
Buf(-1)	0.0560(0.95)	−0.1701(−1.47)	−0.1788(−1.54)
ROA	3.8320(2.41)**	10.6642(3.70)***	10.8067(3.79)***
ROE	0.0349(1.14)	−0.3524(−2.67)***	−0.3655(−2.71)***
NPL	0.4176(1.73)***	−0.1443(−1.92)*	−0.2292(−1.07)
LOAN	−0.0405(−0.53)	−0.2200(−1.00)	−0.1544(−2.07)**
MSKD	0.1431(1.81)*	0.6811(2.89)***	0.7091(2.86)***
GGDP	0.2408(5.19)***	0.3750(1.76)*	0.3140(1.81)*
ASSET	−0.2989(-0.76)	0.0313(0.04)	0.0304(0.04)
LC	0.0289(2.66)***	−0.0381(−2.13)**	−0.0349(−1.93)*
_cons	0.9258(0.11)	2.9427(0.26)	3.3348(0.30)

Note ***, **, and * represent that t values in the table are at 1, 5 and 10 % significance levels

5 Conclusions and Policy Proposals

In this work, we use GMM estimation method to study the relationships between commercial banks' liquidity-creation level and capital buffers. As indicated herein, the capital-buffer level of commercial banks is pro-cyclical. The enhancing liquidity-creation function of large commercial banks helps improve the capital-buffer level. However, the enhancing liquidity-creation function of joint-equity commercial banks will decrease the capital-buffer level, and the liquidity-creation function of the entire financial system is negatively correlated with capital-buffer level. Therefore, based on the above-mentioned theories, first we should adjust the provision of risk capital to cope with different liquidity-creation levels, and increase the ability of commercial banks to deal with liquidity risk. Second, we should strengthen the implementation of dynamic counter-cyclical policies to alleviate the pro-cyclical property of commercial banks' operating activities.

Acknowledgments Supported by the colonel project "Characteristics, channels and effects of China's shadow banks" (2014J12) of Dongbei University of Finance and Economics.

References

1. Allen F, Gale D (2004) Financial fragility, liquidity, and asset prices. J Eur Econ Assoc 2 (6):1015–1048
2. Berger AN, Bouwman CH (2009) Bank liquidity creation. Rev Financ Stud 22(9):3779–3837
3. Diamond DW, Rajan RG (2001) Banks short-term debt and financial crises: theory, policy implications and applications. Carnegie-Rochester Conference Series on Public Policy. North-Holland 54(1):37–71
4. Diamond DW, Rajan RG (2001) Liquidity risk, liquidity creation, and financial fragility: a theory of banking. J Polit Econ 109:287–327
5. Fungacova Z, Turk R, Weill L (2015) High liquidity creation and bank failures. IMF Working Paper: No. 15/103

The Impact of Interest Rates on Bank Risk-Taking: A Comparison Between State-Owned Banks and Joint-Stock Banks in China

Shun-Ho Chu, Wenjie Yang and Jie Hu

Abstract Since the global financial crisis, commercial banks have highlighted concern about risk management. In 2013, People's Bank of China deregulated the limitation on bank lending rates, which implies the emergence of a market-oriented interest rate policy. The study aims to investigate the effect of interest rates on bank risk-taking. The empirical results of this study show that interest rates have a relation with bank risk-taking. The short-term interest rate has a negative relation with total risk, idiosyncratic risk, and default risk, but not systematic risk, whereas the long-term interest rate has a positive relation with total risk, idiosyncratic risk, and default risk, but not systematic risk. This study found that state-owned banks are negatively correlated with total risk, systematic risk, and default risk but have idiosyncratic risk compared with joint-stock banks.

Keywords Total risk · Systematic risk · Idiosyncratic risk

1 Introduction

The major risk commercial banks faced in accepting deposits and extending lending is the interest rate risk resulting from the mismatch n the maturities of the deposits and lending [1]. Bank decisions relating to interest rates rely heavily on monetary policy and supervision of banks [2]. Expansionary monetary policy, implying lower interest rates, will increase bank risk-taking and vice versa [3]. Previous studies of

S.-H. Chu (✉) · W. Yang · J. Hu
School of Business, Macau University of Science and Technology, Macau, China
e-mail: shchu@must.edu.mo

W. Yang
e-mail: yangwenjiecarol@foxmail.com

J. Hu
e-mail: sophia12183636@sina.com

M. Li et al. (eds.), *Proceedings of 2015 2nd International Conference on Industrial Economics System and Industrial Security Engineering*,
DOI 10.1007/978-981-287-655-3_46

Ali and Daly, Delis and Kouretas, and Zribi and Boujelbène showed that interest-rate changes have negatively affected bank risk-taking [2, 4, 5].

After the global financial crisis of 2007 to 2008, economic growth has dramatically increased in China. Macroeconomic factors have significantly affected risk-taking by the bank sector. Some macroeconomic factors positively affect bank risk-taking, whereas some have a negative impact on bank risk-taking. In recent years, China has aggressively put forth the market-oriented interest rate policy. People's Bank of China conducted monetary policy to raise interest rates by an increase in the deposit-reserve ratio for mitigating soaring housing prices and the overheating real estate market. 2013, People's Bank of China deregulated the interest rate of lending to allow banks to decide the lending rate independently. This may improve bank performance in lending markets, but it may result in an increase in risk-taking by banks due to more competition in lending markets. This causes banking sectors to focus on interest-rate risk and well-prepared soundness of bank operations.

The work aims to investigate the impact of interest rates on risk-taking. It also forms the basis of a comparative analysis of two groups of banks, including four state-owned commercial banks and eight joint-stock commercial banks. Thus, the study tries to examine the level of risk-taking affected by interest rates between the two groups: state-owned banks and stock-joint banks. The remainder of this work is organized as follows. Section 2 contains a review of literature and hypothesis. Section 3 presents data and methodology. The model estimation and a discussion of results are explained in Sect. 4. The final section concludes the work.

2 Review of Literature

Interest rate is the proxy of earnings and costs of funding for banks. Lower interest rate reducing profit margins drive banks to search for borrowers with higher credit risk, which results in greater risk-taking [2, 6]. Conversely, a higher interest rate poses an incentive to banks to extend much more lending, which may lead to an increase in credit risk due to a problem of asymmetric information [7]. In general, interest-rate risk is incurred by banks as the maturities of assets and liabilities are mismatched and the interest rate changes [8].

Previous empirical studies show that interest rate is negatively related to bank risks. Zribi and Boujelbène found that interest rate has a significantly negative relation with bank credit risks in the case of Tunisia [5]. Vogiazas and Nikolaidou found that short-term interest rate has a significant and negative relation with bank credit risks in the case of the Romanian banking system [9]. Delis and Kouretas found that the effect of interest rate on risk-taking for banks with higher equity capital is diminished but is amplified for banks with higher off balance sheet business [2]. Some studies have argued that interest rate is related to bank risks. Castro found that the real interest rate has a significant and positive relation with bank credit risks in the case of GIPSI [7]. However, Poudel stated that interbank rate does not have a significantly positive relation with credit risk in the Nepalese banking industry [10].

3 Data and Methodology

Data were collected from the websites of commercial banks in China listed on the Shanghai Stock Exchange for analysis based on the criteria of a -panel data approach. Annual data from 2008 to 2013 were used and yielded a total of 72 observations. The study selects 12 banks from 14 listed commercial banks, except for China Everbright Bank and Agricultural Bank of China due to their stock market data commencing from the year of 2010. The study classifies 12 banks as 2 groups of for state-owned commercial banks and 8 joint-stock commercial banks.

Referred to the study of Lepetit et al. this study employs market-based data to measure bank risks and insolvency risk [11]. The capital asset pricing model (CAPM) is frequently employed to measure risk-taking in empirical studies [11, 12]. Following the single-factor CAPM approach, introduced by Sharpe, is shown in Eq. (1) as follows [13]:

$$R_t = \alpha + \beta_m(R_m) + \varepsilon_t \tag{1}$$

where t expresses time, and α is a constant that varies across banks. Equation (1) regresses the return on each bank, R_t and on a market return, R_m. The beta coefficient of R_m is used as a proxy for market risk. The standard deviation of the monthly return of stock on each bank is expressed as a proxy of total risk where idiosyncratic risk is the difference between total risk and systematic risk [14]. The additional risk is insolvency risk as a proxy of Z-score (Z value) calculated by the following Eq. (2), which used to measure the possibility of bank default [11].

$$Z_t = \frac{\bar{R}_t + 1}{\sigma_t} \tag{2}$$

where Z_t is to measure the possibility of failure on each bank at time t, while $\overline{R_t}$ and σ_t are the mean and standard deviation, respectively, of monthly stock return on each bank at time t.

Taking reference to the empirical-regression model used in several studies by Gallo, Apilado, and Kolari, Lepetit et al. and Haq and Heaney, the study proxies total risk (TR), market risk (MR), idiosyncratic risk (IR), and Z-score (Z) to measure bank risk-taking using the CAPM approach and Z-score method [11, 12, 14]. Second, this study employs short-, medium, and long-term interest rate as a proxy of the independent variable with reference to the study of Delis and Kouretas [2].

In addition to the dependent and independent variables, this study also adds micro-level and macro-level control variables. The micro-level variables are capital adequacy ratio, core capital-adequacy, and ratio of loans to deposits, which are related to capital regulations performed by the banking authority. The micro-level variables consist of GDP, CPI, foreign exchange rate, and money supply.

The cross-section time-series regression model of this study, shown in Eq. (3) to examine the impact of interest rates on bank risk-taking, is as follows:

$$\begin{aligned} Risk_{it} = & \alpha_{it} + \beta_1(SI_t) + \beta_2(MI_t) + \beta_3(LI_t) + \beta_4(CAR_{it}) + \beta_5(CCAR_{it}) + \beta_6(LDR_{it}) \\ & + \beta_7(GDP_t) + \beta_8(CPI_t) + \beta_9(MS_t) + \beta_{10}(D_{it}) + \varepsilon_{it} \end{aligned} \quad (3)$$

where the dependent variables of $Risk_{it}$ express total risk, market risk, idiosyncratic risk, and insolvency risk of each bank I at time t, which are denoted in TR_{it}, MR_{u}, IR_{it} and Z_{it}, respectively. Short-term interest rate, SI_t, is the average monthly rates of the 90-day interbank rate. Medium-term interest rate, MI_t, is denoted in the average monthly yield rates of 5-year government bonds, while long-term interest rate, LI_t, is denoted in the average monthly yield rates of 10-year government bonds. CAR_{it} is a proxy of the capital-adequacy ratio; $CCAR_{it}$ is the core capital-adequacy ratio; and LDR_{it} is the ratio of loans to deposits. GDP_t denotes the growth rate of GDP; CPL_t denotes the growth rate of the consumer price index; and MS_tMS denotes changes in the money supply in terms of M2. D_{it} is a dummy variable. $D_{it} = 1$ if they are state-owned banks; ε_{it} denotes value of error.

4 Empirical Results

To understand the basic characteristics of all of the variables, the study outlines descriptive statistics for the dependent, independent, and control variables as shown on Table 1. Correlation analysis among all variables of the study shows that the term interest rate variable is highly correlated to short- and long-term interest-rate variables. Second, the variables of capital-adequacy ratio and core capital-adequacy ratio have high correlation. Hence, the study tends to delete two variables, medium-term interest rate and core capital-adequacy ratio, in the estimation equation.

According to the results of Hausman test, the study selects a random effect model in all regression models. Table 2 presents the empirical results of OSL regression analysis for four models.

Through statistically empirical analysis, the short-term interest rate has a significantly negative relationship with total risk and idiosyncratic risk, whereas it has a significantly positive relationship with total risk. That short-term interest rate has positive correlation with Z-value implying that the short-term interest rate has a negative correlation with default risk. Second, the long-term interest rate is

Table 1 Descriptive statistics for selected variables

Variables	Maximum	Minimum	Mean	St. deviation
Total risk (TR)	0.176	0.033	0.097	0.038
Systematic risk (MR)	1.857	−0.034	0.860	0.507
Idiosyncratic risk (IR)	0.126	−1.716	−0.763	0.495
Z-score (Z)	30.101	5.954	12.423	6.129
Short-term interest rate (SI)	0.053	0.017	0.039	0.013
Medium-term interest rate (MI)	0.037	0.026	0.031	0.004
Long-term interest rate (LI)	0.039	0.034	0.036	0.002
Capital-adequacy ratio (CAR)	0.241	0.091	0.124	0.021
Core capital-adequacy ratio (CCAR)	0.207	0.050	0.096	0.022
Ratio of loans to deposits (LDR)	0.800	0.564	0.684	0.056
Gross domestic production (GDP)	0.104	0.077	0.090	0.010
Consumer price index (CPI)	0.059	−0.007	0.032	0.022
Money supply (MS)	0.284	0.136	0.184	0.049
Dummy (D)	1.000	0.000	0.333	0.474

Table 2 Empirical results of OSL regression

Variables	TR	SR	IR	Z
C	0.682*** (6.686)	−4.137*** (−2.898)	4.824*** (3.564)	−100.639*** (−5.851)
SI	−8.732*** (−5.544)	108.189*** (4.844)	−116.948*** (−5.507)	1566.537*** (5.868)
LI	9.648*** (3.016)	−135.417*** (−2.984)	144.987*** (3.361)	−1647.900*** (−3.039)
CAR	−0.167 (−1.119)	−1.840 (−0.912)	1.630 (0.858)	39.291 (1.581)
LDR	−0.085 (−1.294)	−0.627 (−0.761)	0.550 (0.717)	22.465** (2.119)
GDP	−7.717*** (−6.000)	110.082*** (6.038)	−117.842*** (−6.799)	1283.782*** (5.890)
CPI	4.091*** (5.878)	−73.897*** (−7.504)	78.022*** (8.337)	−751.108*** (−6.068)
MS	0.351*** (4.547)	−6.092*** (−4.340)	6.446*** (4.831)	−23.0109 (−1.372)
D	−0.040*** (−3.568)	−0.373*** (−3.358)	0.334*** (3.332)	7.292*** (4.465)
R-squared	0.689	0.731	0.753	0.648
Adjusted R-squared	0.650	0.697	0.722	0.604
F statistic	17.470***	21.448***	24.022***	14.513***
Durbin-Watson stat	2.191	2.033	2.008	1.396

Note t-statistics are reported in parenthesis. * Denotes significance at the level of 10 %; ** denotes significance at the level of 5 %; *** denotes significance at the level of 1 %

positively and significantly correlated to total risk and idiosyncratic risk, whereas it is negatively and significantly correlated to systematic risk. The negative correlation between long-term interest rate and Z-value shows that the long-term interest rate is positively and significantly correlated to default risk.

The study also examines the impact of regulatory-level variables on bank risk-taking in the estimation equation. The ratio of loans to deposits has only a positive correlation with Z-value, which indicates that the ratio of loans to deposits has a negative impact on default risk.

Furthermore, the study investigates the effect of macroeconomic-level variables on bank risk-taking in the estimation equation. GDP is negatively and significantly correlated to total risk, idiosyncratic risk, and default risk, whereas it is positively and significantly correlated to systematic risk. CPI and MS have a positive and significant correlation with systematic risk, whereas they have a negative and significant correlation with total risk, idiosyncratic risk, and default risk.

In addition, the study analyzes the influence of interest rate on risk-taking based on a comparison between state-owned and joint-stock banks. State-owned banks have significantly negative correlation with total risk, systematic risk, and default risk, whereas they have a significantly negative correlation with idiosyncratic risk.

Finally, the empirical results find that state-owned banks are more sensitive to interest rate than joint-stock banks. State-owned banks have large-sized asset portfolios. Moreover, they may have off balance sheet activities, including credit, interest rate, and foreign exchange derivatives, which may potentially heighten their risk exposure.

5 Conclusion

The purpose of this study was to investigate the impact of interest rates on bank risk-taking by using a sample of 12 listed commercial banks listed on the Shanghai Stock Exchange spanning the time period from 2008 to 2013. The empirical results of this study show that interest rates have a relation with bank risk-taking. The short-term interest rate has a negative relation with total risk, idiosyncratic risk, and default risk, but not systematic risk, whereas the long-term interest rate has a positive relation with total risk, idiosyncratic risk, and default risk, but not systematic risk. This implies that an increase in the short-term interest rate and a decrease in the long-term interest rate will increase systematic risk.

After the deregulation of the bank-lending interest rate conducted by People's Bank of China, commercial banks should pay more attention on interest-rate risk management. Thus, banks should have a strong capability to measure and manage risks derived from volatility of the interest rate.

References

1. Mishkin FS (2014) The economics of money, banking, and financial markets, 10th global edition. Pearson Education Limited, Harlow, pp 285–286 (English)
2. Delis MD, Kouretas GP (2011) Interest rates and bank risk taking. J Bank Financ 35:840–855
3. Bario C, Zhu H (2008) Capital regulation, risk-taking and monetary policy: a missing link in the transmission mechanism? BIS Working Paper No. 268
4. Ali A, Daly K (2010) Macroeconomic determinants of credit risk: recent evidence from a cross country study. Int Rev Financ Anal 19(3):165–171
5. Zribi N, Boujebène Y (2011) The factors influencing bank credit risk: the case of Tunisia. J Account Tax 3(4):70–78
6. Apel M, Claussen CA (2012) Monetary Policy, Interest Rates and Risk-taking. Sveriges Riksbank Econ Rev 2:68–83
7. Castro V (2013) Macroeconomic determinants of the credit risk in the banking system: the case of the GIPSI. Econ Model 31:672–683
8. Saunders A, Cornett MM (2012) Financial markets and institutions, 5th edn. McGraw-Hill Companies, Inc., New York, pp 580–582
9. Vogiazas SD, Nikolaidou E (2011) Investigating the determinants of nonperforming loans in the romanian banking system: an empirical study with reference to the Greek crisis. Econ Res Int 85:1–13
10. Poudel R (2013) Macroeconomic Determinants of credit risk in nepalese banking industry. In: 21st international business research conference
11. Lepetit L, Nys E, Rous P, Tarazi A (2008) Bank income structure and risk: an empirical analysis of European banks. J Bank Financ 32:1452–1467
12. Haq M, Heaney R (2012) Factors determining European bank risk. J Int Financ Mark Inst Money 22(4):696–718
13. Sharpe WF (1964) Capital asset prices: a theory of market equilibrium under conditions of risk. J Financ 19(3):425–442
14. Gallo JG, Apilado VP, Kolari JW (1996) Commercial bank mutual fund activities: implications for bank risk and profitability. J Bank Financ 20:1775–1791

Potential and Challenges of Carbon Finance Development in China

Xiaoming Hu

Abstract The Chinese central government determined that the focus of future economic work should be on the adjustment of economic structure and the transformation of development mode. The comprehensive transition to a low-carbon economy would be the main direction of China's future economic development. Finance, an important means and approach for the optimization of resource allocation and redistribution of funds, will definitely play a major role in the development of a low-carbon economy in China.

Keywords Carbon finance · Low carbon · Challenges

1 Introduction

The Chinese government attaches great importance to the development of a low-carbon economy. At the Central Economic Working Conference held in December 2009, the central government determined that the focus of future economic work should be on the adjustment of economic structure, and the transformation of development mode. The comprehensive transition to a low-carbon economy would be the main direction of China's future economic development. As the "low carbon" concept gradually takes root, the low-carbon economy has begun to produce benefits. However, the development and rapid growth of the low-carbon economy and industry face barriers that prevent the flow of finances and technology into this sector. As technology could not be developed and applied without the support of funds, funds form the core for the development of a low-carbon economy in China. Finance, an important tool for the optimization of resource allocation and

X. Hu (✉)
Beijing Jiaotong University China Center for Industrial Security Research, Beijing 100044, China
e-mail: huxiaoing@sino-carbon.cn; 48516100@qq.com

M. Li et al. (eds.), *Proceedings of 2015 2nd International Conference on Industrial Economics System and Industrial Security Engineering*,
DOI 10.1007/978-981-287-655-3_47

redistribution of funds, will definitely play a major role in the development of a low-carbon economy in China.

2 Development Potential of Carbon Market

China attaches great importance to addressing climate change, and has incorporated it into the national economic and social development plan as a major national strategy. It has also taken greenhouse gas emissions control as an important handle in its economic restructuring and transformation of development pattern. A full-fledged mechanism for addressing climate change, and a policy system with specific objectives, is gradually taking shape.

At present, China has set a greenhouse gas emissions target according to which, by 2020, the carbon dioxide emissions per unit of GDP will see a 40–45 % reduction from that of 2005 [1, 2]. In the 12th Five-Year Plan, it has been further clarified that, by 2015, the phased target of a 17 % reduction will be achieved. To achieve these goals, the Chinese government is trying to use the invisible hand of market, and to bring into play the fundamental role of the market mechanism in the resource allocation of the greenhouse gas control, instead of solely relying on administrative leverage.

In addition to facilitating the registration of domestic CDM projects before 2012, and their issuance after the beginning of 2012 while also encouraging voluntary carbon emissions reduction trading, China is making great efforts to promote the building of a unified national carbon market. Local pilot projects of carbon trading have been initiated by the National Development and Reform Commission, and a lot of preparatory work has been done on the national scale.

In October 2011, the National Development and Reform Commission (the NDRC) released the Notice on Launching Pilot Emissions Trading System (NDRC Climate Office [2011] No. 2601), according to which Beijing, Tianjin, Shanghai, Chongqing, Guangdong, Hubei and Shenzhen would launch pilot systems from 2013 through 2015. It asserts that China, as the world's largest GHG emitter, has officially kicked off its Emissions Trading System (ETS). The Chinese government hopes to get its GHG emissions under control through building a market scheme that sets a price on emissions, and incentivizes emissions reduction. The pilot ETS in seven regions, namely Shenzhen, Shanghai, Beijing, Guangdong, Tianjin, Hubei and Chongqing, became operational during June 2013 and June 2014. The total emissions of the seven pilot schemes combined exceed 1 billion tons, which makes China the largest country in terms of covered emissions, and second only to EU ETS in terms of market size. The initiation of ETS in China has had a positive effect on the development of the global emissions market and the efforts to cut GHG emissions. China is now speeding up the building of a national-level emissions trading market based on its pilot schemes.

3 Development Potential of Carbon Finance Infrastructure

In practice, it is very important to build the carbon finance infrastructure. Generally speaking, the financial infrastructure refers to the hardware facilities and institutions in the financial operation, which are all essential to its sustainable development. The carbon finance infrastructure includes three main systems, namely the market trading system, the payment and settlement system, and the support and safeguard system. As for China, all of the systems have been gradually taking shape in its initial phase of carbon finance infrastructure development.

3.1 Carbon Trading System

With the looming of the deadline of the first commitment period of the Kyoto Protocol, which serves as the legal basis of CDM, the destiny of CDM is still uncertain. The EU became the first one to announce that it would only accept CDM projects from the least developed countries, casting great uncertainty on the existing CDM mechanism. As the largest developer of global CDM projects, and the largest supplier of CERs, China must blaze a new trail for the existing and future CDM projects, and the establishment of a domestic carbon trading system is undoubtedly one of the most feasible options [3, 4].

In recent years, the Chinese government has developed clearer concepts of the carbon market and its trading methods. The current status of carbon trading systems in developed countries has also helped China set a goal for understanding and planning its own trading system. The upcoming carbon trading system in China may take the experiences and lessons of existing foreign carbon trading systems, but they can be based on China's own national conditions. Relying first on the carbon spot trading based on total quantity control and allowance allocation, China would also possibly develop other trading products, such as carbon futures and derivatives, and would eventually establish a carbon finance market with deep involvement of its mainstream financial institutions and emission entities, connecting it with the international carbon market.

3.2 Payment and Settlement System

In China, although a mandatory carbon emissions trading system is yet to come, the voluntary carbon trading mechanism and the dedicated platform was launched as early as the 2010 World Expo. The World Expo Voluntary Emissions Reduction (VER) trading platform, created by the Shanghai Environment and Energy Exchange, is the first of its kind in China. It consists of multiple systems, including login, information publishing and search, trading, payment, settlement, and registration. It also provides integrated services, such as information publishing, event marketing, emissions reduction trading, clearing, and settlement. Funds from the purchasing enterprises can be transferred through a special settlement account and be paid to the owner of the voluntary carbon emissions reduction.

In comparison with the carbon trading and settlement systems in developed countries, especially the European Union, the trading system in China has yet to develop, and it is far from perfect. However, China is actively working to establish a national carbon trading registration system, and is likely to further build a central settlement system similar to that of the stock market, and a supporting central registered custodian bank for carbon trading. In the future, the carbon trading payment system in China is likely to fall in line with those of the EU and other countries, and its overall structure would be similar to the payment system of the stock market, with banks and clearing institutions playing a role similar to those in other carbon trading systems.

3.3 Support and Safeguard System

Support and safeguard systems provide guarantees for financial security and efficient cash flow, and consist of a credit system, science and technology, background support, information availability, rules, and regulations. Rules and regulations are not only an important part of the support and safeguard system, but also provide a legal basis for the establishment of a carbon market and policy certainty to market participants. Thus, the establishment and improvement of relevant rules, regulations, and policies remain a top priority for China in the development of its carbon market and carbon finance.

Currently, China has yet to pass a law or a regulation to establish the national carbon market. All seven pilots regions are preparing to pass local legislations or regulations for advancing local pilot ETS programs. In addition, China has already passed many relevant laws, regulations and policy documents that can also contribute to the creation of an ambient environment for the establishment of a carbon market.

Further, a series of local legislative initiatives and administrative measures in support of establishing carbon-trading mechanisms have also been taking place in the seven pilot provinces and cities. These might include basic rules for the establishment and management of emission trading systems, supportive technical rules, such as emission monitoring and reporting guidelines, trading rules, and market supervision rules, etc.

The carbon market in China is still under construction, and not even in its infancy, so the rules and regulations in this respect would have to be gradually formulated and improved together, along with the development of the market. In the future, national law or regulations establishing a unified carbon emissions trading program would be in place providing both a legal basis and a guarantee for the normal and smooth operation of carbon market and carbon finance in China.

4 Challenges

4.1 China Is Still to Develop a National Carbon Market

Currently, China only gets its limited experience with the carbon market in two ways. The first one is through participation in the international carbon market through CDM as the host country, and by selling CERs to developed countries to help them meet their emissions reduction obligations under the Kyoto Protocol. The second one is the voluntary emissions reduction trading. Enterprises, driven by the need to fulfill corporate social responsibilities or promote public image, voluntarily buy carbon-offsetting credits to offset the carbon emissions produced by their own production and business activities. In 2011, China decided to establish its own domestic carbon market, starting from the seven pilots programs. Preparatory work for the launching of the pilots is still under way, and although all the pilots would start their early operation in 2013, there is still a lot of work to be finished.

Due to limited experience with the carbon market, public awareness of the concept of carbon market in China is weak. Enterprises often mistakenly think that a domestic carbon market is equivalent to a CDM Market. This is due to the fact that China mainly gets its experience with the carbon market from CDM, and about 90 % of China's CDM projects have been sold to international buyers. A voluntary carbon trading market is emerging in China, but it is still a very small and immature market. In addition, due to the lack of knowledge of the carbon market, including the project-based trading process and strategic positions, only a few commercial banks get involved in carbon finance and relevant financial activities, with few other Chinese financial institution participating.

Taking China as a whole, the invisible hand of market rarely plays a role in addressing environmental problems, such as energy conservation and emissions reduction. Its financial entities, such as investment banks, private equity funds, exchanges, etc., are not really participators. Due to the current state of the domestic carbon market and the lack of experience of the government, enterprises and financial institutions, China faces a great challenge in building its carbon finance system based on the carbon market. The future carbon market in China will directly depend on the performance and experiences of pilot projects. There is still a long way to go in developing a national unified carbon market in China.

Based on international experience, the development of carbon finance cannot be separated from a mature carbon market. Thus, the construction and development of a national carbon trading market in China is the precondition for carbon finance development.

4.2 The Domestic Financial System Is yet to Develop

The financial system in China is not mature. The development of carbon finance is inevitably based on the financial system, and a wide range of financial instruments and derivatives are needed in carbon finance to develop a complete financial environment. But the development of carbon finance in China is especially blocked by insufficient innovation and rigid supervision of the Chinese financial system.

China seems to be a financial powerhouse in terms of "hardware", such as its capital strength, the number of financial institutions and practitioners, its foreign currency reserve, etc. However, when referring to financial innovation, the development level of the financial system, financial institutions governance mechanism, financial social environment, financial supervision capacity, degree of financial internationalization, and the financial talents allocation, China is still far from a "financial powerhouse" China is especially short of financial innovation that can promote the healthy and rapid development of the financial sector and the economy. Financial innovation is closely linked to the demand of economic development, the ever-changing external environment, and its own advantages, in order to promote the optimal allocation of financial resources, and enhance the core competitiveness of financial institutions. However, China's current financial institution system is dominated primarily by financial institutions that are large and all-inclusive, making it difficult to respond in a timely manner to the financial services in emerging industries, such as carbon trading, that are flexible, diverse and constantly changing. In addition, the financial supervision in every country has somehow been intensified since the recent global financial crisis, which also constrains the financial innovation in reality.

4.3 China Is Still to Develop a National Carbon Market

Carbon finance is new because it has only come into China since the emergence of the international carbon trading market in 2005. Therefore, due to lack of practice and experience, many domestic financial institutions do not fully understand the concept of carbon finance. They are unfamiliar with the profit margins of carbon finance, its mode of operation, risk management, operating methods, as well as its underlying carbon market operation and project development procedures. Commercial banks normally face not only the risks of the traditional financial market, but also the uncertainty of the carbon market and unfamiliar business models, so financial institutions are wary of engaging without a full understanding of it. Except for a small number of commercial banks, other financial institutions rarely set foot in carbon finance. Due to the requirements for financial institutions by national policy, and also the relatively simple business modes, Chinese financial industry is mainly starting to get focused on "Green-credit" without going deeper. So far, there has not been a mature carbon trading system, regulations, or a platform in China, and it is also short of the supports of carbon-swap trading, carbon securities, carbon futures, carbon fund, and other carbon finance derivatives and services.

References

1. X Yu (2010) The confusion and ideas of the low carbon economy investment and financing in China. Seeker 7:64–65
2. Wagner G (2009) Docking into a global carbon market: clean investment budgets to finance low-carbon economic development (March 17' 2009). The economics and politics of climate change. Oxford University Press, Oxford
3. Pacudan B (2005) The clean development mechanism: new instrument in financing renewable energy technologies. Springer, Netherlands
4. Najam A (2002) Financing sustainable development: crises of legitimacy. Prog Dev Stud 2:153–160

Part VII
Special Session on International Economics and Trade

A Comparative Study on International Competitiveness of Iron and Steel Industry of Four Asian Countries

Weida He and Zhifeng Lin

Abstract The iron and steel industry occupies a significant part of the national economy. In this paper, multi-level grey correlation analysis method is applied to evaluate and analyze the international competitiveness of the iron and steel industry among four Asian iron and steel producing countries: China, Japan, South Korea, and India, by cross-sectional data of the years 2003, 2007, and 2012. The research results show that Japan has the strongest international competitiveness among the four countries, followed by China and South Korea, while India is the least competitive country. Production capacity and the economic environment of China are in a state of competitive advantage, while market performance and the technology level of China are in a state of competitive disadvantage.

Keywords Industrial economy · Iron and steel industry · International competitiveness

1 Introduction

Since 1996, China has become the largest exporter of iron and steel in the world market, and this industry has maintained high-speed development for many years. But, with respect to international competitiveness, China falls into a dilemma. On the one hand, added value of export products is not improved with increasing exports, and the whole industry is trapped in a low level of international division of labor. On the other hand, China depends heavily on imported iron ore, thus it never masters the pricing power in international iron ore trades. Meanwhile, China's

W. He · Z. Lin (✉)
Donlinks School of Economics and Management, University of Science and Technology Beijing, Beijing 100083, China
e-mail: linzhifeng22@126.com

W. He
e-mail: hewd@ustb.edu.cn

M. Li et al. (eds.), *Proceedings of 2015 2nd International Conference on Industrial Economics System and Industrial Security Engineering*,
DOI 10.1007/978-981-287-655-3_48

neighbors, Japan, South Korea, and India are also superpowers and strong competitors of China in the international iron and steel production-market.

In the field of research on international competitiveness, Porter [1] proposed the famous Diamond model that analyzed international competitive advantage from six aspects. British scholar Dunning [2] and Chinese scholar Rui [3] respectively added "multinational business" and "knowledge absorption and innovation ability" into Porter's diamond model. Represented by Schwab [4] and Garelli [5], the World Economic Forum and International Institute for Management Development together put forward a competitiveness equation considering both competitive assets and processes. In the research on the competitiveness of iron and steel industry, Sahoo [6] used DEA method and translog production function to calculate scale efficiency and technical efficiency of iron and steel enterprises. Movsguk [7] adopted panel data and a stochastic frontier model to assess performance of China's iron and steel companies. He and others [8] framed a three-double model to evaluate the competitiveness of China's iron and steel industry. But, so far, there is little literature adopting comprehensive evaluation methods to comparatively study international competitiveness of major iron and steel countries in Asia. Thus, this paper adopts multi-level grey correlation analysis method to evaluate international competitiveness of the iron and steel industry of four Asian countries.

2 Evaluation Index System and Model

Referring to the existing research and considering industrial characteristics, we establish an evaluation index system as shown in Table 1. It consists of four level 1 indexes and twelve level 2 indexes.

Table 1 Evaluation index system

Level 1 index	Level 2 index
Production capacity A1	Annual crude steel output B1
	Labor costs B2
	Industrial labor productivity B3
	Industrial concentration B4
Technology level A2	Continuous casting ratio B5
	Annual patent application quantity B6
Market performance A3	International market share B7
	Revealed comparative advantage (RCA) B8
	Trade competitiveness B9
	Industrial sales profit margin B10
Economic environment A4	Per capita GDP B11
	DGP annual growth rate B12

In this paper, grey correlation analysis method is adopted for the construction of a comprehensive evaluation model. First of all, reference sequence V_0 and comparison sequence V_i should be determined. V_0 consists of the best observed value of each index, and V_i is composed of all index values of a country, shown as formula (1) and (2). i represents evaluated country, $i = 1, 2...n$ and k represents evaluation index, $k = 1, 2...m$. Thus V_{ki} is the observed value of country i at index k. All observed values of n countries at m evaluated indexes constitute a matrix V, as shown in formula (3).

$$V_0 = (V_{10}, V_{20}, ...V_{m0}),\ V_{k0} = optimum(V_{ki}) \quad (1)$$

$$V_i = (V_{1i}, V_{2i}..., V_{mi}) \quad (2)$$

$$V = (V_{ki})_{m*n} = \begin{bmatrix} V_{11} & V_{12} & \cdots & V_{1n} \\ V_{21} & V_{22} & \cdots & V_{2n} \\ \cdots & \cdots & \cdots & \cdots \\ V_{m1} & V_{m2} & \cdots & V_{mn} \end{bmatrix} \quad (3)$$

We choose min-max normalization to transform all data into dimensionless as formula (4) and (5) show. Then we can get normalized reference sequence X_0, comparison sequence X_i, and a data value matrix X.

$$X_{ki} = \frac{V_{ki} - \min\limits_i (V_{ki})}{\max\limits_i (V_{ki}) - \min\limits_i (V_{ki})} \quad (4)$$

$$X_{ki} = \frac{\max\limits_i (V_{ki}) - V_{ki}}{\max\limits_i (V_{ki}) - \min\limits_i (V_{ki})} \quad (5)$$

After all data have been normalized, correlation coefficient can be calculated by formula (6), where ρ is a distinguishing coefficient, and it is usually chosen of 0.5. Then we can obtain correlation coefficients matrix E.

$$\gamma_{ki} = \frac{\min\limits_k \min\limits_i |X_{k0} - X_{ki}| + \rho \max\limits_k \max\limits_i |X_{k0} - X_{ki}|}{\left|X_{k0} - X_{ki}\right| + \rho \max\limits_k \max\limits_i |X_{k0} - X_{ki}|} \quad (6)$$

Variation coefficient method is used to determine index weight. First, we calculate the mean $\overline{Z_k}$, standard deviation S_k, and variation coefficient I_k of each evaluation index. I_k can be calculated by formula (7). Finally, evaluation index weight W_k is equal to the proportion I_k taking up in the sum of all index variation coefficients, as shown in formula (8).

$$I_k = \frac{S_k}{\overline{\overline{Z_k}}} \tag{7}$$

$$W_k = \frac{I_k}{\sum_{k=1}^{m} I_k} \tag{8}$$

Index weight multiplied by variation coefficient is equal to grey correlation degree, as formula (9) presents. For multiplied-level index system in this paper, repeat the above processes to get the final grey correlation degree.

$$R = (r_1, r_2, \ldots r_t)_{1*n} = WE \tag{9}$$

3 Results and Analysis

In this paper, we calculate grey correlation degree by cross-sectional data of the years 2003, 2007, and 2012 and then rank international competitiveness among China, Japan, South Korea, and India by correlation degree value. The following Tables 2, 3 and 4, show the evaluation results. We find that the ranking of the four countries' international competitiveness of iron and steel industry fluctuated slightly in 2003, 2007, and 2012. The first place was always occupied by Japan. India was always in the last position. Since 2007, China's ranking rose and overtook South

Table 2 Grey correlation degree and ranking in 2003

Index	China		Japan		South Korea		India	
	Correlation degree	Ranking	Correlation degree	Ranking	Correlation degree	Ranking	Correlation degree	Ranking
A1	0.6860	2	0.5512	3	0.6932	1	0.5225	4
A2	0.5053	3	0.9341	2	1.0000	1	0.3333	4
A3	0.3452	4	0.8023	1	0.6495	2	0.6172	3
A4	0.6269	2	0.7091	1	0.4152	4	0.4798	3
Final	0.5096	3	0.8048	1	0.7472	2	0.4634	4

Table 3 Grey correlation degree and ranking in 2007

Index	China		Japan		South Korea		India	
	Correlation degree	Ranking	Correlation degree	Ranking	Correlation degree	Ranking	Correlation degree	Ranking
A1	0.7531	1	0.4850	4	0.6017	2	0.5799	3
A2	0.5790	3	1.0000	1	0.6602	2	0.3333	4
A3	0.5656	4	0.6039	1	0.5910	2	0.5664	3
A4	0.6392	2	0.7001	1	0.4962	3	0.4430	4
Final	0.6307	2	0.8005	1	0.6046	3	0.4221	4

Table 4 Grey correlation degree and ranking in 2012

Index	China		Japan		South Korea		India	
	Correlation degree	Ranking	Correlation degree	Ranking	Correlation degree	Ranking	Correlation degree	Ranking
A1	0.7480	1	0.4734	4	0.6248	2	0.4928	3
A2	0.6921	3	0.7146	2	0.9778	1	0.3333	4
A3	0.5599	3	0.9186	1	0.6172	2	0.3956	4
A4	0.7235	1	0.6201	2	0.4020	3	0.3661	4
Final	0.6735	2	0.7053	1	0.6732	3	0.3858	4

Korea, taking second place. Competitiveness rankings were stable in 2007 and 2012 as follows: Japan, China, South Korea, and India.

In production capacity, China and South Korea dominated the first place in turn. But it is necessary to note that China is in the bottom of the level 2 index industrial concentration. China's CR4 was only 21.52 % in 2012, which is behind Japan's 61.50 %. Small enterprises and scattered layout lead to low industrial concentration, and great difficulty in achieving economies of scale.

In technology level, the ranking was stable. Japan and South Korea alternately occupied the first place among the four countries. China always ranked third. India's technical level was in the last position of the four. Continuous casting technology is primary technology of iron and steel production. Continuous casting ratio of China, Japan, and South Korea in 2012 was 98.40, 98.80, and 98.30 % respectively, almost realizing full continuous casting production and reaching the advanced level in the world. India's was only 79.90 %, revealing a deep gap with the other three countries.

In market performance, China's low industrial sales profit margin and RCA explained its bad behavior. China had the lowest sales profit rate, and even fell to 0.04 % in 2012, reflecting almost no profit. Japan's sales profit percentage was steady. In 2012, the other three countries suffered different degrees of sales profit margin decreasing, while Japan's was still 9.18 %.

4 Conclusion

In this paper we evaluate international competitiveness of the iron and steel industry of the four Asian countries, and adopt grey correlation degree analysis method to carry out empirical research. We obtain the following basic conclusion. Japan has the strongest international competitiveness in the iron and steel industry among the four Asian countries. China and South Korea have roughly the same comprehensive advantage. India is at the bottom of the four countries in the final correlation degree ranking in 2003, 2007, and 2012, which proves that it is not at the same competitive level as Japan, South Korea and China. China is just ahead of India in technology level. Low sales profit margin and poor RCA result in China having a relative disadvantage in market performance. Meanwhile, low industrial concentration severely threatens its comprehensive competitiveness.

Acknowledgments The research work is supported by National Social Science Foundation of China under Grant No. 14ZDA088 and Social Science Foundation of Beijing under Grant No. 14JGA014 and the Fundamental Research Funds for the Central Universities under Grant No. 06106117.

References

1. Porter ME (1990) The competitive advantage of nations. Free Press, New York
2. Dunning JH (1992) The competitive advantage of countries and the activities of transnational corporations. Transnatl Corp 1(1):135–168
3. Rui MJ (2006) New diamond model of industry competitiveness. J Soc Sci 4:68–73
4. Schwab K (1996) The global competitiveness report 1996. Oxford University Press, New York
5. Garelli S (1996) IMD world competitiveness yearbook 1996. International institute for management development. Lausanne
6. Sahoo BK (1999) A comparative application of data envelopment analysis and frontier translog production function for estimating returns to scale and efficiencies. Int J Syst Sci 30(4):379–394
7. Movshuk O (2004) Restructuring, productivity and technical efficiency in China's iron and steel industry, 1988–2000. J Asian Econ 15(1):135–151
8. He WD, Wan XJ, Wu YB (2009) Study on competitiveness of China's iron and steel industry based on the strategic capability-based view. China Ind Econ 11:56–65

Dynamic Evolution of the Influence Factors of the Development of Electronic Banking Business in Local Commercial Banks of Underdeveloped Areas

Wei Chen, Wenli Li and Jian Zhang

Abstract Under a context that electronic banking gradually become the main direction of bank development, a study on the motive factors of electronic banking business of underdeveloped areas is what we need to draw out experiences and references, and to encourage the majority of domestic commercial banks to explore an appropriate mode of Internet finance. In this paper, placing the electronic banking development of Bank A as the main thread, using method of longitudinal dynamic single case study, studies on three aspects, namely, consciousness of leaders, influences from outside and internal management mechanism, so as to open the "black box" that covers the influence factors of the development of electronic banking business. The result indicates that given the regional particularity, development of electronic banking in underdeveloped areas is with hysteretic and phased nature, and the influence factors are evolving dynamically in different phases, which cause an uneven distribution of leading factor.

Keywords Electronic banking · Influence factors · Dynamic evolution

1 Introduction

Along with the rapid development of IT, as well as increases in the volume and complexity of banking business, a trend of electronic banking that is electronic-technology-oriented, has been born. By 2014, Chinese electronic banking users have amounted to 400 million, and the conversion rate of electronic

W. Chen (✉) · W. Li
Faculty of Management and Economics, Dalian University of Technology, Dalian 116024, China
e-mail: chenwei5960@sina.com; zjtom919@163.com

J. Zhang
Guanghua School of Management, Peking University, Beijing 100871, China

M. Li et al. (eds.), *Proceedings of 2015 2nd International Conference on Industrial Economics System and Industrial Security Engineering*,
DOI 10.1007/978-981-287-655-3_49

banking business has exceeded 60 %. In the same time, small and medium banks have more advantages in developing electronic banking than big banks, since they are in small in scale, have concentrated users, and flat institution setting. While in a regional viewpoint, thanks to the support of favorable national policies, rapid economic development of underdeveloped areas requires more in the electronic banking service.

2 Literature Review

Many scholars believe that electronic banking has become an indispensable part of the banking industry in the Internet era. Some of them regard electronic banking as an important tool to upgrade the banking industry and raise the satisfaction rate. Khan held that the success of the modern banking industry lies mostly on the amount of electronic banking products it provides, and the quality of their service [1].

In terms of the advantages small and medium size banks possess in the electronic banking business, Xing held that, compared with those small and medium size banks, the Four Great Banks of China are inefficient in electronic banking business, thus it is inadvisable to expand electronic banking business blindly [2]. Hu held that intermediate banking business might bring certain risks to the banks' profits (i.e., instability of earning), while for small and medium sized banks, such a risk is not significant [3].

As to how electronic banking should yield an innovative development, Liu put forward that electronic banking should change the marketing and service methods, as well as operation mechanism, since it shoulders the duty to make innovation in banking products and management systems [4]. In the meantime, financial services for residents' daily life, as well as supply chain finance and industry cluster finance, should be given more importance. Tang studied college students and reached the conclusion that electronic banking should offer a fine service on its customers, and adopt different marketing programs for different users [5].

In conclusion, studies on electronic banking are currently mostly on the direction that deems electronic banking as an effective supplementation and transformation option, and the suggestions they put forward are mostly for the operation and innovation of electronic banking of large commercial banks; however, the factors that influence electronic banking business development are still stay in a "black box." Therefore, this paper, based on study of the small local commercial banks in underdeveloped areas, as well as analysis on phases, will discuss the evolvement and mechanism of the influence factors.

3 Case Study

The study adopted a method of dynamic analysis on longitudinal single case to uncover the "black box", i.e., the influence and mechanism of electronic banking development of commercial banks in underdeveloped areas.

In this study, Bank A, a local commercial bank in the mid-east region of China, is chosen according to the typical principle. Bank A, reorganized from a rural credit association, is the first local cooperative bank. After the reorganization, Bank A focused on those financial products that support local "three-agriculture (namely, agriculture, farmer and rural area)" work and local industry chain, as well as on electronic banking business. What is worth mentioning is that while the local economy is ranked behind in the province, electronic banking business of Bank A ranks in the top three in all the 83 rural credit association reorganized banks.

Data analysis and in-depth interview indicate significant phases of the development process of Bank A. Furthermore, factors such as leaders' consciousness, external environment, and internal management system influence each and every phase. Leaders' consciousness mainly refers to the importance leaders attach to electronic banking, and the strategic position at which leaders place electronic banking. Given the small scale of local commercial banking, compared with national commercial banking, initial development of such new services are mostly decided by leaders' consciousness. External environment refers to the demand of society on electronic banking business. With the special situations of underdeveloped areas, we'd be wise to see the external environment in two aspects: (1) demand of the country or the whole society, and (2) demand of local companies and residents. Internal management system refers to the distribution of management functions concerning electronic banking business.

Even more noteworthy is that the three factors are not isolated, but interrelated: external environment affects so heavily that it can shake leaders' consciousness and internal management system; in fact, the last two interplay as well.

To probe deeper into the mechanism of such influences, we divide the development process of Bank A into four phases: initial, infancy, integration, and innovation.

(1) Initial Phase (2000–2004)

In this period, electronic banking business has developed like a raging fire in western banking industry. In China, China Merchants Bank has already launched the first domestic electronic banking business. The three influence factors were inoculating: environments, national or local, were pervaded with a conservative attitude toward electronic banking business. As a local small commercial bank, most of its clients are farmers or prefecture-level residents who have no demand on electronic banking business and as such, the leaders were sitting on the fence. Without the external environment and leaders' consciousness, internal management system stood still.

(2) Infancy Phase (2004–2007)

This is the period of development during which the domestic electronic banking business made leaps. National commercial banks and large stock-holding banks presented ample types of electronic banking products. Bank A launched its electronic banking business program, which is, to be frank, promoting the electronic banking products developed by local rural credit associations. The three influence factors in this phase are analyzed and below are the results.

Leaders' consciousness: president of Bank A was so insightful that he pushed electronic banking business forward, and attached importance to it in the board meeting, when other banks were still focusing on traditional business. External environment: a relatively huge domestic demand on electronic banking business has formed. Electronic banking business has reached consumption field in big cities. However, given the underdevelopment and skeptical attitude toward safety of electronic banking business in the city A, such demand was small. Domestic environment and local environment were in adverse effect and weakening each other, so there was only a tiny effect of external environment. Internal management system: given the influence of leaders' consciousness, and in spite of the small effect of external environment, internal management system changed a little: technology department was appointed to develop an electronic banking business.

It is obvious that the leading factor that influences development of electronic banking business in this phase is leaders' consciousness, which drove the change of internal management system.

(3) Integration Phase (2007–2013)

This phase features a nice momentum of electronic banking business in Bank A. Types of electronic banking products that are suitable to local clients were developed, and the number of clients and amount of card balance ranked in the top three in the province. In particular, given the advantages of traditional businesses, large number of clients and branches, Bank A integrated traditional businesses with electronic banking business and promoted them through peer-to-peer propaganda and special events.

In this phase, leaders attached even more importance to electronic banking businesses, and deemed them a critical complement to traditional businesses. Along with the birth and growth of domestic electronic banking businesses, manifested by the popularity of the ATM and POS, and the development of safety technology in particular, demand of residents of city A increased gradually. In terms of internal management system, which witnessed the development of electronic banking, which included product design and marketing role were gaining in weight, while the technology department, which used to take charge of electronic banking business, turned to Internet and information construction, since electronic banking businesses are "too demanding." At the request of related mangers, a resolution about establishing an electronic banking department was passed, which brought new fuel for Bank A: the new department, combining local demand and traditional business, developed Social Security Card (for social security payment), Jin Nong Card (for agricultural aid), and Citizen Card (for utility bills, telephone bills, bus and taxi fees, local consumption accounts, etc.). In addition, according to local situations, it held events such as "Peer-to-Peer" and "POS Machine in Rural Supermarket." Its excellent work helped the electronic banking business volume of Bank A exceed that of great national banks and ranks top in the Province. One should note that, in this phase, the influence of local environment, which turned from negative to positive, was affected by both internal management system and domestic environment. In the integration phase, internal management system served as the leading factor and reacted upon the leaders' consciousness and external environment. Three in one, Bank A gained leaps in electronic banking businesses.

(4) Innovation Phase (2013–)
This phase features the combination of Internet and finance. With the rapid development of 3G, optical fiber technology, and wireless network, the concept of "Internet +" becomes more and more popular, and Internet finance becomes the trend of history. Bank A puts forward that it should place electronic banking business as the direction of transformation and upgrading and are planning programs such as Local Specialty E-mart Program and Whole Supply Chain Internet Finance Program for Special Industries of City A. In this phase, leaders no longer treat electronic banking business as the complementary to traditional finance, but rather as an equally important strategy, or even priority. In terms of external environment, with technology becoming mature, online consumption demand thriving, and government support and guidance becoming more solid, the domestic environment is playing an unprecedented positive role with the strongest momentum. Therefore, it could be seen as a positive momentum of influence exerted by geographical environment. In terms of internal management system, after the concept of Internet finance is raised, electronic banking business is facing the task of integrating with traditional businesses, and the relation between electronic banking department and technology department is turning into amalgamation and collaboration. Currently, electronic banking product development project groups of Bank A are often composed of both departments, which shows a benign communication and coordination mechanism.

In this phase, the external environment is the leading influence factor. The effect of it exceeds all other leading factors above-mentioned—geographical environment, leaders' consciousness, and internal management system are all affected badly by and together with domestic environment, which brings a huge force for the rapid and innovative development of electronic banking businesses of Bank A (Fig. 1).

4 Results and Expectations

In the study, a dynamic evolving analysis on influence factors of the electronic banking business of Bank A, a local bank of an underdeveloped city, is conducted, and the results are as follows: (1) The influence factor of electronic banking

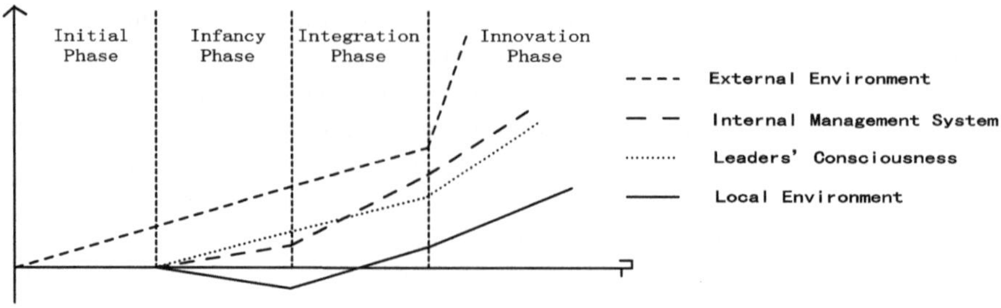

Fig. 1 Changes in force and trend of every phase

business of small-sized local banks in underdeveloped areas evolves dynamically. (2) The distribution of leading influence factor of electronic banking business of Bank A is uneven in different phases.

Despite the work that has been done on study and analysis, as well as results that have been reached, shortcomings are inevitable. The first one is the limitations of the sample. The second one is that we simplified the internal management system into the function distribution of technology department and electronic bank department of Bank A. In spite of the reasonability in electronic banking business research, is would be complicated in reality. Therefore, the job has to be done in the next step to yield more systematic and many-sided conclusions.

References

1. Khan (2011) Customers' perception of e-banking service. Int J Bus Econ Manag Res 2 (3):156–164
2. Xing Y (2014) Research in management of electronic banking business in the bank of China. Shandong normal university Press, Jinan
3. Hu H (2015) Intermediary business and banking risks. Financ Re 2015(11):62–65
4. Liu X (2012) Trend and challenges of electronic banking in China. New Financ 2012(10):26–30
5. Tang L (2011) Campus marketing strategy of electronic banking products. New Mark 2011 (5):34–36

Research on the Effects of Trade Openness on Employment in China

Wenjing Wang

Abstract This paper uses provincial panel data of 1995–2013, and adopts static fixed effect panel data model from two angles of the time comparative and regional comparative analysis of our country's trade openness contribution to employment growth. The results show that the employment effects of foreign trade, after the WTO accession, is greater than before, and there is a significant difference in the employment effects on the three regions of the eastern, the central and the western. The negative role of employment growth caused by imports was ranked as eastern, central and western. The positive role of employment growth caused by exports in the western is greater than in the central. Accordingly, the article puts forward some policy recommendations.

Keywords Trade openness · Employment · Provincial panel data

1 Introduction

Since China joined the WTO in late 2001, total foreign trade volume increased year by year from \$509.65 billion in 2001 to \$4158.9 billion in 2013, an increase of 7.16 times, far more than the total amount of the early period of the reform and open policy. However, with the deepening of China's opening up and adjustment of industrial structure, the issue of domestic employment has been one of the problems during the process of economic and social development in China that cannot be ignored. Although China's registered urban unemployment rate is leveling out at around 4 %, the actual unemployment rate in China is far greater than this value. There are currently three types of unemployed workers in China: (1) the registered urban unemployed; (2) laid-off workers; (3) disguised unemployment, which mainly includes the redundant personnel in state-owned enterprises and the labor

W. Wang (✉)
School of Economics and Management, Beijing Jiaotong University, Beijing 100044, China
e-mail: 14120556@bjtu.edu.cn

M. Li et al. (eds.), *Proceedings of 2015 2nd International Conference on Industrial Economics System and Industrial Security Engineering*,
DOI 10.1007/978-981-287-655-3_50

force in rural areas. The unemployment rate announced by the Statistics Department only includes the urban registered unemployed people, while the actual scale of unemployment is considerably larger. Admittedly, with the further expansion of openness, it is limited in supply and demand at home, and the expansion of the international market will increase employment and promote economic growth. This is the original intention of this paper.

2 The Related Literature Review

Smith first proposed the "surplus product export" theory. He believed that with the development of foreign trade, the employment opportunities in backward countries would increase. Thus, the problem of unemployment would lessen, and, ultimately, the increased labor and other idle resources investment would make the economy grow in backward countries. Metgerler and Harburger, the followers of Keynes, proposed the theory of trade multiplier. This theory believed that a country's export, like new investment, could also play a role in increasing employment. When goods and services export, the corresponding department's employment can be increased, and, due to the correlation effect, the unit can also lead to employment increase in other sectors. Subsequently, the H-O-S theory, which was based on the hypothesis that elements can fully flow thought that wages can be adjusted freely, however, the labor market is always in a state of full employment; the trade between countries will cause a fall in import sector output and a rise in export sector output; and the domestic labor force will also flow to the export sector from the import sector [1].

In recent years, the literature on the relationship between trade and employment is gradually increasing in our country. Yang found that China's export-oriented policies contribute to increasing employment [2]. Zhou and Li believes that changes in trade structure of manufactured goods in China is not conducive to the increase of employment, and trade with developed countries will have a positive effect on employment [3]. Jiang found that trade liberalization can promote China's employment, but this effect is crumbling. On the contrary, in the foreign-funded enterprises, the promotion effect on employment is strengthened [4].

In summary, the empirical studies have a rich reputation at home and abroad, but the studies about interprovincial are still relatively rare. Most of the studies show that trade has a positive effect on employment, while there are no consistent conclusions for what the effect of export or import is respectively. The following article will use the 1995–2013 annual inter-provincial panel data, and adopt static fixed effect panel data model, from two angles of the time comparative and regional comparative analysis of our country's trade openness contribution to employment growth.

3 Empirical Model and Data

3.1 Models

This article uses Hine and Wright's method to study the employment effect of trade [5]. First, function is assumed to be the Cobb–Douglas production function:

$$Q_{it} = A^{\gamma} K_{it}^{\alpha} N_{it}^{\beta} \tag{1}$$

where Q_{it} represents the actual output of province i in period t, K_{it} is the capital stock of province i in the period t, N_{it} is the number of labor, α and β represent the output elasticity of capital and the output elasticity of labor, and A is the factor of technical progress. In the process of production, technology will continue to progress over time, and trade will affect the technology. Therefore, we assume that the parameter A varies with time, and satisfies the following equation:

$$A_{it} = e^{\delta_o T_i} M_{it}^{\delta_1} X_{it}^{\delta_2}, \quad \delta_0, \delta_1, \delta_2 > 0. \tag{2}$$

where T represents the time trend, M is the extent of import reliance, X is the extent of export reliance. Putting them into the working equation, we get:

$$\text{In}\, N_{it} = \varphi_0^* + \mu_0 T + \mu_1 \text{In} M_{it} + \mu_2 \text{In} X_{it} + \varphi_1 \text{In} K_{it} + \varphi_2 \text{In} Q_{it} \tag{3}$$

Accordingly, the panel data model is:

$$\text{In}\, L_{it} = C_t + a_1 \text{In} M_{it} + a_2 \text{In} X_{it} + a_3 \text{In} K_{it} + a_4 \text{In} Y_{it} + \varepsilon_{it} \tag{4}$$

3.2 Data Sets

This paper uses provincial panel data (except Chongqing) for empirical analysis. In 1997, Chongqing was carved out of Sichuan Province into a municipality. Its data will be incorporated into Sichuan. All data comes from the "China Statistical Yearbook." L is the level of employment, reflecting the total regional employment. Y is the level of output, and the paper takes GDP (100 million yuan). X and M represent the extent of export reliance and the extent of import reliance, respectively. The extent of export reliance is the ratio of total exports to GDP in the region, and the extent of import reliance is the ratio of total imports to GDP in the region. The total imports and exports converted into RMB, according to the RMB exchange rate against the dollar, and calculated through RMB-denominated GDP. K, is the capital stock (100 million yuan). This paper uses the data estimation and calculation method of Zhang [6]:

$$K_{it} = K_{it-1}(1-\delta_{it}) + I_{it} \quad (5)$$

where δ is the depreciation rate of the capital stock, taking 9.5 %; I is the fixed investments.

4 Empirical Findings

4.1 Time Comparison of the Impact of Trade on Employment

This section will be divided into two time periods, 1995–2001 and 2002–2013, in order to compare the effect of trade openness on employment before and after China's joining the WTO. The regression results are shown in Table 1.

The coefficient of the extent of export reliance in two periods are both positive, indicating that exports has a role in promoting employment. The latter coefficient is greater than the former, indicating that after entering the WTO and the expansion of trade openness, the leading role of exports on employment is growing. The positive impact of exports on employment is due to the fact that our labor force is relatively abundant and, through exporting labor-intensive products after opening, the employment of our country has been vigorously promoted. As the foreign trade multiplier theory proposed by Metgerler and Harburger suggested, a country's export has a multiplier effect on national income and employment.

The coefficient of the extent of import reliance in the two periods are both negative, indicating that with the increase of trade openness and the extent of import

Table 1 The analysis of regional foreign trade (US $ 100 million)

Variables	1995–2001	2002–2013
Constant C	7.251120*** (174.3586)	6.108619*** (114.1218)
Output LNY	0.038800*** (3.546741)	0.196010*** (7.884998)
Extent of export reliance LNX	0.011528** (1.997413)	0.026928*** (4.274566)
Extent of import reliance LNM	–0.019560*** (–5.391287)	–0.022362*** (–3.635294)
Capital stock LNK	–0.035353*** (–3.581702)	–0.040056** (–2.033431)
R^2	0.999692	0.998502
Redundant FE test value	2370.728970	473.969873
Prob(F-statistic)	0.0000	0.0000
Hausman test value	257.261229	158.208795
Prob(Hauseman)	0.0000	0.0000
total sample	210	300
Remark	FE	FE

T statistics are in parentheses. ***, **, * represent 1, 5, and 10 % levels significantly. FE represents the fixed effects

reliance, China's employment has been inhibited. The larger the absolute value of the coefficient of the extent of import reliance, the more obvious is the impact of the effect of imports on employment.

The Regression coefficients of GDP are significantly positive, and the absolute value of the coefficient is far greater than the extent of import and export reliance, which verifies the theory that economic growth can increase employment, and is the main power for promoting employment.

4.2 *Regional Comparison of the Impact of Trade Openness on Employment*

In this paper, our 30 provinces are divided into three areas: the eastern, the central, and the western. The regression results are shown in Table 2.

From the coefficient of the extent of import reliance, the eastern is larger than central and the western in the impact of imports on employment. Compared to the central and the western, the level of economic development and labor productivity in the eastern is higher, indicating that with the increase of trade openness and the extent of import reliance, the substitution effect of the employment is more obvious.

For the extent of import reliance, the coefficient of the eastern is not significant at the level 5 %, and the coefficient in the western is larger than the central. This is

Table 2 Regional comparison of the impact of trade openness on employment

Variables	Eastern	Central	Western
Constant C	5.839069*** (159.1869)	7.088970*** (249.4413)	6.253586*** (524.1764)
Output LNY	0.294463*** (38.72031)	0.072310*** (10.92795)	0.058563*** (9.868098)
Extent of export reliance LNX	−0.007698* (−1.888428)	0.018032*** (4.180410)	0.029915*** (15.30176)
Extent of import reliance LNM	−0.044160*** (−10.49690)	−0.011546*** (−3.170879)	−0.001802** (−2.002438)
Capital stock LNK	−0.110032*** (−17.86034)	0.012305** (2.204509)	0.042851*** (9.022369)
R^2	0.999686	0.997960	0.999890
Redundant FE test value	10172.734775	6556.476794	20693.841884
Prob(F-statistic)	0.0000	0.0000	0.0000
Hausman test value	156.093954	21.685620	94.047271
Prob(Hauseman)	0.0000	0.0000	0.0000
total sample	187	136	187

t statistics are in parentheses. ***, **, * represent 1, 5, and 10 % levels significantly

mainly because the openness, trade and economic development levels in the western are far below that of the eastern and central. And the lower labor costs make the western more dominant on the export of labor-intensive industry. The effect of export in the eastern is negative. This is partly because the trade strategy inclined too much toward the exports. Total exports have increased, while the contradiction of the regional resource is also on the rise. This means that the export of the scarce goods can, to some extent, offset the positive employment effect of labor-intensive industry. Moreover, our trade mainly concentrates on the second industry, and does not give full play to the pulling effect of the third industry on employment. Exports increase year by year, but the effect on employment is limited. The impact of exports on employment tends to be in the opposite direction.

As for GDP, the regression coefficients of the three regions are significantly positive, and in descending order they are the eastern, the central, and the western. The results are consistent with the regional development and regional economic level in China.

5 Conclusions and Policy Recommendations

The analysis from the time perspective shows that the extent of export reliance has a significant positive effect on employment growth in the two periods, while the extent of import reliance has a significant negative effect on employment growth. Since our country joined the WTO, trade openness increases year by year. And the effect of export on employment has been enhanced, while the inhibitory effect of imports on domestic employment is also enhanced. The analysis from the regional perspective shows that the pull effect of the extent of export reliance on employment is not significant, and the contribution rate of the extent of export reliance on employment growth in the western is larger than the central. The extent of import reliance has a negative effect on employment growth, and the impact in descending order is the eastern, central and the western.

The findings of this research suggest that our government should participate in the international market actively and increase the total foreign trade. In the context of globalization, our country should seize the opportunity to further increase the degree of openness, and give full play to the positive role of foreign trade in economic growth and employment.

Authorities should expand exports, and play the comparative advantage of abundant labor force to promote job growth. We should give full play to the advantage of labor-intensive industries and, at the same time, change the way of boosting trade positively. Simultaneously we should commit to increasing high-tech exports. Furthermore, we should improve the technological content of export products to enhance value-added and optimize the structure of export products so that we can create more employment channels and opportunities.

Finally, the government should actively promote the development of foreign trade in the western and central, and use the positive role of trade liberalization for employment growth. The trade volume in the eastern accounts for a larger proportion, attracting a lot of labor. But the effect of the eastern trade growth on employment is limited. On the contrary, the foreign trade is developing fast in the western and central, and the employment effect is also very significant. Therefore, we should make full use of the advantages of the employment effects of the foreign trade in central and the western to increase employment and alleviate the employment pressure in the east.

References

1. Samuelson PA (1948) International trade and the equalization of factor price. Econ J 6:165–184
2. Yang W (2013) An empirical analysis of the effect on employment of foreign trade in our country, based on provincial panel datal from 1995 to 2011. Econ Forum 20:64–65
3. Zhou S, Li C (2006) Impact of changes in the industrial structure of employment in China trade. Number Tech Econ 7:137–151
4. Jiang H (2007) The empirical study of effect on employment of foreign trade in China, taking the foreign capital enterprises for example. Int Trade Issues 10
5. Hine R, Wright P (1998) Trade with low wage economies, employment and productivity in UK manufacturing. Econ J 108(450):1500–1510
6. Zhang W (2004) Phase transitions and backbones of the asymmetric traveling salesman problem. J Artif Intell Res 21(4):471–497

[illegible]

Part VIII
Others

The Concept of Defining Home Care for the Aged Services Based on the Connotation and Extension in China

Mingxiu Cao

Abstract With the advent of an aging society, having a large number of elderly people creates increasingly significant aging trends. The content of home-based care services has been gradually enriched and perfected, but it is not yet standardized. Service management and service staff lack professional training, and there is conflict between service staff and clients as well as other issues. In order to better regulate home-care service, the author participated in the development process of China's domestic service standards and redefined the concept combined with the actual situation. In this process, the author will regulate the specification of the service sector and play a positive role for the whole family industry.

Keywords Home care for the aged service · Concept of definition · Connotation · Denotation

1 Introduction

Many researchers have defined the concept of home-care services for the aged in China. But with the development of domestic service, concepts need to be unified and standardized for the whole industry. On the basis of home-care services connotation and extension-analysis studies, the author redefined the concept of home-care services, not only to regulate home-care service agencies and management staff, but work also to have a positive effect on the overall development of the domestic service industry.

M. Cao (✉)
China Center for Industrial Security Research, Beijing Jiaotong University, Beijing 100044, China
e-mail: caomingxiu2008@163.com

M. Li et al. (eds.), *Proceedings of 2015 2nd International Conference on Industrial Economics System and Industrial Security Engineering*,
DOI 10.1007/978-981-287-655-3_51

2 Relevant Scholar's Research

Home care was first proposed by Western countries, but different scholars have different views of the basic definition. In present literature, most scholars believe that home care involves elder people living at home, but the way to enjoy a social pension that provides the elderly nursing home and community-based services means family-focused, community pension service network perimeter for the protection to the pension system, also known as "the nursing home has no walls."

An earlier scholar of theoretical research on home-care services in China is Professor Yuan [1]. He proposed the concept of home care from the retirement place aspect. More scientific research system and the concept began in the early 2000s. Professor Yuan thinks family pension and social pension are relative to the pension fund sources. The family pension is the pension funds from family and family members. It is institutional pension if pension funds from the community provide the community through social pension or pension benefits and other forms. Home care and hospitalization workplace pension is relative to pension places. If the family is the main place of pension, it is home care; if nursing homes or apartments for the elderly are the main place of life, it is called admission pension. Jincheng Wang thinks that home care is a combination of family pension and social pension [2]. In Guangzong's viewpoint [3], the home care for the aged is live at home, receive community care services and government support for the protection of the family, and the social pension which is combined with the old pension system.

Most current official research institutions and scholars believe that over a period of time, home care for the aged will be a combination of the national pension model and the social and family model. But the family-based pension model is the core of home-care services and social services or the community as auxiliary pension model [3]. Some scholars purpose funding sources in several major pension comparisons with the family pension and old-age social pension, the home care and institutional care places for the purpose of pension. If the family is the main place where older people live, it is home care. If nursing homes or apartments are the main place of life for the elderly, it is an institutional pension [4].

Peng [5], Xiang [6] emphasized that home care for the aged service is a system in which the elderly live at home, but enjoy nursing home services for the elderly in the community—the kind of social pension model in which the family is the core. Luo [7] from content services that home-care services in the community are to create a support family pension social service system, the elderly living at home decentralized, community service nonprofit agencies provide public services to solve elderly care, housekeeping services, and mental comfort problem.

Academia currently defines the concept of home-care services primarily by the way of services items. In January 2008, the National Aging Committee issued [8] "comprehensively promote community home care services work" which clearly defined home care for the aged services refer to all levels of government agencies and community organizations rely on the power of communities to choose home care for the aged to provide daily care, health care, spiritual solace, and in the form of other aspects of pension services. Some researchers considered home care to encompass medical services as well as other care, including sports activities, cultural engagement, emotional and psychological benefits, legal rights, and other various services [9].

3 The Connotation and Extension of Home Care for the Aged Services

3.1 The Connotation of Home Care for the Aged Services

At present, the study of home care for the aged services concepts, ideas, and definition are different; it is difficult to reach a consensus. The author summed up the views of home care for the aged service concept according to the information literature.

(1) The combination home care for the aged mode: the combination of family endowment and community endowment
Home-based care for the aged- - the choice of urban pension model in the paper, Chen said [4]: "the family endowment is the organic combination of family endowment and community endowment. Its basic content is: Housework shall be borne by the society, the spiritual life pension shall be borne by the family, the material aspects of pension shall be borne by the three aspects: the state, the collective and the individual." In April 1998, at the closing ceremony of the national family old-age social pension service seminar, Zhang, president, put forward: [8] "In the general home-based care for the aged is given priority to family pension, social pension is complementary pension mode, all social forces should actively mobilize to form accord with the will of the elderly pension, and to maintain and strengthen the elderly independent ability, the most practical and most efficient endowment security system, to establish a pattern contribute to the development of the society endowment."
(2) The service style and service content: the old people should receive service from agencies and community to provide comprehensive service mode and service content.
In 2008 [10], The office on aging issued "The advice of comprehensively advancing home endowment service work" no. 4 (2008), home-based care for the aged was defined: relying on the community, government, and social forces to provide pension for the elderly life care, domestic service, rehabilitation care, and spiritual comfort service as a form of service. In addition to the traditional family pension mode, developing community service in our country to establish endowment service system is an important concept. Liu [11], in the context of the present situation and countermeasures of urban pension, also

mentioned "home-based care just within the scope of the old people, the various social services, providing shopping, cleaning and care to the elderly comprehensive care for daily life."

(3) The living place and the environment: The family is the main carrier of family pension. Living place and the environment have a very close relationship to pension.
If the family is the main place for old-age life, it is home-based care. If nursing homes or elderly apartments are the main place of living, it is called a hospital endowment. In the context of home-based care and pension mode in China: Peng [12] said "the family endowment concept accurately reflects the living of the elderly pension form." House is old people scattered in their own family pension rather than concentrated on old-age pension institutions. So home-based care means the home is pension platform that based on the endowment relatively fixed in the community environment, government, and society. Family and several aspects have different effect on this platform.

(4) Funding sources: mainly on the basis of pension funds.
Jihui Yuan proposed the "family pension and social pension are relative to the pension funding sources. If pension funds derived from the family, provided by family members, is the family pension; if pension funds from the society, the social pension or social benefits provided by social endowment, it is pension agency.

All in all, the concept of home care for the aged services should be defined from the combined mode, service means and service content, living place and the environment and sources of funds.

Understand this concept needs to pay attention to the properties of the service product. From the point of view of product attributes, home care for the aged services endowment service should belong to the mix product. On the one hand, meeting the needs of old people's basic life services is clearly the duty of the government This part belongs to the government need to provide "public products", a sort of a non-profit "meet the old person program. On the other hand, in addition to the basic requirements of personalized needs, or even a higher level demand service, the old person needs to purchase through the market to meet their own needs. This part of the service, provided by the market "Commodity", is a non-public product; it is a for-profit product.

3.2 The Extension of Home Care for the Aged Services Concept

If we want to scientifically define home endowment service, we must first make clear the object of service; home care for the aged service object is to choose home endowment patterns of the elderly. Second, to focus on the purpose of the service, family endowment service aims to meet various pension needs for service object and to improve the quality of the elder life. Third, we must make clear the content of service. The home endowment service content should include elderly life care, domestic services, rehabilitation nursing, spiritual comfort, legal aid, charitable aid, etc. Fourth, the place of service: home care for the aged service takes place is the home of the elderly family, as well as in the within l the life in the community. So the home care for the aged and family at home, there are differences between institution endowment.

Compared with the family pension, home care for the aged service has an emphasis on pension service provider from social workers with no family blood relationship; compared with the institutional pension, more emphasis on placed on the family and the social relations of old people.

The contract between the home care for the aged and family endowment is that they cannot leave the family and accept professional pension services as institutional endowment.

Compared with other forms of pension, home care for the aged service has irreplaceable advantages:

(1) Conform to the old people psychology
This way can make the elderly feel security and belongingness in their familiar home environment and retirement community environment. Most old people are not willing to leave their familiar place of life and institutions at home, and they enjoy pension services, whether from face-to-face or psychological. Although the institutions endowment, will provide access to medical care and life care, older people need family care and spiritual communication. This cannot be replaced by endowment institutions.
(2) The uses of idle resources
Home care for the aged service also reduces pension costs and alleviates the impact ageing has on arrival. On the one hand, to carry out the home care for the aged service can make full use of family resources, solve the practical difficulties, and implements service mode to alleviate the current rich-family pressure is very effective. On the other hand, it can also provide jobs. Many laid-off job hunters can find job opportunities to mobilize the community of idle resources and achieves the goal: a win-win.
(3) The deployment of forces
Home care for the aged also allocates the community volunteers, the people of organizations, enterprises, and institutions and other forces, cultivate the spirit of cooperation between the organization and cooperative spirit, are more likely to produce a sense of belonging and identity. In addition, the elderly are both by the service object and service providers. Young men take care of the elderly by means of pairing a mutual support network among old people, and it is a positive role will effectively develop the talent to work with old people.
(4) Broaden the endowment service content
The main target of home care for the aged services is to meet the needs of the elderly's daily care and to provide spiritual comfort, as its main function is to reduce the pressure of family care. Daily care, primarily through door-to-door

services and daycare centers, relieves the pressure. Spiritual solace relies mainly on improving the cultural entertainment education system. Relying on community platform, based on the tendencies of the elderly to carry out all kinds of mass cultural, sports, and entertainment activities open to old people, elderly universities cultivate the learning interest and the hobbies, all of which enrich the spiritual life and protect the physical and mental health of the elderly. The various team activities and the elderly university study will decrease the sense of loneliness in the aging. It makes them feel that they are still active members of the society, and it allows them to help each other and creates positive energy through their contributions to society.

4 Redefining the Concept

On the whole, the concept of home care for aged should be redefined: Home care for the aged service refers to a pension model which combines family support and social pension. It is guided by the government, but based on community. Family is core, door-to-door service and community daycare is the main form. It will provide housekeeping services, rehabilitation care, spiritual comfort, legal aid, charity, and other services for old people.

Therefore, although there sis weakening of the family-pension Function, the government does not yet have abundant conditions to construct agencies for old people, Home care for the aged concentrates the advantages of home care, and institutional pension is a complete harmonization of laws, morality, and traditional pension model which is a realistic choice in line with China's national conditions.

References

1. Yuan J (1996) The pension of Theory and Practice. Chin J Gerontol 10:300
2. Wang J (2000) Home care: an inevitable choice for China's urban. Elder Popul Sci 4:19–22
3. Guangzong M (2000) Reform and Prospect of traditional pension mode in China. Renmin Univ China J 5:41
4. Chen J (2001) Home care:Selection of a city pension model, social, (9): 22–24
5. Peng X (2006) Hong Liang: urban pension service system. Shanghai people's Publishing House, Shanghai, pp 55–76
6. Xiang L (2010) Home care for the aged and services Research Review. Technology Wind 2:58
7. Luo X (2008) Urban communities: Exploration and Inspiration for home care services. Jiangxi Admin Inst 4:63
8. Zhang W (2002) China's aging population and strategic choice. Urb Plan 2:68
9. Huang Y, Liu H (2006) empirical research needs of urban elderly services. Bei hua Univ J (Soc Sci Edition) 4:90–93

10. National Ageing Office issued (2008) comprehensively promote community home care for the aged services work (4)
11. Liu F (2009) Actively promote the home care for the aged:-new social pension model [J], Mod Econ Res (8):48–51
12. Peng J (2013) Social endowment service supply relation analysis—Based on the perspective of agency theory. Soc Sci Front 6:184–189

Research on Controlling Measures of Overcapacity

Mei Feng and Kehui Yu

Abstract The contradiction of overcapacity is more and more prominent; this article from the perspective of government, industry associations, and enterprises reviews the control measures of overcapacity. Deepening reform, eliminating backward production capacity, overcoming local protectionism, limiting market access, and expanding domestic demand are the control measures of government; industry associations should play an active role to control overcapacity. As for enterprises themselves, they should improve the technical level, apply a "going out" strategy actively, and speed up mergers and acquisitions.

Keywords Overcapacity · Industry associations · Control measures

1 Control Measures from a Government Perspective

In recent years, with the sustained and rapid development of the economy, the phenomenon of overcapacity has been particularly evident; the government also takes sweeping measures to control overcapacity in some industries.

1.1 Deepening Reform

Firstly, the government should advance the reform of the investment system, improve the approval and filing system of corporate investment, and allow various types of capital, especially private capital, to enter freely [1]. Ding Chunxiang and Wang Xin think that the excess capacity is the result of imperfections in China's

M. Feng · K. Yu (✉)
Donlinks School of Economic and Management, University of Science and Technology Beijing, Beijing 100083, China
e-mail: yukehui1991@163.com

M. Li et al. (eds.), *Proceedings of 2015 2nd International Conference on Industrial Economics System and Industrial Security Engineering*,
DOI 10.1007/978-981-287-655-3_52

market economic system. In order to fundamentally solve the problem of excess production capacity, it should deepen the reform of China's investment and realize the diversification of investors, actively encourage private business investment, liberalize access restrictions on the field of investment, improve the financing system of the private economy, and promote private economic broadening of the financing channels, thus enabling enterprises to become autonomous, decision-making, self-investing, and self-restraining market subjects.

Secondly, it should promote the reform of the government management system; faced with excess capacity, all levels of government should adapt to the market economy, and weaken government functions, under the guidance of scientific development concepts to promote the transformation of government functions [2]. Yang Yingjie stated that the excess capacity in our country is due largely to administrative interference, and that to resolve the deep-seated problems in the system, the government should gradually relax financial resources and the control of land resources, to create a good institutional environment. For those who invest blindly and bring adverse consequences constraining the policy, if necessary the government should take legal action to control or eliminate the behavior of those protective of excess capacity.

1.2 In Accordance with Relevant Laws and Regulations, Eliminate Backward Production Capacity

Our industries with excess capacity represent a serious phenomenon of backward production capacity in most industries; the production capacity of many companies is basically just building, not declining [3]. Various policies that the government has introduced have also mentioned the elimination of backward production capacity, which would have significant influence in easing overcapacity in some industries.

The government should speed up the elimination of backward production capacity, resolutely stop the construction of any simple project to expand production capacity, strictly restrict any existing low level of backward production capacity, rigorously clean up the current industry overcapacity, and strike a better balance between supply and demand [4]. At the same time, it should be giving appropriate incentives in land, capital, taxes, and so on, so that corporate officers will have appropriate arrangements. Subsidies to eliminate backward production capacity of enterprises should also be taken into account, to reduce social pressures. For those areas where small- and medium-sized iron and steel enterprises are concentrated, through the introduction of high-tech industry, the government can fill the gap after eliminating backward production capacity, solve the problem of unemployment stemming from the elimination of jobs in backward industries, and help maintain social stability. In addition, the government can use economic policy to give some financial compensation to the companies experiencing job elimination, bolstering the smoothness of these transitions.

Cheng Xinwei indicated that the elimination of backward production capacity should also take into account the region and development affected; due to different levels of economic development, different regions of backward production capacity should be defined differently. For the developed areas, their equipment and technology are significantly better than in the less-developed regions; enterprises have expectations of eliminating backward equipment and purchasing advanced equipment. But because of the blanket governmental policy implications of financial subsidies, companies will wait for elimination until the state includes current equipment under rejection criteria.

1.3 Overcome Local Protectionism

With a number of industry overcapacity problems getting serious, the phenomenon has become a prominent issue in economic development; in order to effectively solve the problem of overcapacity, the national government should take strict precautions against local governments' protectionism [5] Yi Peiqiang pointed out that governing the problem of excess capacity should use economic, legal, technology, and environmental protection standards, as well as necessary administrative means, and that all levels of government should guard against and overcome local protectionism; in terms of eliminating backward overcapacity, every province should practice local government responsibility, and resolutely avoid maintaining local performance and GDP growth to protect backward industries.

Wang Xiaoguang pointed out that the process of managing overcapacity should resolutely avoid local protectionism, as local protectionism will make a lot of cross-sectoral, cross-industry, and cross-ownership mergers and acquisitions difficult to achieve. The national government should take measures to eliminate local protectionism, encourage cross-regional and cross-sector mergers and acquisitions, eliminate backward industries that do not meet effective production capacity, and close businesses that waste resources and pollute the environment.

2 Control Measures from a Guild Perspective

The last administrative reform withdrew most of the administrative department for the mining industry, weakening some regulatory functions [6]. In response to excess capacity, Lu Baifu pointed out the need to improve the guild system, strengthening supervision and discipline functions, thus making it a powerful assistant for government regulation. At the same time there is a need to strengthen industrial planning; guide enterprises to invest reasonably; release timely industry, production, marketing, technology, and international trends and other information to enterprises, as well as to monitor the development of industry in a healthy, orderly, and well-coordinated manner, under the guidance of the industry associations.

Du Feilun also pointed out that industry associations should play an active role in the governance of overcapacity. Industry associations should issue industry production capacity and market demand forecasting information in a timely way, announcing capacity utilization and other indicators to the public through regular follow-up analysis, so that enterprises have timely understanding of the industry status quo. The relevant industry associations should also be improved to strengthen statistical work regularly, and to publish information promptly to remind of industry overcapacity risk. This will give full play to the role of industry associations in dealing with related industries' overcapacity.

3 Control Measures from an Enterprise Perspective

Overcapacity is normal in the market economy; it is related not only to the economic cycle and economic reform, but also to the development of enterprises. Exploring ways to solve the problem of overcapacity should involve not only government and industry associations, but also enterprises themselves. These should adopt certain strategies to cope with the emergence of overcapacity.

3.1 Deepening Reform

Excess capacity in most industries is structural overcapacity. In the steel industry, for example, low value-added products represent severe excess capacity; high value-added products are heavily dependent on imports. In response to the structural overcapacity issues, enterprises themselves need first to improve the technical level. Cheng Wanquan pointed out that for the vast majority of overcapacity businesses, the most realistic option is to upgrade the technological level; in general, overcapacity is most prone to appear where there are low barriers to entry, low-tech products are being produced, energy consumption is high, safety performance is not high, and environmental requirements are not being met. Industries at overcapacity should take a new road to industrialization, improve the technical level, and achieve the development of industry intensively and environmentally. Therefore, the enterprises affected should improve their technical level, promoting the industrial chain from low to high, and from single to comprehensive development.

Wang Liguo pointed out that the direct result of our technological backwardness is "structural surplus"; on the one hand, backward technology reduces barriers to entry, affecting investment behavior, while on the other hand, backward technology will lead to enterprise technology introduction meant to improve production. Technology introduction would, then, undermine the enthusiasm of independent innovation of enterprises; this vicious circle risks throwing enterprises into long-term low levels of technology. A backward technical level promotes overcapacity. Therefore, improving the technical level, and training technical innovators in the ability to alleviate overcapacity problems, is important; companies should pay attention to improving the capability of independent innovation, focus on product innovation, accelerate the technological transformation of enterprises and, with it, technological progress.

3.2 Encourage Enterprises to Adopt the Strategy of Going Out

Responding to excess capacity only by expanding domestic demand is not enough; as many scholars have pointed out, enterprises should vigorously implement a "going out" strategy to ease overcapacity. Among these scholars, Liu Zhongguang has stated that particularly for our country's iron and steel, cement, and textile industries, we should actively promote the relevant enterprises' engaging in going out. There have been "going out" seminars held to raise awareness of excess capacity, and to help analyze the situation, exchange information, and develop practical going out programs. Enterprises must solve the problem of overcapacity through the external investment allocation of resources in the global market. At present, China's economy develops steadily and fast, and the strength of social development has been at a certain level, for the development of foreign investment has had a good foundation. Only vigorously implementing going out strategies can help make up for the lack of domestic energy resources, ease domestic overcapacity in some industries, and promote sustained and steady growth. Actively promoting going out strategies, we can also handle foreign investment and attract foreign relations to avoid large-scale industrial transfer abroad and aggravation of unemployment.

Yin Baoda reckons that the country now has a good basis for foreign investment, and it should vigorously implement a going out strategy, improve the system of tax relief, reduce the burden on enterprises' overseas investment, and enhance their international competitiveness. The government should encourage enterprises to develop overseas resources for international cooperation, to participate in international economic cooperation and development, and to invest in and set up factories in the appropriate states to reduce overcapacity in some industries. Meanwhile, industries at overcapacity should take advantage of export trade to absorb excess capacity; this can be done by the government's increasing the export tax rebate rate, changing the export structure, opening up new international markets to promote export growth, and then using these international markets to absorb the excess capacity. Thus, increasing foreign investment and expanding exports can have some positive effect in easing domestic overcapacity.

3.3 Accelerate the Pace of Mergers and Acquisitions

In the long run, one important reason that overcapacity occurs is that many companies produce similar products, to the point that supply exceeds demand. Solving the problem of excess production capacity must also involve addressing this issue [7]. The State Council has issued a notification on accelerating the restructuring of overcapacity industries. It declares that in accordance with market principles, powerful conglomerates are encouraged to take assets, resources, brands, and markets as a link to implement cross-regional, cross-industry mergers and acquisitions, promoting large-scale and base industrial centralization. For example, large-scale iron, steel, and cement overcapacity enterprises and others within the region are encouraged to carry out joint reorganization. In this regard, there are many scholars expressing their support. Cheng Wanquan pointed out that for the vast majority of enterprises, through mergers and acquisitions, resources can be relocated to achieve optimization of volume and structure. For small and medium enterprises (SMEs), merger and acquisition can enhance their own strength; for large enterprises, either absorbing SMEs or moving into other industries can cultivate new growth points and ease China's overcapacity in some industries.

Zhu Chunsheng also believes that to solve the problem of excess production capacity, the relevant industries' merger and acquisition activities should be increased; this will make the advantaged enterprises bigger and stronger, improve industrial concentration, stabilize the market order and prices, and promote related industries development. Enterprises should make full use of market mechanisms to control production capacity, improve efficiency, speed up their own reform and restructuring, and help improve industry concentration and the overall level of technology. Low market concentration has often led to investment fragmentation and lack of market rationality, increasing the risk of excessive entry in some industries. Enterprises should fully respect the market, actively carry out mergers and acquisitions, and improve their competitiveness, the larger ones meanwhile helping SMEs to achieve modern technology upgrades and improved management rationalization. In this way, all should promote the common development of all types of enterprises, forming a good industrial order.

With overcapacity appearing, the study of control measures only gradually increased. On the whole, the governance of overcapacity will further rely on the government to formulate relevant macroeconomic policy; based on this, various industry associations should also play their role well, helping overcapacity industries to regulate their own production scale and promoting the development of self-regulation. The enterprises themselves should regulate their own production behavior under the guidance of national policy, paying attention to the information of relevant state departments and industry associations, adjusting in a timely way (and avoiding the expansion of) their own production scale.

4 Conclusion

Through the analysis of this paper, we can find that to resolve the current situation of overcapacity, we should take steps in three directions. The government should guide the production structure effectively, lead industries to eliminate poor production capacity, and promote industrial upgrading. It can also establish an industry information platform and a guidance system to supervise the production of the enterprise and control production quantity. Enterprises should actively carry out mergers and acquisitions, and accelerate the going out strategy. Industry associations should do their part as intermediaries.

References

1. Erturk KA (2001) Overcapacity and the East Asian crisis. J Post Keynes Econ 24(2):253–255
2. Henderson J, Cool K (2003) Corporate governance, investment bandwagons and overcapacity: an analysis of the worldwide petrochemical industry, 1975–1995. Strateg Manag J 24 (4):349–373
3. Sumaila UR, Teh L, Watson R et al (2008) Fuel price increase, subsidies, overcapacity, and resource sustainability. ICES J Mar Sci J du Cons 65(6):832–840
4. Clark CW, Munro GR (2002) The problem of overcapacity. Bull Mar Sci 70(2):473–483
5. Asche F, Eggert H, Gudmundsson E et al (2008) Fisher's behaviour with individual vessel quotas—over-capacity and potential rent: five case studies. Mar Policy 32(6):920–927
6. Ward J, Mace P, Thunberg E (2005) The relationship of fish harvesting capacity to excess capacity and overcapacity. Mar Res Econ 19(2):525–529
7. Innes G, Grafstein E, Stenstrom R et al (2007) Impact of an overcapacity care protocol on emergency department overcrowding. Acad Emerg Med 14(5):85–86

Optimal Types of Traffic Sensors Located in a Stochastic Network: A Bi-Level Programming Model

Qiubo Zhang

Abstract This paper addresses the optimization model of traffic sensor location considering drivers' route choice behaviors. Based on the idea of bi-level programming, a mathematical model with an objective of maximizing total observed traffic flow, it is first formulated to maximize the benefit game between traffic managers and drivers. A hybrid GA-MSA algorithm is proposed to obtain the optimal or near-optimal solution of the above model, in which GA is utilized to solve the upper-level mixed integer nonlinear programming and MSA is adopted to get the link flow pattern in a stochastic user equilibrium state under different traffic sensor location schemes.

Keywords Sensor location · Bi-level programming · Stochastic user equilibrium · Genetic algorithm

1 Introduction

The increasing need for real and immediate information triggers more and more attention to the optimal deployment of sensors. The deployment of sensors should not only consider the effectiveness of the system, but also depend on the real constraints of city conditions, such as weather conditions and limited budget. The budget constraint exerts a lot of influence on how sensors are deployed. Zhou focuses on locating a limited set of sensors in a network to maximize the expected information [1]. Danczyk developed a mixed-integer linear program for optimally allocating sensors, by minimizing the total surveillance error [2]. Li proposed a reliable facility location model to optimize traffic surveillance [3].

There are various types of sensors. Intuitively, people accept to varying degrees monitoring through different kinds of sensors. Little research has addressed the

Q. Zhang (✉)
School of Traffic and Transportation, Beijing Jiaotong University, Beijing 100044, China
e-mail: 14120926@bjtu.edu.cn

M. Li et al. (eds.), *Proceedings of 2015 2nd International Conference on Industrial Economics System and Industrial Security Engineering*,
DOI 10.1007/978-981-287-655-3_53

deployment in the traffic network of different types of sensors under the condition of limited budget.

This paper aims to fill this gap. It builds a bi-level programming model for traffic network surveillance. The upper level program model takes the road managers (or governments) as the leader in decision-making, and regards the upper limit of investment as the constraint to achieve maximal observed flow in the road network. The lower level program builds a multi-sensors-type multi-acceptance-degree stochastic user equilibrium model for a logit-based consideration of the differences in route choice behaviors.

The remainder of the paper has the following layout. The notation and the bi-level programming model are proposed in Sect. 2. Section 3 proposes customized algorithms for this problem. Section 4 conducts numerical experiments to draw managerial insights. Finally, conclusions and future study are presented in Sect. 5.

2 Bi-Level Program Model for Sensor Location Problem

2.1 Notation

Consider a network $G = (N, A)$; N expresses the node of network, A expresses the link of network. R is a set of origin node, and S is a set of destination node. The following notations apply.

r	Origin node, $r \in R$
s	Destination node, $s \in S$
M	Classification of sensors, m belonging to a specific type, $m = 1,2,\ldots,M$
K_{rs}	Set of paths between r and s
q_{rs}	Travel demand from origin r to destination s, $r \in R, s \in S$
x_a	Flow on link a, $a \in A$
$\boldsymbol{x}$	Vector(…, x_a, …), $a \in A$
C_a	Traffic capability on link a, $a \in A$
$t_a(\boldsymbol{x})$	Travel cost function of link a, $a \in A$
$f_{rs}^{k,m}$	Flow of sensor class m on path k connecting r and s, $k \in K_{rs}, r \in R, s \in S, m \in M$
c_{mk}^{rs}	Travel cost on path k connecting r and s under sensor class m, $k \in K_{rs}, r \in R, s \in S, m \in M$
$\delta_{rs}^{k,a}$	0-1 indicator variable; it equals 1 if link a is on path k connecting r and s, and 0 otherwise
y_a	0-1 decision variable; if a sensor is installed on link a, then $y_a = 1$; otherwise $y_a = 0$
$\boldsymbol{y}$	Vector of sensor location variables (…, y_a, …), $a \in A$
β^m	Additional cost for sensor class m, $m \in M$
Q_{max}	The maximum number sensors that can be deployed in the network

2.2 The Upper Level Optimization Problem

The network planners expect surveillance of more traffic flow; thus the location of sensors is critical to them.

The sensor location problem is a discrete problem. Due to the limited budget, the maximum number of sensors which can be deployed in the network is fixed.

Upper level:

$$\max G(\boldsymbol{xy}) = \sum\nolimits_a x_a(\boldsymbol{y}) \cdot y_a \quad (1)$$

$$s.t. \sum\nolimits_a y_a \le Q_{\max} \quad (2)$$

$$y_a = \{0, 1\}, \forall a \in A \quad (3)$$

The objective (1) is to maximize the total observed traffic flow using different types of sensors. Constraint (2) denotes the maximal number of sensors that can be installed in the network. Constraint (3) is the binary decision variable for a sensor deployment.

2.3 The Lower Level Logit-Based Stochastic User Equilibrium Assignment Problem

The sensor location scheme will influence the assignment of traffic flow. Based on random factors, the logit-based SUE model is built to account for the random behaviors of travelers.

The sensor types are divided into M classes. Travelers have different acceptances of each type, $\beta^m(\beta^m \ge 0)$. The higher a sensor class is, the higher a value β^m represents.

Lower level:

The flow pattern that minimizes Eq. (4) satisfies the SUE conditions.

$$\min Z(\boldsymbol{x}) = \sum\nolimits_a x_a \cdot t_a(x_a) - \sum\nolimits_{rs} q_{rs} \cdot S_{rs}[\beta^m] - \sum\nolimits_a \int_0^{x_a} t_a(w)dw \quad (4)$$

$S_{rs}[\beta^m]$ represents the expected minimized travel cost from r to s, and can be formulated by:

$$S_{rs}[\beta^m] = -\frac{1}{\theta} \ln \sum\nolimits_k \exp(-\theta \cdot c_{mk}^{rs}) \quad (5)$$

The route travel cost, which is related to the acceptance by drivers of different types of sensors, can be expressed as below:

$$c_{mk}^{rs} = \sum_a (t_a(\boldsymbol{x}) + y_a \cdot \beta^m) \cdot \delta_{rs}^{k,a} \tag{6}$$

Considering the road conditions where congestion may be a factor, all the links are assumed to have BPR functions as below:

$$t_a(\boldsymbol{x}) = t_a^0 \cdot \left(1 + 0.15 \cdot \left(\frac{x_a}{C_a}\right)^4\right) \tag{7}$$

3 Solution Algorithm

A GA-MSA approach will be proposed to solve the sensor location optimization model [4–8].

3.1 GA Algorithm for the Upper Level Optimization Problem

The specific procedure can be described as follows:

GA-Step 0: Initialization. In our study, the population of chromosomes is M, and the size is |Qmax| (the number of candidate sensors). Set 1 to present the deployment of a sensor; otherwise set 0.

GA-Step 1: Computation of objective and fitness function values. With the specific chromosomes, the travel cost can be calculated. Then the MSA method will be used to obtain the balance flow and the value of objective function $G(\boldsymbol{x})$. Apply the ranking method to obtain the fitness value objective, which is derived from the upper level objective function.

GA-Step 2: Gene operations.

Sub-Step 2.1: Selection. According to the calculation of the fitness value objective, use a roulette wheel (a fitness-proportional approach in essence) to obtain a new population.

Sub-Step 2.2: Crossover. Given the probability of crossover pc as one of the parameters of a genetic system, crossover with the one-cut-point method is selected to generate the new offspring.

Sub-Step 2.3: Mutation. With the probability being equal to the mutation rate pm, every bit in the whole population has a chance to undergo mutation (i.e., change from 0 to 1 or vice versa), and thus it will be mutated.

GA-Step 3: Convergence Testing. After a certain number of generations, if no further improvement is observed, the current best chromosome that has been kept represents an optimal (possibly global) solution. Or one can also stop running the algorithm after a fixed number of iterations; the decision to do so depends on speed and resource criteria. Otherwise, go back to Step 2.

3.2 *MSA Algorithm for the Lower Level SUE Problem*

The path flow and route flow can be expressed as follows:

$$x_a = \sum_{rs} \sum_{k} P_{mk}^{rs} \cdot q_{rs} \cdot \delta_{ak}^{rs}, \quad f_{rs}^{k,m} = P_{mk}^{rs} \cdot q_{rs}, \tag{8}$$

According to the logit formula, it is based on the assumption that the utilities of all the alternatives in the choice set are identically and independently distributed Gumbel variates. The specification of the utility function implies that:

$$P_{mk}^{rs} = \frac{\exp(-\theta \cdot c_{mk}^{rs})}{\sum_{l} \exp(-\theta \cdot c_{mk}^{rs})} \tag{9}$$

The descent direction: $d_a = \sum_{rs} \sum_{k} P_{mk}^{rs} \cdot q_{rs} \cdot \delta_{ak}^{rs} - x_a$;

The algorithm described in this section is known as the method of successive averages (MSA). The MSA algorithm can be summarized as follows:

MSA-Step 0: Initialization. According to the given O-D demands q_{rs}, apply $P_{mk}^{rs} = \frac{\exp(-\theta \cdot c_{mk}^{rs})}{\sum_{l} \exp(-\theta \cdot c_{mk}^{rs})}$ to calculate the possibility of route choice by free-flow travel cost. Use the functions $x_a = \sum_{rs} \sum_{k} P_{mk}^{rs} \cdot q_{rs} \cdot \delta_{ak}^{rs}$ and $f_{rs}^{k,m} = P_{mk}^{rs} \cdot q_{rs}$ to calculate x^1, which expresses the initial link flow. Set $n = 1, \varepsilon > 0$;

MSA-Step 1: Update the link travel time and the path travel cost. According to the current link flow x^n, update the link travel time $t^n = [t(\boldsymbol{x})^n]$ and path travel cost $c_k^{rs\,n}$;

MSA-Step 2: Update descent direction. Calculate descent direction d^n;

MSA-Step 3: Update link flow. Let $x^{n+1} = x^n + \frac{d^n}{n+1}$;

MSA-Step 4: Convergence checking. If $\|d^n\| \leq \varepsilon$, stop. Otherwise, let $n = n + 1$; go back to MSA-Step 1.

4 Case Study

To analyze the model, we present a small scale network using the structure of Fig. 1. The text network includes 5 links and 1 O-D pair. Each link cost can be expressed by the parameters in Table 1, using the BRP function. The whole demand of O-D pairs is 200. Let route 1 be 1-2,route 2 be 3, and route 3 be 4-5.

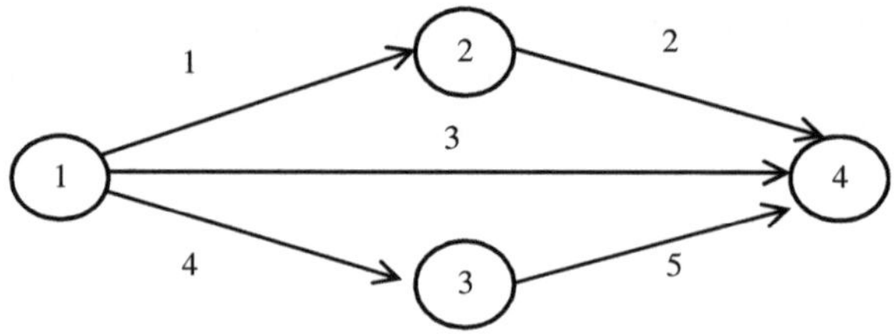

Fig. 1 Text network

Table 1 Parameters of network

Link	1	2	3	4	5
t_a^0	6	6	13	5	7
c_a^0	200	200	250	200	200

4.1 Deployment of Higher-Acceptance Sensors

When higher-acceptance sensors are deployed in the network, these serving to collect the flow of the road and not to gather information about drivers, the sensors will be ignored, adding additional cost $\beta^1 = 0$. Thus the link flow can be calculated as in Table 2; the sensors can be deployed in link 1 or link 2 to obtain more traffic flow.

4.2 Deployment of Lower-Acceptance Sensors

While accurate sensors deployed in the network can obtain more detailed information about drivers (including their vehicle registrations) and about their behavior, vehicle drivers will impose a huge additional cost on road choice. In this condition, we assume that the additional cost is $\beta^2 = 0.5$, so the link flow can be expressed as in Table 3.

Several changes are shown relative to the deployment of higher-acceptance sensors. Now, the sensors should be assigned to link 3, rather than link 1 or link 2.

Table 2 Flow result when deploying higher-acceptance sensors

θ	β	f_1^{rs}	f_2^{rs}	f_3^{rs}
0.1	0	75.2501	67.9919	56.1376

Table 3 Flow result when deploying lower-acceptance sensors

θ	β	f_1^{rs}	f_2^{rs}	f_3^{rs}
0.1	0.5	71.3497	72.1185	57.376

5 Conclusion

In this study, we have shown the possibilities for a sensor location model integrated with a bi-level programming model, which balances the traveler's route choice behavior and the network managers' needs. An algorithm based on GA-MSA was developed to solve the model, applying the GA algorithm to solve the upper level, as well as focusing on the MSA algorithm to solve the lower level. The results offered by the proposed model are of great use for different types of sensor location. Nevertheless, it is also obvious there are some limitations. The model needs to be applied in a complex network for its capabilities to be analyzed, and for the parameters of GA to be modified. With future study, through changing the objectives in the upper level, it will have a considerable influence on managers' decisions.

References

1. Zhou X, George F (2010) An information-theoretic sensor location model for traffic origin-destination demand estimation applications. Transp Sci 44(2):254–273
2. Danczyk A, Liu H (2011) A mixed-integer linear program for optimizing sensor locations along freeway corridors. Transp Res Part B 45(1):208–217
3. Li X, Ouyang Y (2011) Reliable sensor deployment for network traffic surveillance. Transp Res Part B 45(1):218–231
4. Murayama A, Kanasugi A (2010) A novel coding method for genetic algorithms based on redundant binary numbers. Artif Life Robot 15(3):306–308
5. Kuoa RJ, Han YS (2011) A hybrid of genetic algorithm and particle swarm optimization for solving bi-level linear programming problem-a case study on supply chain model. Appl Math Model 35(8):3905–3917
6. Yang J, Zhang M, He B, Yang C (2009) Bi-level programming model and hybrid genetic algorithm for flow interception problem with customer choice. Comput Math Appl 57(11–12):1985–1994
7. Bar-Gera H, Boyce D (2006) Solving a non-convex combined travel forecasting model by the method of successive averages with constant step sizes. Transp Res Part B 40(5):351–367
8. Sbayti H, Lu C, Mahmassani H (2007) Efficient implementation of method of successive averages in simulation-based dynamic traffic assignment models for large-scale network applications. Transp Res Rec: J Transp Res Board 2029:22–30

The Characteristics of Traffic Congestion from the Perspective of Spatial Structure, Urbanization, and Motorization in Chongqing

Chengfeng Huang, Yin Ding, Yuanyuan Zhang and Zheng Shen

Abstract This work analyzes the particularities on present urbanization, urban spatial structure with "multi-centre and cluster-mode," and motorization of the main urban area in Chongqing. In addition, the characteristics of traffic congestion are analyzed from the perspective of time and space evolution. The characteristics show that the congestion sources are combined with congestion points and regionalized; the congested points locate in connections between clusters, and transportation supply is insufficient; traffic congestion occurs intensively during the morning and evening peak hours and is in the stage of duration extension. The characteristics indicate that the traffic congestion in the inner-ring urban areas of Chongqing is getting worse with more congested regions and last longer than peak-hour duration. Finally, the ideology on congestion management is proposed briefly in the aspects of urban spatial structure, public transit, traffic-demand management, and travel civilization.

Keywords Urban spatial structure · Urbanization · Motorization · Traffic congestion · Characteristics

C. Huang (✉)
Western China Transportation-Economy-Society Development Study Center,
Chongqing Jiaotong University, Chongqing 400074, China
e-mail: zhenwushanxia@126.com

Y. Ding
Dalian Auto Vocational and Technical College, Dalian 116031, China
e-mail: 3804941589@163.com

Y. Zhang
ShangHai WSP Consulting LTD, Shanghai 200041, China
e-mail: 281388738@qq.com

Z. Shen
Jiangsu Transportation Institute, Nanjing 210017, China
e-mail: 673111937@qq.com

M. Li et al. (eds.), *Proceedings of 2015 2nd International Conference on Industrial Economics System and Industrial Security Engineering*,
DOI 10.1007/978-981-287-655-3_54

1 Introduction

As China's enter the 21st century, there is an expansion of the traffic demand and deterioration of the urban traffic–running environment with rapid development of urbanization, motorization, and restructured urban spatial form. While enjoying the social and economic prosperity, people in cities are forced to bear the negative effects caused by traffic congestion. In face of the severe urban traffic status, it is essential to rethink the generation and development of traffic congestion in different urban development stages.

Taylor, a scholar of transportation planning and policy in the United States, argued that cities exist because they promote social interactions and economic transactions. Traffic congestion occurs where many people pursue these ends simultaneously in limited spaces [1]. However, the characteristics of traffic congestion at a particular level of social and economic development are not demonstrated by the results of his further studies.

He and Xie [2], traffic community scholars in China, argued that traffic congestion is not only a technical issue and economic issue but also a social problem. Traffic congestion in cities of China is associated with urbanization and economic development. By researching the relationship between traffic congestion and urban spatial expansion, Xu [3] concluded that simple spatial expansion (for example, a single center of urban sprawl and suburbanization) leads to traffic congestion. The mitigation of traffic congestion needs construction of employment centers and complex spatial extension to coordinate industrial conversion. An effective construction of a multicenter layout must be supported by effective public transport systems. Xu et al. [4] pointed out that polycentric urban structure is a sustainable strategy of alleviating traffic congestion in large cities, but the study lacks of empirical data.

This work analyzes the characteristics of urban traffic congestion from the perspective of the particularities of urbanization, urban spatial structure, and motorization of Chongqing based on an empirical study of the inner ring urban area in Chongqing. This study aims to conduct a basic theoretical study for further research of effective congestion-control strategies.

2 Particularities of Urbanization, Urban Spatial Structure, and Motorization

2.1 *Definition of the Basic Unit*

The main urban area of Chongqing is divided into several clusters. These clusters compose the basic units of the urban spatial structure with a "multi-center, cluster-mode". This work takes administrative district as the basic unit because its regionalism is positively correlated with cluster division as well as due to the difficulty accessibility of the data.

2.2 *Particularities of Urbanization*

Urbanization is an important symbol of human civilization as well as an important stage in the development of social productive forces. In brief, urbanization is a natural historic process in which a traditional village society develops into a modern urban society [5]. Many research data show that there are dramatic changes in economy, society, politics, and culture of the city when its urbanization rate reaches 50 %. Urban problems, such as traffic congestion, environmental pollution, and housing stress, are particularly prominent.

(1) Rapid progress of urbanization

The rapid progress of urbanization makes high-level requirements to urban transport infrastructures. The "logistic" curve of the development of urbanization states that an urbanization rate rising from 30 to 70 % reflects an accelerated development period of urbanization. And the rate at the later stage of this period from 50 to 70 %. The urbanization rate of Chongqing is 53 %, showing that Chongqing has entered the later stage of the accelerated development period.

The average annual growth rate of urbanization in Chongqing was 1.68 % from 1996 to 2010, which is 0.29 % higher than the national average increase rate of 1.39 % during the same period. The annual increasing % age of Chongqing exceeds that of Europe, America, Japan, and Korea. The comparison is shown in Table 1.

Table 1 The comparison of the speed of urbanization between developed countries and Chongqing

City/country	Accelerated development period of urbanization	Times used (year)	Urbanization Rate (%)	Annual increase (%)
Chongqing	1996–2010	14	29.5–53.0	1.68
China	1996–2010	14	30.48–49.95	1.39
UK	1801–1890	90	26–72	0.51
France	1851–1970	120	25.5–70	0.37
U.S.A	1870–1960	90	25.7–70	0.49
Japan	1950–1980	30	37.3–76.2	1.30
Korea	1960–1988	28	27.71–70.24	1.52

Sources The data for Chongqing comes from Chongqing statistical yearbooks of relevant years. The data for China comes from the 2011 China Statistical Yearbook. The data of the UK, France, and the U.S.A comes from Peiyi Gao's literature of Comparative Study of Chinese and Foreign Urbanization. The data of Japan and Korea derives from Yong Liu's literature of The Strategy Research of China's Urbanization

(2) Increase in the size and density of population

The urbanization of Chongqing is characterized by population urbanization, which directly generates an urban-population explosion. Because the size and density of population in Chongqing are increasing rapidly, it inevitably results in a drastic rise in the volume and intensity of the urban traffic demand. These finally lead to enormous pressures on urban traffic systems.

The resident population in the main urban area increased from 6.1783 to 7.4576 million from 2003 to 2010, and the average annual growth rate was 2.9 % with a resident-population density of 1,400 persons/km. The daily travel volume of residents in the main urban area of Chongqing has increased continuously by 10.8 % from 2007 to 2010 and was up to 13.39 million in 2010. By 2020, the citizen population of citizens will be up to 12 million [6], and the travel volume will continue to maintain a rapid growth in the future because approximately 2.25 times of travel will be generated from every person. Therefore, the traffic supply is under serious pressures. Urban traffic congestion becomes the costs of urbanization development during this period.

(3) Greater urban agglomeration effect

Agglomeration and dispersion effects are the main driving forces promoting the urbanization process. Each effect plays a leading role in urban development during different stages of urbanization. Experience from developed countries and regions demonstrate that the agglomeration effect is greater than the dispersion effect when urbanization rate reaches 50 %. The strong attraction of city residents to populations and industries makes the city crowded and induces traffic congestion. Where there is a high level of agglomeration effect on economic factors, the traffic flow is busier, and it naturally will generate conflict between transportation supply and traffic demand.

Therefore, industry density, an indicator for measuring economic development, indicates the characteristics of the agglomeration and dispersion effects during the process of urbanization. Thereby, the characteristic of traffic congestion can be explained. Based on comparative data analysis between the area of administrative districts and the total GDP of the main urban area of Chongqing in 2010, the industry density of each district is shown in Fig. 1.

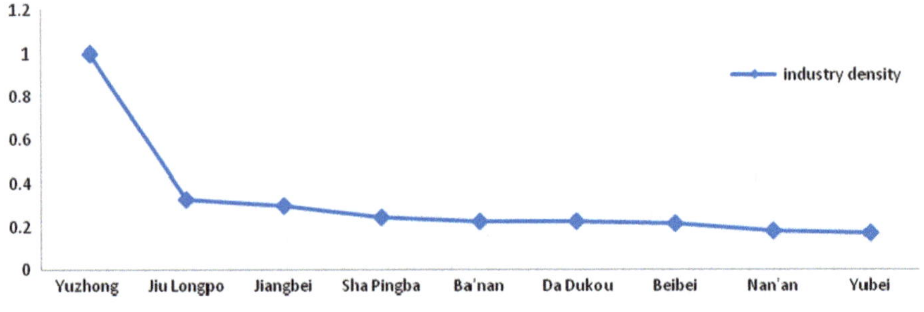

Fig. 1 The comparison of industry densities of administrative districts of Chongqing

The industry density of Yuzhong district is higher than that of other districts, and the data show an incomparable advantage in the level of economic development. This demonstrates that the center cluster (Yuzhong cluster) has an obviously strong agglomeration effect, whereas the economic agglomeration capacity of the four subcenter clusters is weaker. Therefore, large-scale profit gaps exist in the four subcenter clusters and the center cluster. The subcenter clusters also still rely highly the center cluster with strong economic connections. As a result, it strengthens the agglomeration effect of the center cluster and induces over-concentrated traffic flow in terms of both me and space (mainly in rush hours on roads between the center clusters and within the circle road around CBDs). The contradiction between the agglomeration effect and the limited urban space intensify urban traffic congestion.

2.3 Particularities of Urban Spatial Structure

Urban spatial structure and urban land use are the origins of traffic demand. The city size, development intensity, and functional layout of urban spatial structure determine the total traffic demand, the spatial distribution, and the distribution of traffic flow intensity. Then these affect the characteristics of congestion such as congestion severity, congestion spatial distribution, and duration of urban traffic congestion.

The main urban area of Chongqing originates from Yuzhong peninsula and gradually evolved from a group-shaped city into an urban form with a decentralized urban level concentrated in a clusters level and point-like under the segmentation of natural barriers such as rivers and mountains. The city eventually formed an urban spatial structure with "multi-center and cluster-mode" (Fig. 2). This differs from the urban-development mode of some polycentric cities both at home and abroad. The urban spatial structure with "multi-center and cluster-mode" is corroded by rapid urban construction with the development of urbanization, the expansion of urban population, and increasing land demand.

(1) Vertical growth of clusters and higher intensity of urban construction

The urban construction area increased from 300.09 km^2 in 2003 to approximately 620 km^2 2010. However, urban spatial growth has focused on the vertical growth of clusters, and the major constructions are concentrated in the main urban areas within mature clusters.

There is cluster sprawl, i.e., erosion of the green space between clusters and adhesion between clusters frontiers [7]. The clusters integrate with each other and evolve into larger urban "regions." Facilities of commerce, entertainment, business, science and technology, culture, and education institutes are centralized in the regional center. The jobs housing balance, which reflects cluster's functions, is emphasized in a larger scope of regions [7].

The city center cluster Yuzhong cluster constitutes implemented inner-aggregated growth. The building density per m^2 of commerce as well as housing is steadily

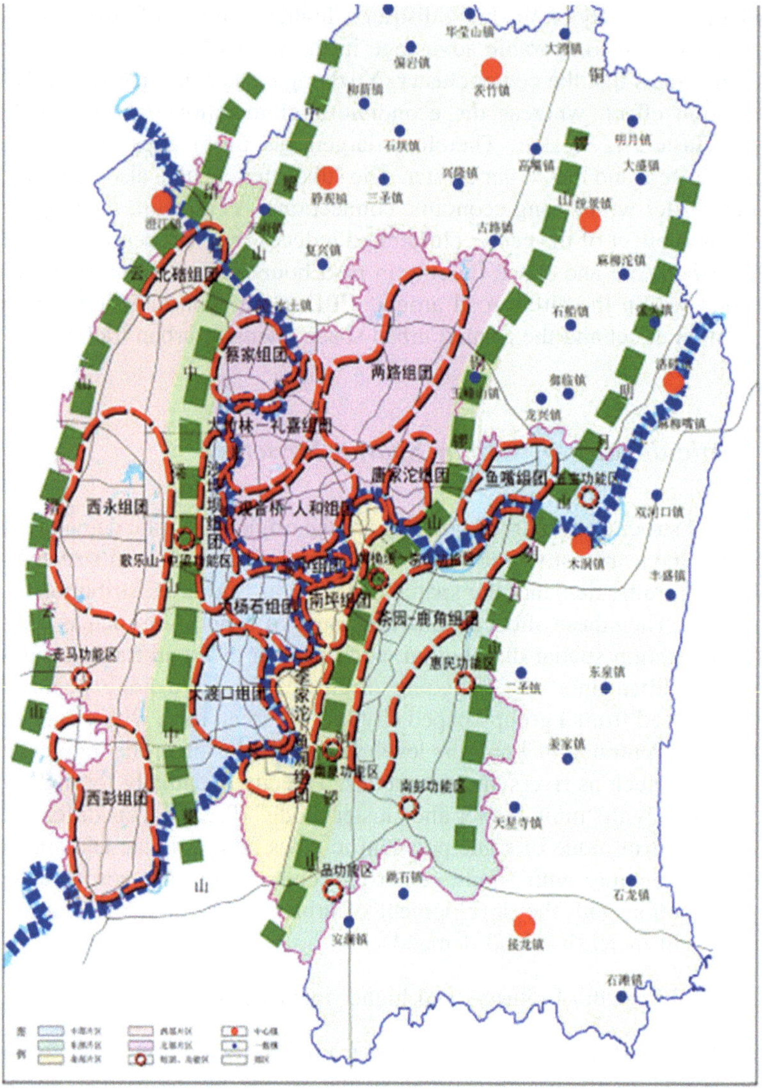

Fig. 2 Spatial structure of the main urban area of Chongqing [6]

increasing, which results in a large total construction and traffic volume exceeding the capacity of transportation facilities.

The development of urban road traffic is lagging behind the vertical growth of clusters. By 2010, the per-capita area of roads in the main urban in Chongqing was only 5.9 m^2, which is far below that of other major cities of China (Shanghai is nearly 16 m^2, Guangzhou is 12.8 m^2, Tokyo 13.5 m^2, and London 24.5 m^2) [8]. In the face of the huge traffic demand brought by urban development, sufficient road-traffic supply is lacking.

(2) A weak coupling of urban spatial structure and traffic structure

Studies in the literature have indicated that there is a dynamic and reciprocal-feed interaction relationship between urban spatial structure and traffic structure. It can provide various levels of accessibility for the urban spatial system and maximize both land-use efficiency and traffic operational efficiency by establishing a transport structure with "multi-level and three-dimensional mode" and coupling it with an urban spatial structure of "multi-center and cluster-mode." Therefore, traffic congestion can be mitigated effectively.

The system of "city center, subcenter, and cluster center" in the main urban areas of Chongqing has not been formed, and it is still in the stage of "city center and sub-center." Public-transport resources of center clusters are intensive, whereas the public transport service of noncenter clusters is very limited. On one hand, there are a large number of bus lines in the city center, but public transit has the problem of high double-line coefficient with low punctuality rate and low running speed due to traffic congestion. It lacks bus lines and has low level of service in the fringe area of the urban area within the inner ring. On the other hand, the city MRT/LRT are centered in the Yuzhong cluster and radiate to other clusters mainly connecting clusters and regions where there is a city center or subcenter. However, the city MRT/LRT network and its connection with buses have not been formed and lacks of large-capacity transport system due to the low density and limited coverage of transit lines.

To a large extent, a weak coupling of urban spatial structure and traffic structure reduces the attractiveness of public transport; on the contrary, it stimulates the growth of private-car traffic and increases the physical pressure to roads.

2.4 *Particularities of Motorization*

Motorization can be viewed as derivatives of urbanization. Because of the high degree of mobility, it has compressed time and distance and expanded the accessibility and urban scale during the process in changing from a traditional backward rural society into a modern urban society. It facilitates communication in economy, society, and culture and promotes the development of urbanization. However, the particularities of motorization in the main urban area of Chongqing at this stage have become a main pushing force toward traffic congestion.

(1) Automotive consumer culture infected resulting in traffic-consumption upgrade and surge in traffic demand

The private car is the major growth point of rapid developing motorization, which increases the pressure to road transport. With the increase of the disposable incomes of urban residents in Chongqing, there is a sharp rise in people's requirements for accessibility, flexibility, and comfort to transport. The private car is able to meet these requirements. Therefore, the ownership and use of private cars has become the main way to increase levels of private motorization.

Automobile consumption has become the major activity of residents in Chongqing. The ownership of motor vehicles in Chongqing increased significantly compared with other cities and is apparently in a "burst" phase. In addition, it was been higher than that in Hong Kong in 2010 (Table 2).

The per-capita vehicle ownership of the main city of Chongqing is approximately 117/1 thousand people, whereas the per-capita vehicle ownership of Hong Kong and Shanghai is 97.5 and 110/1 thousand people, respectively. This means that the main city of Chongqing has 129,000 vehicles more than Hong Kong and 44,000 vehicles more than Shanghai (excessive consumption is even more shocking if it is compared with Beijing, Chengdu, and Guangzhou in accordance with the equivalent). The growth in automobile consumption will lead to a continue increase in traffic-travel vehicles is urban areas.

In 2011, the number of motor vehicles in the main urban area of Chongqing reached 812,000, which is 145,000 more vehicles compared with 2010. The number of private vehicles is 377,000, which accounted for 46.4 % of motor vehicle ownership. The annual increase in the rate of private cars increase by >25 % within 7 years, and this plays the main role in the increase of motor vehicles [9] (Fig. 3).

Table 2 Car ownership in the city of Beijing, Shanghai, Guangzhou, Hong Kong, the central area of Chengdu, and the main city of Chongqing

Year	Beijing		Shanghai		Guangzhou	
	Ownership of motor vehicles	Rate of increase (%)	Ownership of motor vehicles	Rate of increase (%)	Ownership of motor vehicles	Rate of increase (%)
2005	258.3	12.5	211.5	–	177.4	3
2006	287.6	11.3	213	0.7	183	3.2
2007	319.6	11.1	227	6.6	182.3	0.2
2008	350.4	9.6	234.4	3.3	183.9	0.9
2009	401.9	14.7	243.4	3.8	195.5	6.3
2010	470	16.9	248.8	2.2	214	9.5
2011	498.3	6	251.6	1.1	232.5	8.6
Year	Hong Kong		Central area Chengdu		Main city of Chongqing	
	Ownership of motor vehicles	Rate of increase (%)	Ownership of motor vehicles	Rate of increase (%)	Ownership of motor vehicles	Rate of increase (%)
2005	54.1	1.5	48.3	3.6	30.2	7.5
2006	55.3	2.2	54.6	13	34.3	13.6
2007	56.5	2.2	62.6	14.7	38.9	13.4
2008	57.5	1.8	71.5	14.2	44.1	13.4
2009	58.4	1.6	89.7	25.5	52.8	19.7
2010	66.4	13.7	97	8.1	66.7	26.3
2011	69	3.9	107.9	11.2	81.2	21.7

Data Source 2012 Chongqing City Traffic Annual Reports

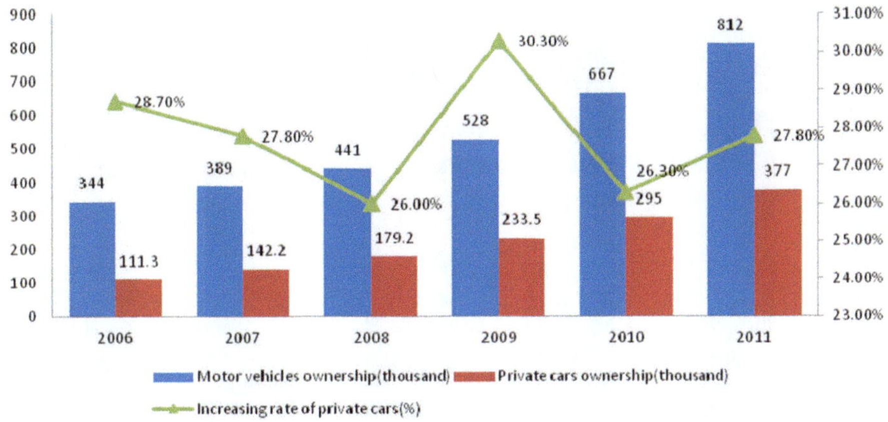

Fig. 3 Ownership of motor vehicles and increasing number of private cars in the main urban area of Chongqing

The direct results of rapid growth in motor vehicles are constantly increasing motorized travels and more pressures to road traffic.

(2) Public transport in passive "competition" with private motorized transport

In 2010, the proportion of public transport in total travel increased by 0.4 %, and the percentage of private cars increased by 1.4 % compared with 2009. It can be seen that the latter is 3.5 times more than that of the former. Therefore, the motorization of travel is characterized by rapid growth of private cars as shown in Fig. 4.

The growth of economic development and increase in incomes support the development of private cars. However, the development of public transport is lagging behind compared with the rapid growth of the private car. The advantage of the public transport system in mitigating traffic congestion is hindered due to the low level of public transport network, single wire nets layer, poor quality in service, and low level of mass rail-transit development. It also causes a preference toward the overuse of the private car, which can aggravate the burden to road traffic and worsen the public transportation environment. Thus, it forms a vicious circle of traffic congestion. As Mumford puts it, the travel speed of cars can be very fast, but a substantial increase in private cars slows down the final speed [10].

Also shown is the degree of dispersion in employment and housing has increased in the main city area of Chongqing after the urban population has become more prosperous. The lifestyle of living in clusters has changed gradually into a lifestyle with car reliance due to more frequent traffic connections between clusters. The development of motorization causes the decrease of impedance factors in the distance of space and time. It also causes the irrational demand for motorized travel, which has played a key role of increasing urban traffic congestion.

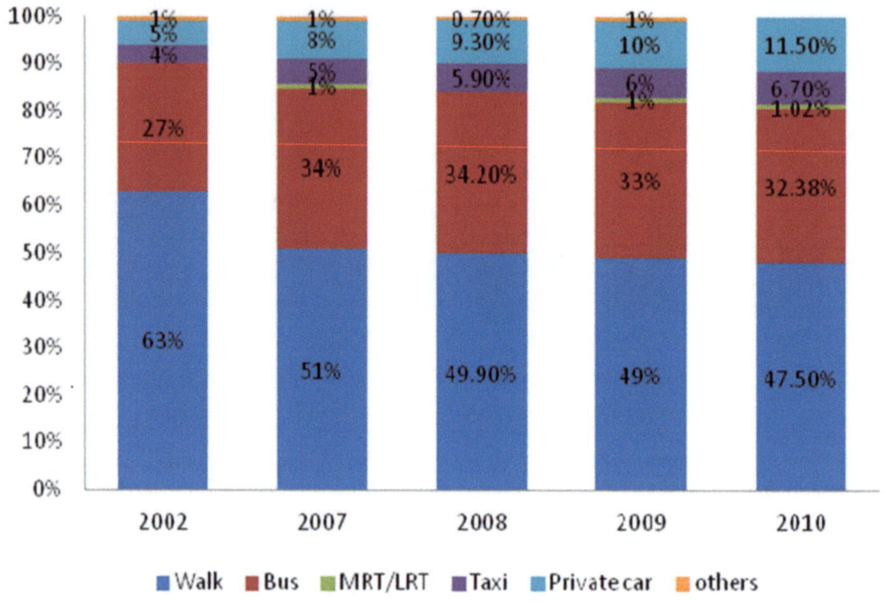

Fig. 4 Travel structure of residents in the main urban area of Chongqing from 2002 to 2010

3 The Characteristics of Urban Traffic Congestion in the Inner Ring Urban Area of Chongqing

Due to the limitation of data acquisition, the characteristics of traffic congestion within the inner ring urban area of Chongqing were studied in this work. The characteristics are analyzed in this section based on the above-mentioned particularities of urban spatial structure, urbanization, and motorization.

3.1 The Characteristics of Congestion Spatial Evolution

Characteristics of congestion spatial evolution, congestion sources regionalized and combined with congestion points, regional congestion in the center of the city, and circle-layer congestion in subcenters.

Traffic congestion in the monocentric cities concentrates on the city center and spreads along the road network shaped with "ring and radiation" to the suburbs. Take Beijing as an example. The congestion is called "pie-style" congestion. Under the spatial structure of "multi-center and cluster-mode," traffic congestion in the inner ring urban area of Chongqing shows the characteristics of the combination of congestion sources and congested points: Regional congestion occurs in the city center (Yuzhong district), and circle-layer congestion occurs in the CBDs of sub-centers (Fig. 5).

"Congestion sources" refers to places that attract and create huge volumes of traffic, which induce conflicts between supply and demand. Examples of places are large shopping malls and office buildings. Congestion points, such as road intersections, bridges, and tunnels, do not produce traffic demands, but congestion occurs there due to the volume of traffic exceeding their capacity. "Regional congestion" refers to the condition in which congestion allocates in interrelated segments of roads, and the intersections between segments coincide. "Circle-layer congestion" represents that the volume of traffic on the ring road in a certain region exceeds its capacity, resulting in disruptions to traffic and congestion on the ring road.

The rate of urbanization in the city center (Yuzhong district) has reached 100 %. It has the largest resident population density, >26,600 people/sq km, which thus induces huge travel demands. Yuzhong district is the financial center, commercial center, information center and culture center of Chongqing. It also embraces high-quality education institutions and medical and health institutions and is the hub of waterborne- and road-passenger transportation. Therefore, this district has multiple centric functions with strong central effects. It is the main area by which north south traffic flow passes and carries a large amount of cross-cluster traffic.

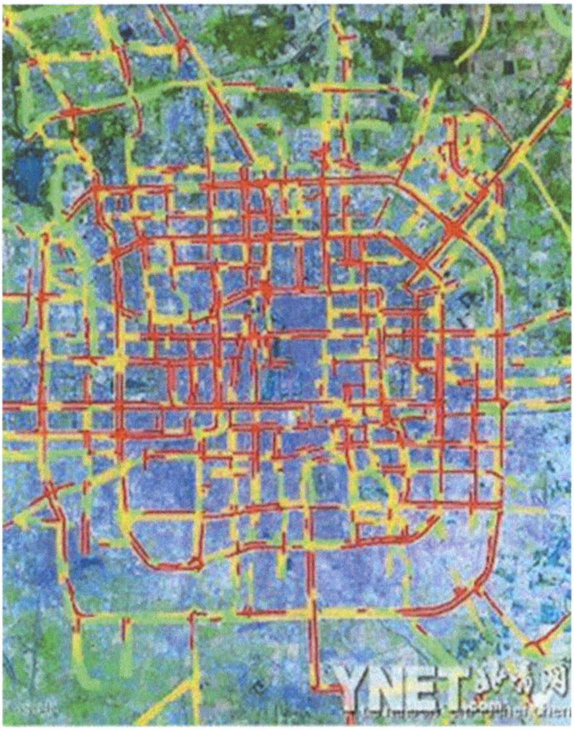

Fig. 5 Real-time traffic chart of Beijing

It undertakes huge traffic pressures both on the internal road networks and the channels between clusters.

Therefore, congested points interweave with congested roads and generate regional traffic congestion. The urbanization rate for all four subcenters is all >90 %, and the population density is second to that of Yuzhong district. CBDs are the regional economic concentration points of the subcenters. However, road traffic enters into the ring roads of CBDs and is mixed with the service traffic flow of the CBDs, which leads to excessive concentration of traffic flow on the ring roads, which become the main congested areas (Fig. 6).

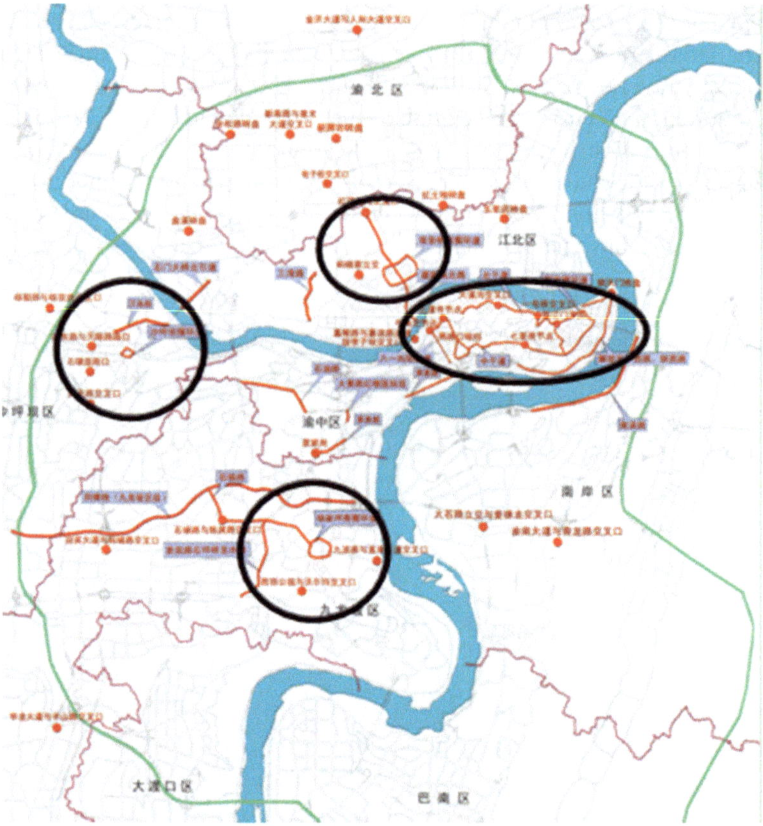

Fig. 6 Congested intersections and segments of the main urban area of Chongqing [11]

3.2 Characteristics of Congestion Spatial Evolution

Characteristics of congestion spatial evolution include congested points locating in connections between clusters. Supply is insufficient compared with demand. Bridges and tunnels become the "bottlenecks" of traffic.

The groups are formed due to the main area of Chongqing being divided by the Yangtze and Jialing rivers with the natural barrier of Zhongliang mountain. Thus, the transportation linkages in cities with spatial structure of "multi-center and cluster-mode" are clearly divided as inner-cluster transportation, trans-cluster transportation, and external transport, which differ from those of noncluster cities. For Chongqing, bridges and tunnels are the main arteries to achieve traffic links between clusters.

The congestion points are collected together as congestion lines. These congestion lines are mainly in the important transport links between groups including the three north south lines that connect Nan'an, Yuzhong, Jiangbei, and YuBei districts (Huanghuayuan bridge and Huanghuayuan tunnel, Caiyuanba, Jialing River, and Yu'ao bridges, Bayi and Xiangyang tunnels, and Er'gongyan and Jiahua bridges); the east west line of Er'gongyan bridge connecting Nan'an, Yuzhong, and Jiulongpo districts; two lines connecting Jiangbei and Yubei districts through Shimen and Gaojia Huayuan bridges; as well as Zhongliang Mountain tunnel, which connects the main city with the university town of Shapingba district. These lines connect the centers of important groups with heavy traffic such that the transport lines become bottlenecks.

According to the relevant data, the daily volume of traffic on 22 bridges across the Yangtze and Jialing rivers was 1.521 million PCU in 2011, which is increased by 9.1 % compared with that of 2010. The volumes of traffic on some bridges during peak hours tend to exceed the capacities. Bridges are the main channels between clusters from north to south where the volume of traffic is >10,000 PCU.

The daily traffic flow of 11 tunnels crossing the Zhongliang Mountain and Tongluo Mountain tunnels in the main urban area reached 420,000 PCU in 2011, which increased by 11.1 % compared with compared 2010, and it will continue to increase in the coming year. The Tieshanping tunnel of Yuyi highway is the connection between the Eastern Area and the central city of Chongqing, and the volume of its traffic has increased by 73 %. The University City tunnel is the channel between the Eastern Area and the central city of Chongqing with an 8.0 % increase in the volume of traffic.

The total daily flow crossing Zhongliang Mountain and Tongluo Mountain tunnels was 479,000 in 2012 with a 14.0 % increase compared with 2011 [12]. The daily flow through Zhongliang Mountain tunnel is 270,800, representing an increase of 16.4 % compared with 2011.

The daily flow through Tongluoshan Mountain tunnel is 208,000, representing an increase of 12.2 % compared with 2011 [12]. Zhenwu Mountain tunnel of Yuqian expressway, University City tunnel of Yusui expressway, and Zhongliang Mountain tunnel of Chengyu expressway served the highest volume of traffic at

>80,000 [12]. Compared with 2011, the daily flow of all tunnels increased in 2012. Beibei, Yufeng Mountain, Yipin, and Beibei tunnels of the Yuwu expressway have increased considerably in terms of the volume of traffic [12].

The congestion bottleneck zones are not the congestion source, just the congestion points. However, the increase in traffic relationships between the city center and the main subcenters, with the shortage in the traffic supplies, resulted in the imbalance between supply and demand, which in turn led to the production of congestion points in large areas and the widely spread of congestion.

3.3 Characteristics of Congestion Duration Evolution

Characteristics of congestion duration evolution include traffic congestion occurring intensively during the morning and evening peak hours, and they in the stage of duration extension.

In contrast to the terrible traffic congestion in cities like Beijing, Shanghai, Guangzhou, New York, and Los Angeles, the spatial and temporal distribution of traffic congestion in the inner ring urban area of Chongqing are characterized by "double peak-hours in working days" and "extension duration of peak hours" (Table 3).

This demonstrates that the traffic peak–hour durations showed a growth in the urban areas of Chongqing from 2007 to 2010. The start times of morning and evening peak hours in 2010 were 1 h earlier than those of 2007. The morning rush-hour duration increased from 1 h in 2007 to 2 h in 2010. The evening peak-hour duration increased from 1 h in 2007 to 3.5 h in 2010. Therefore, the lengthening of the duration during morning and evening rush hours shows extended traffic congestion duration at that time.

4 Traffic Congestion Management Ideology

Based on the preceding analysis, Chongqing—with the layout of "multi-center and cluster-mode"—is in the stage of rapid developing urbanization. Therefore, the traffic congestion cannot be alleviated simply by increasing the limited transportation

Table 3 The peak-hour durations of the main urban area of Chongqing from 2007 to 2010

Year	Morning peak hours	Evening peak hours
2007	8:00–9:00	17:00–18:00
2008	8:00–9:30	17:00–18:00
2009	8:00–9:30	17:30–19:30
2010	7:00–9:00	16:00–19:30

supplies to meet the expansion of unlimited traffic demands. Chongqing must establish and improve multicenter structure according to the law of economic development in the process of urbanization. It must also enhance the core competitiveness of the centers to form new growth poles that are able to compete against those in the city center. Then it will be possible to fundamentally achieve decentralization and equalization of traffic flow.

At the same time, traffic-demand management and creation of a reasonable traveling civilization should be introduced in the face of the irrational expansion on travel demands and a rising proportion of private car under the rapid development of motorization. These measurements should be based on the implementation of public transport priority policies with the aim of creating a rational use of transport tool, internalizing the costs of private cars, controlling irrational traffic demands, and adjusting the structure of urban transport.

5 Conclusion

This work studied urban traffic congestion in the socioeconomic-development environment. It started from the analysis of the particularities of urbanization, urban spatial structure, and motorization, analyzed the present characteristics of traffic congestion in the main urban area of Chongqing, and finally put forward a brief idea of countermeasures to traffic congestion.

The characteristics of traffic congestion in Chongqing demonstrate that as a late-developing city, Chongqing still has the opportunity to make improvements in traffic congestion through a persistent mixture of functional inner clusters and the promotion of job housing balance under the urban layout of the "multi-center and cluster-mode." In addition, Chongqing should pay more attention to the coordinated development of public transportation and nonmotorized traffic with urban functions. Additionally, it should make a change in the preference of excessive reliance on private cars and promote public transport. Given these suggestions, it will certainly not be easy to alleviate traffic congestion.

Acknowledgments The work is supported by a research project, National Think Tank Project of Research in Urban Traffic Congestion and Countermeasures of Chongqing subsidized by Chongqing Association for Science and Technology. The project number is 2011-78.

References

1. Taylor BD (2002) Rethinking traffic congestion. access magazine. Access Number 21(21):8–16
2. He Y, Xie F (2011) System, policy and idea: routing selection of urban traffic congestion management. Jiangxi Soc Sci 9:209–215

3. Xu D (2009) The relationship of urban traffic congestion and urban spatial expansion. PhD thesis, Beijing Jiaotong University, Beijing
4. Xu D, Ji M, Gui-ying C (2009) Research on urban traffic congestion management in the perspective of urban structure. Compr Transp 05:60–64
5. Gao P (2006) Comparative study of urbanization in China and abroad, Tianjing
6. Chongqing Planning Bureau (2007) The urban and rural general planning of Chongqing (2007–2020). Chongqing
7. Fu Y, Zhou T, Gao Z (2009) On urban structure and urban traffic in Chongqing. Shanxi Archit 35(05):20–22
8. Ye Z, Fang B, Li J (2012) The practice of "multi-center, cluster-mode" urban spatial structure of the main urban area of Chongqing. Chongqing Urban Rural Plan Forum 25:43–47
9. Chongqing Municipal Planning Bureau, Institute of Urban Transportation Planning of Chongqing (2012) 2011 Traffic operation analysis report of Chongqing
10. Mumford L, Song J, Ni W (2004) The city in history: a powerfully incisive and influential look at the development of the urban form through the ages.Beijing
11. Urban and Rural Construction Commission, Municipal Design Institute of Chongqing (2010) Plan of traffic-smooth Chongqing-transport infrastructure building program. Chongqing
12. Chongqing Municipal Planning Bureau, Institute of Urban Transportation Planning of Chongqing (2013) Transportation development annual report on the main city of Chongqing in 2012

Integrated Allocation-Management Evaluation System for Public Housing

Yanxia Zhao and Fei Wang

Abstract This article establishes the seamless joint evaluation system of distribution management in public housing using Analytic Hierarchy Process (AHP). There are significant differences among different modes of public housing, while the entire distribution management mode is obviously better suited to the public housing than others.

Keywords Public housing · Distribution management · The analytic hierarchy process (AHP)

1 Literature Review and Problem Posing

1.1 Overseas and Domestic Research Status Review for Public Housing

Research by Fame (1970) shows that Singapore promoted the central provident fund system and set up the central CPF Board in 1955 to manage and use accumulating and consolidating fund savings. The provident fund system plays a decisive role in solving housing financing problems, and it overcomes the problem of inability to afford residential housing. Jacob (2004) pointed out that the Japanese housing security fund mainly arises from national finance and widely absorbs

Y. Zhao (✉)
Postdoctoral Programme of China Industrial Economic Security Research Center, Beijing Jiaotong University, Beijing 100044, China
e-mail: 996080326@qq.com

Y. Zhao · F. Wang
College of Management, North China University of Science and Technology, Hebei Tangshan 063000, China
e-mail: 913622477@qq.com

M. Li et al. (eds.), *Proceedings of 2015 2nd International Conference on Industrial Economics System and Industrial Security Engineering*,
DOI 10.1007/978-981-287-655-3_55

various elements of social capital. Recently, financing reforms have mainly shown subsidies flowing from housing builders to low-income residents, and financing channels converting from fiscal investment orientation to self-raised funds orientation. The domestic housing research has mainly concentrated on such aspects as land supply, capital rising, and distribution management. Lyu Ping, etc. suggested that it is necessary to cancel the allotment supply of public housing land, change the current public housing supply method by converting implicit subsidy to explicit subsidy, and transform public housing land-free allocation to paid land supply, consequently making various land-supply methods uniform [1]. Zhong Youcha indicated that we must give full play to the leading role of government in the affordable housing financing mechanism; innovate the credit mode; bring commercial banks' initiatives into full play; and innovate new mechanisms to absorb social financial capital [2]. Wei Yanqiu put forward the idea that our country must build a closed financial system to support rental public housing fund-raising [3]. Dong Fan and Wu Ruishu claimed that it is necessary to enrich audit standards such as original family assets, asset incomes, career risk degrees, etc. Based on these, they worked out the basic standards and mode of public housing admittance [4]. Luo Xiaoling proposed solving strategies for avoiding questionable conduct in public housing applications based on game theory. The research result revealed that some applicants make questionable applications because they have compared the costs of dishonesty to their own incomes. When the income is greater than the cost, a questionable or power rent-seeking phenomenon occurs. If we want to raise housing security benefits, we must strictly control the management aspects with applicants and audit departments. As the intensity of punishment increases, dishonest behavior decreases [5].

1.2 Research Status Review and Research the Objectives Arising

This project is based on prior research; however, how local government arranges sustainable public housing development is not studied enough in that prior research, especially in the aspects of land-supply and land-choosing methods, insured objectives screening and quitting, final-period management of public housing, etc. They are lacking integration into local practice. Hebei province is a pivotal subject in the integration of Jing-Jin-Ji. Public housing is significant to speed up the process of coordinating the development of Jing-Jin-Ji. It relieves the functions of the capital and absorbs floating population. Based on these considerations, this work chose the planning and construction, allocation and management, operation and value-added of public housing in Hebei province as the study objective, and it takes construction of sustainable development and solution of the public housing mechanism as research objects, with the purpose of providing a theoretical basis and reference to decision-making fir departments and executive departments.

2 Management Evaluation System for Public Housing in China

The public housing project is a livelihood project and a far-reaching developing project that is connected with residential living. The sustainable development of public housing is an indicator of social fairness and efficiency. However, the current public housing policy stresses construction rather than management, quantity rather than quality, and admittance rather than quitting. These phenomena have a serious impact on the sustainable developing management of Hebei province's public housing. Therefore, this research chose construction of a public housing sustainable developing mechanism and implementation path as research objects. This reflected the project's innovative theoretical significance and its practical guiding significance.

The contradiction between the lack of public housing supply quantity and high requirements from low-income residents is completely obvious now. If the public housing policy lacks a sound security objective admittance and quitting mechanism or management of security objective, there are two serious results: On one hand, residents who have obtained housing security are unwilling to quit the security system even though they have better income conditions; on the other hand, new low-income residents cannot get housing security in time because of limited security houses. This may cause unfair allocation. To solve this problem, we must build a new mode of evaluation system, which is declared by the public, and then audited by the government, tax administration, civil administration, public security, manpower resources, social security, industry and commerce, finance departments, etc. This new mode of evaluation system receives public supervision from television channels, websites, newspapers, etc. In addition, the government must start an information technology tracking system of the insured families' incomes and jobs. We used analytic hierarchy process to construct a seamless, jointly integrated allocation-management evaluation system for public housing. This mode can well reflect openness, justice, and efficiency criteria. The core procedure in this evaluation system is choosing the proper influence factors of each criterion. We chose evaluation factors according to the circumstances in Hebei province; meanwhile, we acquired factors and information materials according to our national indicators.

3 Integrated Allocation-Management Evaluation System

3.1 Research Method

The AHP can intuitively take on the mode and path of decision-making thought and procedural process when people are solving problems. Because the AHP is practical when it utilizes mass data to treat complex decision problems, it can operate highly efficiently under supporting software. Thus, use of the AHP has spread widely given its incorporation of qualitative and quantitative research tools. This research has set up a new evaluation system of public housing to further optimize the policy implementation of public housing allocation and management.

3.2 Integrated Allocation-Management Evaluation System

Most influence factors of allocation and management of public housing cannot be scientifically calculated. We formulated questionnaires and carried out pointed in-depth exploration with experts, representative families, and community cadres to confirm the main influence factors. Firstly, we screened and analyzed 275 Tangshan families' valid questionnaires, and then we studied three influence principles: public audit application principle, coverage security principle, and quit in-time efficiency principle. Second, we determined from this investigation that the influence factors come mainly from public participation, standardized objective, initiative-quitting application, and truthfulness of information. Therefore, we determined nine influence factors and divided them into four layers based on the construction principles of the AHP. They are objective layer A, criterion layer B, element layer C, and project layer D, all as shown in Fig. 1.

3.3 Allocation-Management Evaluation Index System Based on AHP

We constructed each layer of a judgment matrix from different influence factors of public housing and compared the importance of each element with its upper layer.

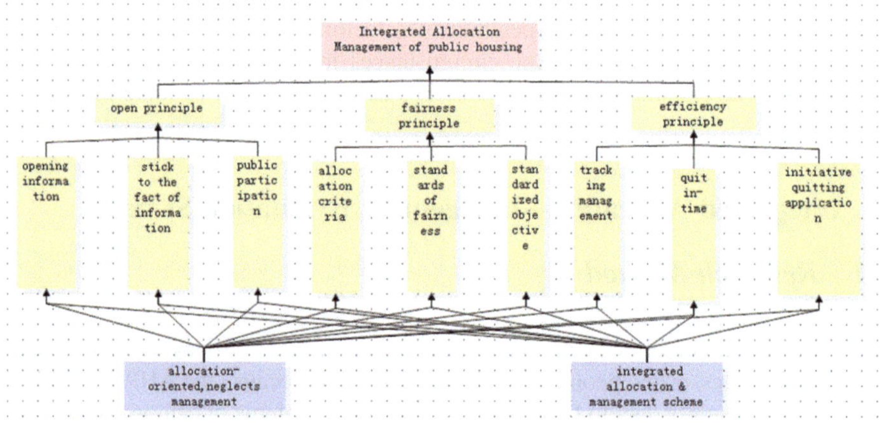

Fig. 1 Diverse AHP patterns of allocation and management of public housing

Then we assigned a value for each layer from 1 to 9. For example, we compared the criterion layer with the general objective layer, e.g.,the efficiency principle with the fairness principle; The former is slightly more important than the latter, so we inputted 2 in the matrix. By analogy, we can construct the matrix of each factor with the general objective and check its consistency.

First, we constructed the judgment matrix with a large number of investigations, surveyed and interviewed data results, and operated the YAAHP7.0 software. The matrix of criterion layer to objective layer and its check consistency show that the weight importance of each criterion layer to the general objective layer are as follows: efficiency criterion (0.4934) > fairness criterion (0.3108) > opening principle (0.1958). For the consistency ratio of the judgment matrix we have: 0.0156 < 0.1. Through the check consistency, we determine the weight of the general objective as 1.0000. The judgment matrix of each element layer and its weight of criterion layer and consistency check are as shown in Table 1.

The target layer A is public housing allocation and management. The criteria layer B1 is opening principle (0.1958); B2 is the fairness principle (0.3108); and B3 is the efficiency principle (0.4934). The element layer C1 is opening information (0.6833); C2 if factual information (0.1168); C3 is public participation (0.1998); C4 is allocation criteria (0.1095); C5 is standards of fairness (0.5816); C6 is standardized objective (0.3090); C7 is tracking management (0.0977); C8 is quit in-time (0.1870); and C9 is initiative quitting application (0.7153). The weight of each C element layer to the general objective A is, respectively, as follows: 0.1338, 0.0229, 0.0391, 0.0340, 0.1808, 0.0960, 0.0482, 0.0923, and 0.3529.

Finally, we constructed diverse modes of allocation and a management-evaluation index system for public housing using our yaahp7.0 computational results. The top five weights in the whole evaluation index system are initiative quitting application 0.3529, standards of fairness 0.1806, opening information 0.1338, standardized objective 0.0960, and quit in-time 0.0923.

It is proven that to allocate and manage public housing more effectively, fairly, and openly, first we should create an integrated scientific-implementation scheme. Meanwhile, we must increase effective guidance and education for applicant families, so that they will get acquainted with the quitting standards and quit when they exceed the threshold. We suggest that we need the mass media to spread public information of application standards for public housing. Only if integrated mass-tracking management follows can we reach the goal of an all-inclusive public housing security network and improve the fairness of the social security policy.

Table 1 The judgment matrix and check consistency of criterion layer to objective layer

Allocation and management of public housing	Opening principle	Efficiency principle	Fairness principle
Opening principle	1.0000	0.5000	0.5000
Efficiency principle	2.0000	1.0000	2.0000
Fairness principle	2.0000	0.5000	1.0000

4 Results of Evaluation

After evaluation with two different modes of allocation and management of public housing, the evaluation results show that in the aspect of allocation and management of public housing, the "integrated allocation management mode" is better than the "distribution-oriented but oversight management mode." The research results can constitute a favorable decision basis for the theory and practice of allocation and management of public housing.

References

1. Ping L, Dapeng X, Shuang L (2013) The theoretical and practical significance of mutual property rights mode used in affordable housing system. Urb Dev Stud 2:20–24
2. Youcha Z (2012) The study of main financing channel of public housing. China Financ 13:78–80
3. Yanqiu W (2014) The international comparison and reference of public housing security and policy optimization in China. Urb Dev Studs 12:14–19
4. Fan D, Ruishu W (2012) The selection of ensured families and the allocation of welfare housing in China. J Cap Univ Econ Bus 4:33–39
5. Xiaoling L (2013) Research on the influence on commercial housing price from affordable housing. Soft Sci 10:1–15

Traffic-Congestion Problem from a Time–Space Conversion Perspective

Shen Yan and Tian Zhang

Abstract This work presents an SEM analysis framework of the interaction mechanism of city traffic congestion and related variables. Instead of the traditional transport economics theories, our work is based on the time–space economy perspective whereby the reasons for and solutions to traffic congestion are reconsidered. The main contribution of our work is to explore Downs Law and advance the following congestion paradox: No matter how traffic develops under the city agglomeration, road congestion will be relieved in the short term but will return to the original conditions in long term. We find that the traffic congestion cannot be avoided but that accessibility can be improved by providing a large-volume time–space tunnel. The conceptual framework and the paradox are empirically tested using the case of the Tokyo metropolitan area. We also recommend relevant authorities to improve the space–time conversion ability by establishing TOD mode.

Keywords SME framework · Accessibility · Metropolitan · Job-housing balance

1 Introduction

Traffic congestion has become a major problem among the metropolis causing a series side effects including traffic safety, waste of resources, environmental pollution, etc. Although government authority, academic fields, or other individuals are highly concerned about the problem of traffic congestion, the fact is that through the global perspective, whether in Los Angeles and New York (car-dominant) or the Tokyo metropolitan area (rail transit dominant), despite the nearly 100 years of development and management of transport network and infrastructure, the problem

S. Yan (✉) · T. Zhang
School of Economics and Management, Beijing Jiaotong University, Beijing 100044, China
e-mail: syan830@gmail.com

T. Zhang
e-mail: zhangtian@bjtu.edu.cn

M. Li et al. (eds.), *Proceedings of 2015 2nd International Conference on Industrial Economics System and Industrial Security Engineering*,
DOI 10.1007/978-981-287-655-3_56

of road congestion has not been solved. Therefore, we must reexamine the validity of traditional transportation economy theory to deal with the congestion problems. Or should we change our way of thinking to reframe the rationality of the congestion problem itself.

Concentrating on the disadvantages of the traditional supply–demand approach, Downs Law pointed out that "the demands of road traffic always have a trend to overwhelm the supply of traffic facilities ability" Downs [1]. However, the law also has many limitations. First of all, Down takes public transportation as a given exogenous variable and discusses traffic congestion problems on the road only. Should we add public transport into the Downs' analysis? After considering public transport, should we consider whether Downs law still applies? Answers to these questions are unknown. Second, the studies of Downs [2] only consider the influence of traffic infrastructure on traffic demands and ignore the important influence of city scale and layout on traffic. Many scholars e.g. Rong [3], Lu et al. [4], Wang and Rong [5] have pointed out that city structure an important impact on traffic accessibility. Therefore, it is necessary to introduce city structure into the analysis framework and to rethink the congestion problem.

2 SME Framework

2.1 Conceptual Framework

All city "diseases"—such as population expansion and industry agglomeration—derive from the negative externalities of the development of the city. Therefore, we cannot therefore deny the role of urban agglomeration. The historical experience of New York, London, and Tokyo tells us Xu et al. [6], Wang et al. [7] that the inhibition of urban development does not work. We must effectively exert the scale and agglomeration effects of the city. To avoid city disease giving up the city's economic development is impractical.

Taking the research of Rong [3] as a logical starting point, based on the theory of space–time contradiction and, our work explore the interaction mechanism of urban traffic congestion and related variables providing urban scale (S), transport mode (M), and a transport efficiency (E) framework (Fig. 1) to investigate the relationship between traffic congestion and urban development.

We suggest that with the agglomeration of urban population and resource, city scale will expand constantly, which means that the scale of the distribution of population, resources, and opportunities has a wider spread. However, when the boundary expands to a certain extent, its negative externalities push the residents living at the edge of the city into a job/housing separation problem. Because people's patience is limited in tolerating job/housing separation, they will ask for a more efficient transport tool or transportation mode. In the Tokyo metropolitan area, >92 % residents' have a commute time of <2 h. In contrast, when transportation

Fig. 1 SME framework

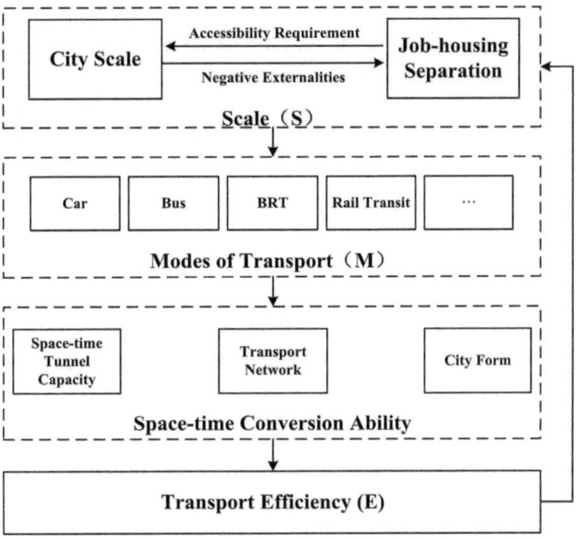

efficiency is improved, or residents have better accessibility, city scope can be further expanded and therefore create new problems of jobs/housing imbalance.

Under the difficult constraints of a 2-h commute time, and with the ascension of the urban agglomeration effect and the expansion of the urban population, the job/housing relationship kept changing. Urban transportation systems also provide different transportation products or modes to meet the higher commuter traffic. In other words, under the constraints of accessibility, a higher number of urban populations require an alternative to existing city-transportation modes. Rong [3] had a good summary of this issue as shown in Fig. 2.

Population (million)	<100	100-200	200-500	500-1000	1000-2000	>2000
City Commuting Area Radius	15Km	20Km	20-30Km	30-40Km	40-50Km	>50Km
Dominant mode 1	Bus	Bus	Bus	Bus	Bus	Bus
Max One-way Time	45min	1.0h	1.8h	2.5h	3.0h	4.0h
Dominant mode 2	Bus	Bus	BRT	BRT	Subway	Subway
Max One-way Time	45min	1.0h	1.5h	2.0h	2.5h	3.0h
Dominant mode 3	Bus	Bus	BRT	Subway+ Railway	Subway+ Railway	Subway+ Railway
Max One-way Time	45min	1.0h	1.5h	<1.5h	<2h	2.5h
Dominant mode 4	Bus+BRT+ TOD	Bus+BRT+ TOD	BRT+Subway +TOD	Subway+Rail +hub+TOD	Subway+Rail +hub+TOD	Subway+Rail +hub+TOD
Max One-way Time	0.5h	45min	1.0h	1.0h	1.5h	2.0h

Fig. 2 City scale and population supported by different dominant public-transport modes

From a longitudinal view, in same population, and with the evolution of the transport mode (the introduction of BRT and commuting rail and subway), the one-way commute time showed a decrease. From a horizontal view, if the transportation mode remains unchanged, the commute will decrease while the population increases. Why does the choice of different transportation modes lead to a change in accessibility? We nest introduce the important concept of space–time conversion efficiency.

2.2 *Traffic-Congestion Paradox and the Role of Space–Time Conversion Efficiency in SME*

First, analyzing the first kind of congestion: commute-based nonrandom demands. Given the transportation-technology circumstances, there is a contradiction between city development and the ease of traffic congestion.

In a given initial city's attraction of commuting approach and transport efficiency, assume that congestion is improved due to some reason or reasons (for example, the road is broadened or new subway line is added). By that time, the congestion is at a minimum level, which extends the 2-h commuting area. It is possible to cover more settlement that departsfrom a given job position in 2 h' commute time, which results in an increase of traffic-demand attraction or leads to an increase in road-traffic flow. Because different people have different sensitivities to the value of their time, people with low time sensitivity people will accept more commuting time to get a job offer. After an increase in the degree of congestion, the person with greater time sensitivity will effectively "squeezed out" of that 2-h commute. Therefore, during expansion of the 2-h commuting-area, traffic congestion becomes increasingly worse until the situation returns to the initial level as does the benefit of post-accessibility improvement. When the commuting area stops expanding, the congestion situation will reach equilibrium. That is why New York and Tokyo, after much effort, are still stuck with the problems of traffic and congestion, to which the world currently does not have a solution. Under the congestion paradox, it is in vain to evacuate the urban population and its traffic flow by way of administrative guidance or mandatory policy.

However, different transport modes have different degrees of efficiency. While the population of Beijing is less than the number of rail-transit commuters in Tokyo, the crowding on the road is almost the same as that in Beijing, possibly even worse than crowding in Tokyo. Why are the traffic conditions of Beijing and Tokyo so different? This work argues that the difference is the embodiment of the difference in transport efficiency, and the reason for the difference can be attributed to the two cities' different space–time industrial agglomeration conversion efficiency.

A city under the double conditions of expansion and tremendous tides of traffic flow requires an urban traffic system that can provide a large-capacity of transport

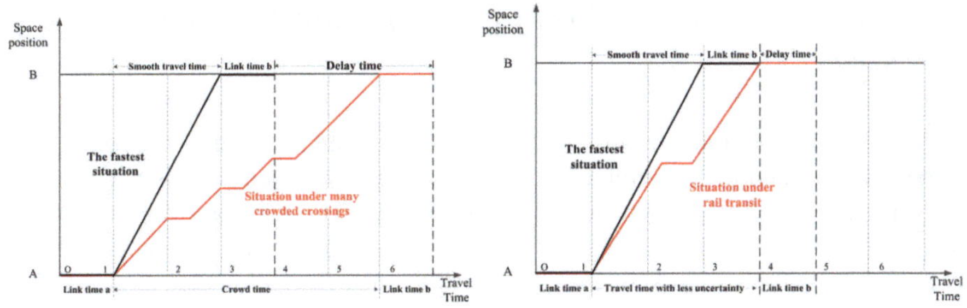

Fig. 3 Displacement time chain of different congestion situations

corridor. How should the mode be selected, what mode should be selected, and what mode is most dependable?

There are two concepts we must emphasize. First, to overcome a certain space distance requires a certain amount of time and money Rong [8]. The ability of transform distance into the time is called "time–space conversion efficiency." To what extent can time–distance be compressed in the scope of commuting area is the main index of the space–time conversion ability of a city transport system. A space–time tunnel is a traffic corridor that is to overcome spatial distance or other traffic disorder by significantly shortening the time–distance.

By comparing the two graphs in Fig. 3, we can determine that the longer the commuting distance, the more crossing have to be passed; thus, the uncertainty of traffic time is increased, which means the potential opportunity cost loss will increase. Rail transit is one of the modes of transportation that can significantly reduce uncertainty. Although the BRT and highways both have the features of a space–time tunnel and have a high space–time conversion efficiency, only rail transit can minimize the uncertainty of traffic delays. However, when the population grows to a certain extent, one can simply rely on rail transit until adjustment in the transportation system is unsustainable. By that time, a coordinated development of urban form and transportation network is necessary, e.g., introducing a TOD structure.

3 Case Study of Tokyo Metropolis

Tokyo is famous for its high population density, high concentration degree, and high transport efficiency. The case of TOD mode in the Tokyo metropolitan area is recognized as the most successful case of traffic-resource allocation. This work will test an SEM framework through of the Tokyo metropolitan area case by dynamic analysis of the city and its transport-network development.

3.1 Brief Introduction of Tokyo Metropolitan Commuting Area

The Tokyo metropolitan commuting area can be defined as a traffic network based on the core city and mainly connecting of outside rail transport; an inner-cooperate industrial layout, city function, hub distribution, etc. The core city traffic network contacts people with areas lying outside the city through the following groups: (1) core city traffic network based on the BRT, subway, and other transportation modes; (2) regional transport area based on the suburban railway, intercity railway, ground rail, and highway, which connect the city edge and the region of the outlying outer cities. The Tokyo Metropolitan area has a highly transport, which relies on the traffic network. TRAFFIC-oriented development makes the city sustainable.

Scale of metropolis. Tokyo metropolitan area is within a radius of approximately 100 km around Tokyo district. In the 1950s, the region's population was 19.10 million. Tokyo metropolitan area is the largest area in the world, with of the highest urban densities, and the population reached 41.3 million in 2000. Compared with other developed countries in large urban areas, such as New York, London, and Paris, the Tokyo metropolitan area population level is relatively high.

Tokyo is Japan's political, economic, and cultural center and one of the world's most attractive cities. The urban comprehensive function is very powerful. As a financial center, >30 % the bank headquarters and >50 % of large companies' headquarters are set in Tokyo; as a transportation center, Tokyo has Japan's largest port group and aviation network; high-speed rail like Shinkansen, which can travel at ≤200 km/h, and a subway that can almost reach all important areas.

Transportation network and urban form. Due to the advantages of large volume, fast, comfortable, on time, and effective use of the ground and underground space, city rail transportation plays an important role in the Tokyo metropolitan commuting area. The total length of the network of Tokyo city rail traffic circle is within the scope of 2246.4 km, and the city rail transit network density is 222 m/km^2. This figure is the highest in a range such as that of London, New York, and Paris. At present, the Tokyo subway mileage is 312 km, and that of the Tokyo commuting area railway is 1134 km. In 2010, the total Tokyo metropolitan commuter rail transit ticket number was 7.89 million people; the average daily number of rail traffic is 40.74 million passengers (the use of regular tickets accounted for 63); and the average one-way commute time is 68.7 min (one way in 2 h occupies a the proportion of 92.1 %). It is easy to understand that if Tokyo did not have a 2000-km city rail transit network, it could not operate normally!

In addition, the metropolitan form of Tokyo also combines land-use and transportation lines. Comprehensive development of a railway development strategy profit mode provided great success for such areas as the earliest Osaka-rise office buildings, shopping centers, hotels, and entertainment venues, and followed by the railway company, is widely used in Tokyo and other places in Japan. TOD development mode makes the city more reasonable and effective, but it is also an effective solution to the problem of railway-construction funds.

3.2 *The Test of SME Framework Based on the Case the Tokyo*

Trade-Off Between City Development and Its Negative Externality

To solve the large-city "diseases," the Tokyo metropolitan area has experienced five phases of land-development planning, from constrained city development in 1958 at the earliest, to build transportation network under TOD mode. Different methods lead to different results, some successful and some failures. Thanks to these attempts, we can better understand the interaction between the city and its traffic. Therefore, it is a good case for us to empirically test the above-mentioned conceptual framework (Table 1).

We can determine the change process of Tokyo metropolitan planning from the above information as follows:

- Change from constraining city development to improving the agglomeration economy.
- Change from single-core city into a multi-core city administrative division.
- Change from internal planning to cross-administrative boundary planning.
- Change from enhancing the inner city traffic network to development of a finger-like traffic network.

As can be seen, the first two periods of planning ended in failure. Throughout these planning phases, Tokyo did not reach the aim to ease traffic congestion in the city center. The main measures of the first two planning phases intended to limit expansion by setting the green belt around the city center and establishing satellite towns in the suburbs, the intention which intention was to absorb the population and industry. However, these measures did not have any effects; the area of city congestion still expanded, and the crowded node was instead transferred into the city.

However, the third planning phase achieved remarkable results. In this phase, the city depended on transportation, giving full play to the comparative advantages of different regions, to expand the scope of the metropolitan area to rely on transport efficiency so as to give full play to the regional agglomeration effect. Tokyo's population growth rate reduced from 400,000 to 70,000/year. The proportion of second industry decreased by 4.6 %, and the third industry growth rate was 4.6 %. These measures alleviated road congestion in Tokyo, while the rail transportation facilities let people inside the circle quickly go out of the core urban area, which achieved the balance of economic development and traffic.

Table 1 Five phases of land-development planning in the Tokyo metropolitan area

Phase/time period	Background	Content	Result
First (1958)	Population, industry, and infrastructure disorderly, rapid agglomeration. Tokyo disorderly expansion. Core city population = 8.01 million	Green belt was built to prevent expansion of the city. Established satellite cities to absorb the population and industry	Unable to suppress city agglomeration, continued expansion
Second (1968)	Rapid expansion of population. Industrial agglomeration out of control. Core city population = 10.87 million	Reorganized city space. Set green plant zone. Attempt coexistence with space expansion and urban green land. Continued to implement policies for development of satellite cities	Region development unbalanced. Still stuck in traffic. Caused side effects on city economy
Third (1976)	Population and central authority too concentrated. Formation of Tokyo's unipolar pattern. City overcrowded. Core city population = 11.67 million	Strengthened surrounding small city to improve cities' social, cultural functions. Established emanative transport network	Achieved balanced development of core and surrounding cities. Traffic congestion decreased
Fourth (1986)	Industry structure layout not reasonable. Single core pattern unchanged. = 11.83 million	Established metropolitan area based on core business city Surrounded by multilevel regional structure. Provided inner connection by main transport lines	Promoted formation of multipolar pattern of metropolitan area. Exerted city agglomerations
5th 1999	Aging of population and central-cavity problems. Core of urban population = 12.6 million	Sharing, cooperation, and exchange of mutual function by rail traffic trunk channel to form decentralized network structure, improve regional agglomeration economy	Regional competitiveness improved by using traffic network to give full play to agglomeration economies rather than restricting

Effect of Large Tunnel of Time and Space in Tokyo

The Tokyo metropolitan area transport network mainly consists of two circles: (1) the core traffic circle, which is mainly based on the BRT and the subway; (2) the regional traffic circle, which is mainly based on the suburban railway and highway

Table 2 Rail transit passengers in Tokyo core centre

Year	Percentage over the total traffic volume (%)					
	National rail	Subway	Private rail	Rail subtotal	Other modes	Total traffic flow
1960	36	5.7	22.6	64.3	35.7	1512.9
1970	32.8	17.1	26.4	76.3	23.7	2121.6
1975	33.7	21.2	27.5	82.4	17.6	2265.7
1980	30.3	23.8	29	83.1	16.9	2255.8
1982	30.2	24.4	29.5	84.1	15.9	2233.7
1996	31	25.5	30	86.5	13.5	3005.8

and rail traffic. Through these circles our work investigates the role of the time–space tunnel in the Tokyo traffic situation (Table 2).

The core circle, composed of 23 districts of the city, is the center of the Tokyo metropolis. The following table shows the number of rail-transit passengers during the first to forth land-development planning periods.

As we can see from the table, the transport flowing by the first transport circle dramatically increased to the level of 3 times the number of residents in the past 50 years. The proportion of rail traffic over the total volume was >60 %, and it was rising.

Before the third planning phase, to cope with the increasing traffic demand, Tokyo constantly improved its internal stock of urban-traffic facilities. Before World War II, the railway length in Tokyo was 210 km. However as the Downs Law says, the increase in supply will never be able to meet the increased demand. Data shows that before the third planning period, the internal traffic situation of the first circle had not been significantly improved.

The government of Japan in the postwar planning period tried to control Tokyo's population of 3.5 million, but it soon faced a huge pressure of >10 million people surging into the center of Tokyo in 1962. The enormous traffic pressure forced improvement of the transport as well as different way of looking for solutions to traffic jams. Tokyo decided to adopt the rail-transit network, which had lower transportation cost, larger capacity, and faster speed. This greatly enhanced transportation efficiency as well as improved space–time conversion ability.

The second circle is an important circle in Tokyo to disperse traffic pressure outside the city, in which the railway and highway play an important role in the rapid-transit corridor. A fast actinomorphic transport corridor is an important feature of Tokyo metropolitan's transport links between the central city and the peripheral city. The Tokyo actinomorphic rapid-transit corridor has three kinds: highway traffic, JR Shinkansen high-speed railway, and JR common rail. The length of city railway in the second circle is >2800 km and connects all major periphery cities. The railway lines running by JR and private rail are diverse. In the same channel there is an express line, a high-speed line, and a common line. This network effectively connects the center city with the surrounding area.

3.3 The Results of Empirical Study

Tokyo metropolitan traffic circles works in coordination with each other creating a large-capacity space–time network that has high space–time conversion ability. By providing a large-capacity time–space, the Tokyo tunnel enhances the effectiveness of transport,and improves accessibility as well as the job/housing balance so that equilibrium between city agglomeration and traffic congestion is achieved.

Although no city authority has successfully solved the problem of traffic congestion around the world, the transport efficiency of the Tokyo metropolitan area, which has a well-constructed commuter transportation system, is far better than the internal disorder of the metropolitan area.

4 Conclusion

This work has extended Downs Law [2] by bringing public transport into the analysis of congestion and being the first to put forward the congestion paradox, which was used to analyze the congestion problem of the "Scale-Mode-Efficiency" (SME) framework. The findings pointed out that the agglomeration effect lead to the unceasing expansion of the 's scope, which has brought a series of city "diseases" such as the problem of congestion. The need for residents' living balance must be met by more efficient transportation. Due to the different modes of transportation having different time–space conversion efficiency, transportation that can truly resolve the problem of metropolitan area residents' job/housing separation must accord with the characteristics of a time–space tunnel. When the metropolitan area's population increases to a certain extent, it is necessary to design the traffic network and urban structure uniformly. Combining rail transit with TOD mode is a good way to achieve the metropolitan area's job/housing balance. The analysis pointed out that traffic congestion cannot be fundamentally solved. No matter how traffic develops and how urban agglomeration grows, the degree of congestion will be eased in the short term but will return to the original level in the long run, namely, the congestion paradox. Therefore, the policy recommendations for the regulators is that the authorities should not focus on the congestion problem itself but rather give full play to the urban agglomeration economy through the establishment of large-capacity time–space tunnel.

Acknowledgments This work was supported by the Program of the Co-Construction with Beijing Municipal Commission of Education of China and the Fundamental Research Funds for the Central Universities (2015YJS056).

References

1. Downs A (1962) The law of peak-hour expressway congestion. Traffic Q 16(3):393–409
2. Downs A (2005) Still stuck in traffic: coping with peak-hour traffic congestion. Brookings Institution Press, Washington D.C
3. Rong CH (2014) Functions of railway/rail transit in new urbanization and optimization of metropolis time-spacial form. J Beijing Jiaotong Univ (Social Sciences Edition) 13(2):20–28
4. Lu B, Zhang C, Chen TM (2013) Study on changes in job accessibility for the urban low-income: a case study of Beijing. City Plan Rev 37(1):56–63
5. Wang XR, Rong CH (2014) Interaction development research of urbanization and transportation. Inq Econ Issues 1:52–57
6. Xu DY, Zhang L, Jiang XX (2010) Consideration on strong megapolis center and urban traffic congestion. J Beijing Jiaotong Univ (Social Sciences Edition) 9(3):6–11
7. Wang XR, Rong CH, Sheng LF (2013) Enlightenment of yamanote loop line in Tokyo: an important role of rail loop line in metropolitan transportation. Econ Geogr 33(1):54–60
8. Rong CH (2011) The value of transport-logistics time and its applications in time-spatial analysis. Econ Res J 8:133–146

Acknowledgments. This work was supported by the Program of [illegible]

References

[illegible]

Business Model Innovation of Enterprises Based on the Bottom of Pyramid (BoP) Market Strategy

Xudong Fan and Yuduo Lu

Abstract This article is based on the Bottom of Pyramid Strategy (BoP for short) theory, discussing several models of how Chinese enterprises carry out business model innovation based on BoP. After a detailed review of features related to consumption, cognition, ability, and environment in BoP groups and BoP markets, this article found that a BoP market can deliver two kinds of business model innovation, and enterprises should choose best-fit business models to develop market potential according to their own business features and the specific features of the BoP market they are dealing with.

Keywords Features of BoP market · Business model innovation · BoP strategy

1 Introduction

The theory of Bottom of Pyramid Strategy (BoP for short) is a new exploration in an attempt to fuse the perspective of enterprises and the society into sustainable development [1]. Different from traditional perspectives, this theory focuses on innovative activities oriented to low-income groups from a business perspective, reflecting the belief that low-income groups, being located at the bottom of the economic pyramid (the BoP group), represent huge business opportunities. If enterprises are able to meet the needs of this group in an innovative way, they can not only find new growth opportunities but also create both economic and social value by improving the groups' quality of life [2, 3]. The theory changes people's old view toward the BoP group, because of the belief that this huge BoP group can become active market participants; moreover, without the restriction of the existing

X. Fan (✉) · Y. Lu
School of Economics, Dalian University of Technology, Dalian 116024, China
e-mail: endlessfan@sohu.com

Y. Lu
e-mail: luyuduo@163.com

M. Li et al. (eds.), *Proceedings of 2015 2nd International Conference on Industrial Economics System and Industrial Security Engineering*,
DOI 10.1007/978-981-287-655-3_57

technological approaches, this group is glad to use new technologies to improve the quality of life and economic opportunities for its members. It is worth pointing out that though the BoP market does provide development opportunities for enterprises, the market has huge differences from high-end markets, in terms of politics, economy, and culture, which will certainly bring challenges to those enterprises that intend to enter this market [4]. Based on these considerations, this work is devoted to analyzing and summarizing the features of the BoP market, and it comes up with several business models targeting the BoP market according to these features.

2 Features of BoP Group and BoP Market

Based on previous studies, this article systematically summarizes features of the BoP group and BoP market from four dimensions: consumption behaviors, cognitive styles, ability features, and external environment. The following is a presentation of features of the BoP group and BoP market and the opportunities and challenges of enterprises.

1. **Consumption Characteristic**

Features: (a) There is obvious unsatisfied need. (b) Individual purchasing power is small while the group's is huge. (c) "Poverty penalty" effect. (d) Irrational consumption. (e) Recessive consumer demand.

Meaning: (a) Commercial society tends to ignore the requirements with a result that the basic requirements of many BoP groups cannot be satisfied [4]. (b) The overall scale of the BoP group is huge, with the groups' purchasing power amounting to $2.3 trillion [5]. (c) The BoP group must pay higher prices than other groups to get the same products and services [6]. (d) The phenomenon that excessive consumption (such as holiday spending) appears to improve people's social status is obvious [7]. (e) Market information and the relevant knowledge of the characteristics of consumers are unavailable [8].

Opportunity: (a) Non-consumers in the BoP market are ideal targets for disruptive innovation. (b) The market has enormous potential. (c) Once products are accepted, the market will expand at a high speed. (d) Enterprises can help BoP groups overcome the effect of the poverty penalty.

Challenge: (a) The speed of market development is relatively slow. (b) Enterprises must master information and knowledge of the BoP market through trial and error. (c) Enterprises must break through the thresholds of the BoP groups.

2. **Ability Characteristics**

Features: (a) Dependence on irregular and self-sufficient economy. (b) Unstable income. (c) Low educational level and production skills. (d) Lack of negotiating ability.

Meaning: (a) Plenty of deals are conducted through nonmarket channels and self-sufficient consumption and thus lack the assurance of a formal system. (b) The economic system within which enterprises operate is rather delicate and can be easily affected by natural disasters and economic fluctuation and thus lack the ability to resist risk [7]. (c) The limitation of education lowers the productivity of labor and leads to a high unemployment rate [9]. (d) During economic or noneconomic communication, the BoP group is not always respected and stands in a vulnerable position.

Opportunity: (a) Enterprises can establish a bridge between commercial society and the BoP group. (b) Enterprises can help the BoP group to lower risks and raise income. (c) Enterprises can help the BoP group to establish localizability.

Challenge: (a) Learn to do business in the market where the system is rather weak. (b) Help the BoP group to improve its skill levels.

3. **Cognitive Characteristics**

Features: (a) Powerful social orientation. (b) Paying attention to the function of local organization. (c) Strong and fragmented cultural and geographical orientation. (d) General passive attitude toward life.

Meaning: (a) Do not establish a mutual trust relation with strangers rashly, and socialize to ensure reliability before economic communication [10]. (b) Trust the local organization and informal network. The localized social network sets up a buffer space for the instability of the BoP groups' economic ability. (c) Hope to retain its own culture, behavior, and other features. These features maybe different among the Bop groups, with the results that some BoP groups have a strong sense of belonging, while different BoP groups cannot integrate easily. (d) There is a lack of positive expectation toward the future, for example, of improving living standards and finding working passion.

Opportunity: (a) Using the BoP group's social capital with the lever principle. (b) All kinds of economic and noneconomic organizations in the BoP area can be enterprises' partners. (c) Once joining in the local area, a company can gain the trust of BoP groups.

Challenge: (a) Without the support of local organizations, it is difficult to acquire local people's trust. (b) It takes a long time to join in a new local area, so enterprise must bear a longer investment period.

4. **External Environment Characteristics**

Features: (a) Channels are unavailable. (b) Imperfect system framework and market. (c) Underdeveloped basic infrastructure.

Meaning: (a) The deficiency of physical and informative channels limits the ability to obtain resources and knowledge of the BoP group, thus decreasing the range of choices of strategy [6]. (b) There is a lack of sound market regulation. Meanwhile, the rules of competition are dramatically different from those of other markets with ineffective implementation of laws and regulations. (c) Compared with those in the middle and upper levels of the economic pyramid, all kinds of basic infrastructures are relatively underdeveloped (including basic hygiene conditions and productive conditions), and there is poor social security.

Opportunity: (a) There is no effect of infrastructure's locking, which is beneficial to "leapfrog" development. (b) It can help BoP groups build systems and rules in the BoP market. (c) It can be the tie between BoP groups and the external world.

Challenge: (a) There is a need to build a variety of infrastructure from the beginning, the cost of which will be high. (b) The cost of doing business in an environment in which the system is weak may be very high.

From the above it can be seen that the unique features of the BoP market provide enterprises with both opportunities and challenges. BoP groups, living in areas with relatively underdeveloped infrastructures, have limited disposable incomes and lower levels of education. Those features have put forward new challenges for the innovation of enterprises; meanwhile, BoP groups are excluded from most of the existing industries and belong to the "nonconsumer" category. Moreover, the group is not fettered by traditional technology, bringing about low switching costs in the face of new technology and not being affected by the lock-in of technology. Therefore, the group is more willing to accept new technology and have a higher tolerance for the defects of new technology. Those features enable the group to become the ideal customers of disruptive innovation.

3 The Business Model Innovation Oriented BoP Market

The unique features of the BoP groups and the BoP market have brought forward special challenges to enterprises, which always require the enterprises to enter into the market with the reconstruction of the value chain and adopt a business mode which is different from that in the high-end market [1]. Based on previous research, this article defines business model innovation as the process of providing a new value proposition for customers through new integration measures of resources, and new business procedures under the conditions of the changing of enterprises' logic in value creation and value acquisition. Based on the analysis above, this paper reaches the conclusion that enterprises' business modes with a BoP market orientation need innovations in the ways and means of value delivery and value creation. Specifically, the following two business model innovations can be cultivated from the BoP market.

3.1 Market Development with Deeply Embedded System

Formal systems, with the representation of law and contract and related supporting infrastructures, are often deficient in BoP markets. The features of the BoP market determine that development strategies with deep embedded type are more effective than others (Fig. 1). Therefore, enterprises should first eliminate the geographic distance from the BoP groups and improve the perceptibility of the products themselves, through the establishment of organizations in the communities. Simultaneously, the enterprises should also develop the sense of identity of the groups toward their products, through the marketing means of conducting enlightenment education and integrating with local customs for the purpose of improving the acceptability of the products. Figure 1 shows the different development strategies between the BoP markets and traditional markets.

Many solar energy product manufacturers in China adopt development strategies with deep embedded type in the process of developing rural markets.

3.2 Distributed Economic System

Although the purchasing power of the BoP groups, as a whole, is tremendous, they cannot develop into a unified market. The BoP groups assume the characteristics of intraregional centrality and interregional decentralization. The BoP groups within a certain region are inclined to maintain their culture and behavior characteristics, so a single BoP group has a strong sense of belonging and behavioral consistency internally, but different BoP groups perhaps are not likely to mix well with each other because of their different cultural characteristics, which brings about diversity in the value orientation and value preference among different BoP groups. For the purpose of facing these BoP markets with a strong localized value orientation, to

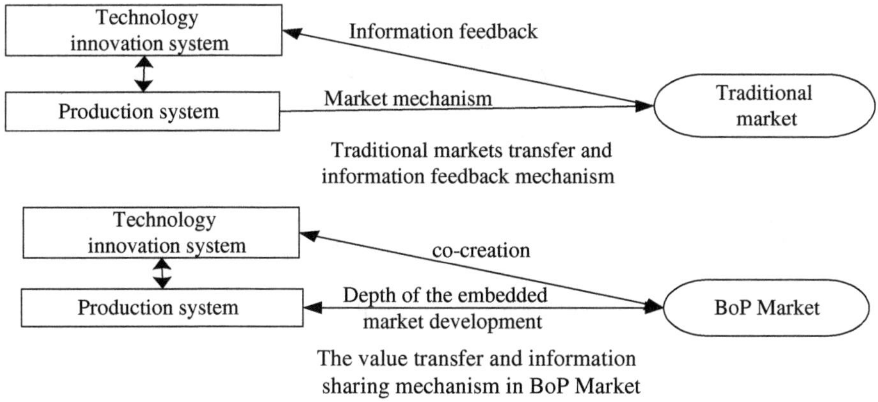

Fig. 1 BoP information sharing mechanism

formulate a series of distributed localized economic systems would be an effective strategy (see Fig. 2).

As Fig. 2 shows that in BoP markets with great heterogeneity, the process of value creation should be transferred to distribute a localized economic system, in order to tackle diverse BoP markets with flexible modes of production and quick responses. As is mentioned above, embedded products and market exploration are effective strategies for companies to develop their BoP business. Compared to giant corporations, a localized economic system would more easily join in a new local area and interact in many ways with the BoP groups. In this way, it can co-create a localized solution together with the local stakeholders. One thing should be pointed out is that establishing this kind of distributed localized economic system requires that it is capable of being involved in the value co-creation. Therefore, local capacity construction is an indispensable process.

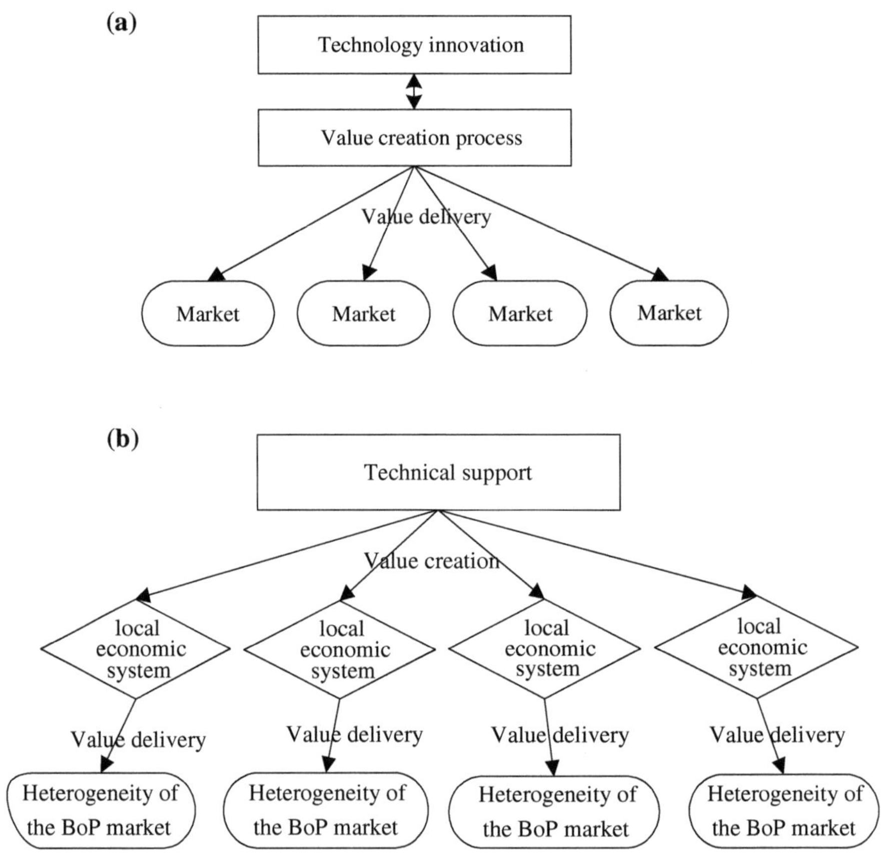

Fig. 2 **a** The traditional market value creation and transmission process. **b** The BoP market Value creation and transmission process

4 Summary

Based on the characteristics of BoP groups and BoP markets, this article made a systematic analysis of two business model innovations facing BoP markets. These business models have challenged the logic system behind the traditional business model, and enterprises should adopt either of them purposefully based on the technical features of their own products and service.

Acknowledgments I would like to express my gratitude to all those who helped me during the writing of this thesis. A special acknowledgement should be shown to Professor Lu, from whose lectures I benefited greatly.

References

1. Prahalad CK, Hart SL (2002) The fortune at the bottom of the pyramid. strategy +. Business. 26:54–67
2. Yunhuan T, Jianghua Z, Xiaoqiang X (2010) Innovation theory for low-income groups (Bop)—review and outlook [J]. Sci Res 28(2):169–176
3. Jianghua Zhou, Yunhuan Tong, Jiheng Li (2010) Enterprise innovation model research for low-income groups. Econ Manag Res 10:12–17
4. Prahalad CK (2004) The fortune at the bottom of the pyramid. Wharton School Publishing, Upper Saddle River
5. World economic forum (2009) The next billions: unleashing business potential in untapped markets. World Economic Forum, Geneva, Switzerland
6. Hammond AL, Kramer WJ, Katz RS, Tran JT, Walker C (2007) The next 4 billion. Market size and business srategy at the base of the pyramid. World Resource Institute, International Finance Corporation, Washington
7. Banerjee AV, Duflo E (2007) The economic lives of the poor[J]. J Econ Perspect 21(1):141–167
8. Grosh M, Glewwe P (1996) Household survey data from developing countries: progress and prospects[J]. Am Econ Rev 86(2):15–19
9. Chesbrough H, Ahern S, Finn M, Guerraz S (2006) Business models for technology in the developing world: the role of non-governmental organizations. Calif Manag Rev 48(3):48–61
10. Chambers R (1997) Whose reality counts?: putting the first last. ITDG Publishing, London

Analyzing Concepts of Crisis Management, Crisis Public Relations, and Crisis Communication Based on the Perspective of Management

Rui Yu

Abstract Based on the perspective of management, this paper reviews and analyzes concepts of crisis management, crisis public relations, and crisis communication. By studying the relationships and differences among these three concepts, this paper draws some conclusions. 1. The function of enterprises' public relations is a perspective of studying crisis management. 2. From the perspective of management, crisis communication is a secondary field of studying crisis management. 3. The study of crisis response strategy in crisis communication is a considerable part of crisis public relations and crisis management.

Keywords Crisis management · Crisis public relations · Crisis communication

1 The Function of Enterprises' Public Relations Is a Perspective of Studying Crisis Management

Management experts always research crisis public relations as a special management activity, and take the strategy of crisis public relations as a management strategy. Coombs, from the U.S., thought of crisis public relations as one area of crisis management, as crisis public relations emphasizes the choice of crisis response strategies to "say something" and "do something" after the crisis has occurred, and the option of choosing crisis response strategies is just the course of management [1].

Crisis public relations emphasize principles of openness, consistency, and timeliness, which are also principles of crisis management. As for the aspects of public relations before a crisis, establishing a crisis warning system, risk control, relationship management, and so on are all the basic content of crisis management.

R. Yu (✉)
Faculty of Management and Economics, Dalian University of Technology, Dalian 116024, China
e-mail: jane7417@126.com

M. Li et al. (eds.), *Proceedings of 2015 2nd International Conference on Industrial Economics System and Industrial Security Engineering*,
DOI 10.1007/978-981-287-655-3_58

Thus from the perspective of management, crisis public relations is a sub-area study of crisis management.

In 1993, Mitroff and Pearson cut into the core part of crisis management to analyze, and they pointed out that crisis managers must undertake tasks to collect, analyze, and disseminate information, and at the same time to find the face, analyze deeply, control loss, and strengthen communication [2].

The function of enterprises' public relations is a perspective of studying crisis management; public relations professionals in this perspective should communicate and dialogue with the public in a timely way, before, during, and after the crisis, to reduce the loss to enterprises, especially image and reputation loss.

Jonathan Bernstein, a scholar from the U.S., thought that crisis public relations can be called crisis management or crisis communication from a professional point of view [3]. The unique role of public relations is that of communicating with the public (internal and external) in a timely and sincere fashion to transmit information and strategy, and to get a stable, coordinated, and harmonious environment.

Professor James E. Grunig, from the University of Maryland, mentioned that a good crisis communication is started before any crisis, and the most effective way to solve the problem and crisis is to communicate with the public before making a decision. If public relations staff can't communicate with the public before the crisis, then the possibility of resolving the conflict will become very small [4].

2 From the Perspective of Management, Crisis Communication Is a Secondary Field of Studying Crisis Management

American scholar Kathleen Feam-Banks defined management communication as follows: "an organization communicates with public before, during and after the crisis" [5]. She suggested that if the response to the crisis is inappropriate, it will damage the credibility of, public confidence in, and reputation of the organization, all built over years. From this point of view, crisis communication is a secondary field of studying crisis management; some aspects of crisis management, like information collection, analysis, processing, and communication belong to crisis communication.

Crisis management aims at managing the crisis and avoiding it as much as possible. Crisis communication aims at informing the public and the media about what happened as much as possible. Risk management serves to identify and predict the risk related to the public; risk communication is how to communicate with the public at every stage of the crisis. In fact, risk communication is crisis communication [6].

Chinese scholar Ruan Zhangqiong (2007) believed that from the perspective of communication, crisis communication is a special form of communication; it focuses on the communication effect, media, audience, and other variables in the process of crisis communication. From the point of view of management, crisis communication is a kind of special management activity; from the point of view of public relations, crisis communication is just crisis public relations [7].

3 The Study of Crisis Response Strategy in Crisis Communication Is a Considerable Part of Crisis Public Relations and Crisis Management

From the presentation of the concept of crisis management, research on the crisis response strategy has been in one of the most concentrated areas. In the field of crisis response strategy, many scholars focus on the study of crisis communication and take it as a part of crisis management.

For example: Renz [8] thought that the study of crisis management should pay attention to communication ethics; Reardon [9] suggests that effective communication can not only relieve the crisis, but also bring a more positive reputation than before [5].

Crisis public relations is very important in public relations, but also an "abnormal" kind of public relations. That is, through the application of public relations during the crisis management process, to achieve effective communication between the organization and the public, it is possible to make up for the loss of image and reputation. Through the crisis management process, the unique role of public relations is to transmit information and strategy in a timely and effective way, by communicating with the public, and to obtain a stable, coordinated, and harmonious environment.

Crisis communication is a special form of communication; crisis communication emphasizes the variables of media, audience, and results in the process of the crisis. Effective crisis communication has the double effect of alleviating or eliminating the crisis, and bringing a positive reputation for an enterprise.

The study of crisis response strategy in crisis communication is a considerable part of crisis public relations and crisis management.

References

1. Coombs WT (2001) Teaching the crisis management\communication course [J]. Public Relat Rev 27:89–101
2. Heath R (1998) Dealing with the complete crisis-the crisis management shell structure [J]. Saf Sci 30:139–150
3. Bernstein J The biggest mistakes in crisis communication [EB\OL]. http://www.Bernsteincrisismanagement.com\articles.html
4. Grunig JE, How to measure your results in a crisis. Guideline for measuring relationships in public relations. The Institute for Public Relations Commission on PR Measurement and Evaluation, 2–3

5. Coombs WT (1995) Choosing the right words: the development of guidelines for the selection of the "appropriate" crisis response strategies [J]. Manag Commun Q (8)
6. Henry RA (2000) You'd better have a hose if you want to put out the fire: the complete guide to crisis and risk communications: Professional tips, tactics, dos, don't and case histories Windsor [M]. Gollywobbler Productions, California
7. Zhangqiong R (2007) The issues of crisis communication—crisis communication research review [J]. Southeast Commun 11:67–69
8. Renz DO, Eddy WB (1996) Organizations, ethics, and health care: building an ethics infrastructure for a new era. Bioeth Forum 12:29–39
9. Reardon KK (1996) The Crossroads of Organizational Communication: Definition or Dichotomy. Manag Commun Q 10(1):106–111

The Opportunities and Challenges that Chinese Traditional Energy Enterprise Confront While Exploiting the Geothermal Resources

Qingxi Li

Abstract With the development of the Chinese economy entering a new normal era, Chinese traditional energy enterprises will meet with some difficulties. As a new kind of energy, geothermal resources have some advantages, such as cleanliness, recycling, and richness. Therefore, we should actively research and exploit geothermal resources. It is an important opportunity for the transformation and upgrading of traditional energy enterprises in China. Start earlier—gain more. The Chinese traditional energy enterprises must convert conceptions, develop new thinking, and solve the problems hindering development if they want to do something great while exploiting geothermal resources. Only this, the transformation of traditional energy enterprises can open up a new world for geothermal development.

Keywords Geothermal resources · Traditional energy enterprises · Opportunities · Challenges

1 Introduction

China is a great powerhouse of traditional energy production and consumption. However, its prospect for energy security is troublesome. Some facts, such as the need to control global climate warming, insufficient reserves of Chinese traditional energy, and a large population, indicate that the traditional energy enterprises in China will be faced with many difficulties. As a new kind of energy, geothermal resources have some incomparable advantages, such as cleanliness, recycling, and richness. Therefore, we should actively research and exploit geothermal resources.

Q. Li (✉)
China Center for Industrial Security Research, Beijing Jiaotong University, Beijing 100044 China
e-mail: liqingxi018@126.com

M. Li et al. (eds.), *Proceedings of 2015 2nd International Conference on Industrial Economics System and Industrial Security Engineering*,
DOI 10.1007/978-981-287-655-3_59

It is an important opportunity for the transformation and upgrading of traditional energy enterprises in China.

2 The Difficulties of Chinese Traditional Energy Enterprises Encountered

With the Chinese economy entering a new normal era, Chinese traditional energy enterprises will meet with the following difficulties in the future.

The first is the contradiction between the high-carbon energy production and the limit on the carbon emissions. In 2014, China and the USA agreed on a set of greenhouse gas emission targets, with Beijing setting a goal for its emissions to peak "around 2030," Carbonaceous energy now accounts for more than 90 % of the current energy consumption in China. This means we will face a real "ceiling" in the future. So the transformation and upgrading of domestic enterprises will face great pressure.

The second is the contradiction between the stable development of enterprises and the fluctuations of energy price. In recent years, the ups and downs of the international energy market brought many difficulties to the stable development of domestic traditional energy enterprises. The profits of some coal enterprises severely declined; some even went bankrupt. Of course, oil enterprises had greater difficulty. In 2014, Sinopec ranked third in the world top 500 companies, but it made a net profit of 46.47 billion Yuan. The enterprise would be under pressure of downsizing staff.

The third is the contradiction between increased competition of enterprises and the lack of traditional energy. President Obama said: "If over a billion Chinese citizens have the same living pattern as Australians and Americans do right now, then all of us will have a very miserable time. The planet just cannot sustain us." We have several interpretations on his remarks. But it is certain we have already realized the necessity of the transformation and upgrading of Chinese energy industry, even if Obama did not tell us. Under the influence of the total of domestic energy, the political factors of international energy market, the limit on the carbon emissions, and other factors, the competition among energy enterprises will become more and more intense. To a certain extent, they will be inevitably in the dilemma of "zero-sum game." So the traditional energy enterprises should actively seek out new roads for their development, and open up a new world.

3 The Several Advantages of Geothermal Resources

Geothermal resources have some advantages when compared with traditional energy.

First, geothermal resources are high-quality resources that have been neglected for a long time. Energy expert Willie said: "We know that we are sitting on an almost inexhaustible source of energy. But not willing to exploit the energy which is beneath our feet; we would like to take oil, gas and coal back from thousands of kilometers away." According to this expert's opinion, if the "hot dry rock process method" can be applied widely, the condition of imbalance of geothermal distribution will make history in the world. Humankind will own an endless wealth of energy.

Second, geothermal resources are clean in essence. Traditional energy, including some new energy, can do harm to the environment in the process of production and use. But geothermal resources generally do not make a great impact on the environment, although they may cause some effect on natural organisms when some geothermal resources enterprises discharge the hot water during exploitation. However, this effect is completely controlled, and the natural organisms can recover quickly. Therefore, at present, exploiting geothermal resources is one of the known ways that belongs to the clean utilization of energy for human beings.

Third, exploiting geothermal resources is helpful in reducing carbon emissions. Owing to the increasing consumption of carbon energy, it is more and more difficult to control global warming. Some people think that the arctic will be ice-free during summers in the future. It could be good news for traffic, but it is terribly bad news for humankind. Under the inflexible standard of carbon emissions as carbon-free, geothermal resources are especially significant.

4 An Important Opportunity for Transformation and Upgrading

Therefore, if we want to break through the limit, we must think in new way and open up a new field. Geothermal resources provide a rare historical opportunity for us to do so.

The first is that exploiting geothermal resources may expand a huge market for development. For example, under the limitations of coal consumption in China, the coal enterprises are finding it more difficult to expand their scale. They are presently suffering from the serious consequences of expansion after the high-speed expansion of many years. Instead of fighting with each other, we could find a new way toward development.

In contrast with other countries, the development of geothermal industry in China has its own characteristics and advantages, yet it has obvious shortcomings. Although we have the largest scale of utilization in the world, Chinese geothermal resources concentrate in the field of low temperature. We lag behind in the field of geothermal power, and our technology strength is weak. The level of exploiting is low.

In 2011 The Chinese Academy of Engineering put forward some development goals, including that geothermal resources be directly utilized to generate electricity. According to these goals, generating electricity by geothermal resources of low-temperature and middle-temperature needs to reach 245 megawatts in 2030.

If we put the high-temperature, middle-temperature, and low-temperature geothermal resources together into a view of the development of traditional energy enterprise, the energy market will be very large. According to the goals of the Chinese Academy of Engineering set in 2011, the direct utilization of geothermal resources should reach 26000 megawatts in 2030, and the direct utilization of geothermal resources should reach 60000 megawatts in 2050 [1]. Start earlier and the gain is more. For the traditional energy enterprises, such a large market is worth looking forward to.

The second is the green development and transformation. At present, some traditional energy enterprises that depend on coal, oil, and natural gas, etc. have the pressure of environmental protection. The petrochemical enterprises caused the serious damage to the local environment repeatedly; coal mining and transport can bring damage to the environment to some extent, too [2]. For example, coal mining may lead to many ecological problems, such as the destruction of water resources, the gas emission, and the surface subsidence. These problems can't be solved completely by simple technical innovation. The traditional energy enterprises may realize transformation and upgrading from the high carbon to carbon-free, when they devote their strength to exploiting geothermal resources. In November 2006, Xin Xing Oil Company, which was a subsidiary of Sino Petroleum Corp, founded its joint venture partner, which is ENEX Company, which is a subsidiary of Iceland Lvyuan Company. They established Shanxi Lvyuan Geothermal Resources Development Limited Company. By the end of 2012, the areas of exploiting geothermal resources of Xin Xing Oil Company had already been expanded to Leling and Shanghe in Shandong, Yuncheng in Shanxi, Changyuan in Henan, and so on. It had 91 geothermal wells. Its heating capacity accounted for 25 % of the normal geothermal heating area in the whole country. It not only replaced 1.4 million tons of coal, but also reduced 3.8 million tons of carbon dioxide emissions a year [3].

The third is to find a stable way for development. In the exploitation of geothermal resources, the use of the bath accounts for 50 %, and the use for direct heating accounts for 30 %. The rest is involved in aquaculture industry, generating electricity accounts for only a small proportion. They all make an important contribution to increasing energy supply and reducing carbon emissions for the country, directly or indirectly. From the perspective of the enterprise, a certain degree of market stability is necessary for development. Owing to factors, such as international politics and economy, the fluctuant range of traditional energy in the market is large, which brought some trouble for enterprise. Because of its own unique advantage, the exploitation of geothermal resources presents the stable tendency. Although it has no profiteering, it does not fall seriously. Investment in the geothermal energy industry can help enterprises grow into comprehensive energy enterprises, and improve their abilities to cope with risks in the market. This mode of the development is especially suitable for some large central energy enterprises in our country. Some of them have also been aware of the problem. For example, Shenhua group is engaged in the development of wind power on the basis of relying on their core business of coal in order to try to become a comprehensive energy enterprise.

The fourth is that Chinese traditional energy enterprises have their own unique advantages in the exploitation of geothermal resources. Central enterprises have a well fund and a technical force. PetroChina and Sinopec have become international energy giants. They rank high among the global top 500 enterprises. Shenhua group has become the largest coal energy enterprise in the world. For one thing, these enterprises are sufficient in funds and have the financial resources for the development of geothermal resources. For another, some large central energy enterprises have strong technical reserves that are relevant to the exploitation of geothermal energy. Moreover, these enterprises have political advantages that can play an important role in organizing and coordinating the exploit of geothermal energy.

For this reason, we can put the development of geothermal resources as a leader, and let it promote the development of science and technology, and guide the downstream equipment manufacturing to upgrade. In this way, we can occupy a strategic place in the global manufacturing. Just as Engels said: "Once the society needs technology, it will push the science forward than 10 universities."

5 Introduction to Several Key Problems to Solve While Developing Geothermal Industry

The purpose of the government and the purpose of enterprises on the development of new energy resources are different. The government focuses on the interests and the coordinated green development of the whole country. But enterprises are in pursuit of profits. In terms of large energy enterprises in China, besides the non-technical factors, there are still several disadvantages in the development of geothermal resources.

The first is the concept. The first adverse idea about geothermal resources development is that the reserve is not enough. Subject to technical progress, people generally think that geothermal resources exploitation is limited to shallow geothermal or specific areas, so its potentiality is not too large. With the development of deep geothermal exploiting technology, this concept has been shaken. "The hot dry rock resources buried in the 3000–10,000 m deep of China's continent is total equivalent to 860 trillion tons of standard coal, which is 260,000 times of China's current annual energy consumption," Xuehua et al. [4] one expert said. The second adverse idea is about cost. This concept mainly concerns the cost of geothermal exploiting: people think that high cost would reduce its competitiveness. Finally, there is the weak idea of environmental protection. Nevertheless, in the era of environment protection concerns, and carbon emission pressure, vigorously developing new energy has moved far beyond the economic perspective, and has been transformed into a value judgment of the world. Of course, all of these changes need time, but the government and academic circles must continue to go ahead.

The second challenge is technology. Relying on governmental transfusion is by no means the right way toward geothermal development. We must reduce the cost through technical revolution for the sake of competitiveness; this is the key to its survival. Geothermal resources now belong to the category of new energy, but it is older than most of the traditional energy. In fact, China's utilization of geothermal resources has a history of more than 2000 years. Therefore, geothermal resources belonging to "new" energy is not because the time since we find it is not long, but because of the slow development of technological innovation. Cost reduction is therefore the key to enhancing the competitiveness of the geothermal industry. At present, the money spent on drilling takes 50–70 % of the total cost of geothermal resources development in EU. A 50 MW typical traditional geothermal power plant installed capacity of 5 MW per well, with a drilling success rate of 67 %, the drilling costs up to 2285 euro's/kWm. Therefore, the main reason for the slow development of geothermal industry is that the technology, especially the drilling technology, which is owned by current enterprises, is not advanced enough, and it leads to a high drilling cost.

The third is nearsightedness. There are more or less incorrect views on political achievements in some companies; these views will cause us to be shortsighted. The traditional energy companies will not be too concerned about geothermal industry which is small in investment, and less output when the energy resources is in a bull markets. The significance of a project for the leadership, not only lies in the rate of return, but is also related to its scale. When the energy resources are in bear markets, due to the low prices, the development of geothermal energy is often unprofitable. As a result, it is difficult to attract the interest of traditional energy giants toward geothermal resources development. They would rather put capital into the large-scale, high-yield traditional energy sector, in order to "obtain prominent achievements" In terms of scientific research, enterprises prefer to focus on the promotion of traditional energy production technology, rather than think too much about the geothermal resources research. It has both directly and indirectly caused the lack of research talents over the last two decades in our country.

The degree of the exploitation of geothermal resources is closely related to the progress of science and technology. With the development of science and technology, and the conversion of concepts held by our government and our citizenry, it is predicted that the exploitation of geothermal resources can have a prosperous future. Geothermal resources will become an important component in the structure of Chinese energy consumption.

References

1. Pang Zhonghe H, Shengbiao WJ (2012) A roadmap to geothermal energy development in China. Sci Technol Rev 30(32):18–24
2. Menggang L (2012) Security report of Chinese energy industry(2011–2012). Social Sciences Academic Press, Beijing, pp 60–61

3. Conglei C, Xiaoxuan X (2013) International energy companies utilization of geothermal energy inspiration for Sinopec. Sino-Glob Energy 10:21–25
4. Xuehua B, Jun S, Qiuxiao L (2014) The perfection of geothermal resources development and utilization. Conserv Utilization Mineral Resour 10:5–8

A New Method of Production: Peer Production

Xiangdong Pan

Abstract This paper presents peer production on the basis of the three industrial revolutions. It briefly introduces the Open Source Software phenomenon and background of peer production, then analyzes the conditions and dynamics which make peer production emerging. The paper points out that *internet communication*, *modularity*, and *customers as innovators* play an important role in peer production. It considers that some external environments such as *hacker culture* and *popularity of knowledge* make peer production lasting. Finally the paper summarizes the characteristics of the peer production method, including *mass collaboration* and *customers as innovators*.

Keywords Peer production · Mass collaboration · Customers as innovators

1 Introduction

The social supply capacity must be gradually improved to fit the growing market demand by industrial upgrading and innovation. Reviewing history, the first two industrial revolutions were both caused by strong demand. Now, the third industrial revolution is coming.

Rifkin [1], who is an American economic and social theorist, published *The Third Industrial Revolution: How Lateral Power Is Transforming Energy, the*

X. Pan (✉)
Postdoctoral Programme of China Industrial Economic Security Research Center,
Beijing Jiaotong University, Beijing 100044, China
e-mail: hbfgwpxd@126.com

X. Pan
Hebi Development and Reform Commission, Hebi 45030, China

M. Li et al. (eds.), *Proceedings of 2015 2nd International Conference on Industrial Economics System and Industrial Security Engineering*,
DOI 10.1007/978-981-287-655-3_60

Table 1 The division of three industrial revolutions

Phases	Time	New technologies	New infrastructures	New industries	Techno-economic paradigm
Ancient times	Before 1760	–	Canals and waterways	–	Household production
The first industrial revolution	1760–1870	Steam engines and machinery; Iron and coal mining	Turnpike roads; Railways	The textile industry; Mining industry; Metallurgy industry	Factory production; Mechanization
The second industrial revolution	1870–1950	Electrical equipment; Heavy chemistry and Petroleum refining	Worldwide shipping; Transcontinental railways; Great bridges and tunnels; Telegraph and Telephone; Electrical networks	Oil industry; Power industry; Automobile industry	Mass production; Flow production; Economies of scale
The third industrial revolution	After 1950	Information and Communications Technologies; Computer-aided biotechnology; 3-D printing; Robots and intelligent manufacturing	Internet; Computers; High-speed multi-modal physical transport links	IT industry; software; New materials; Biotechnology industry	Peer production; Mass collaboration; Mass customization

Economy, and the World in 2011, and Markillie [2], who is an Innovation Editor of *The Economist* and Editor of *Technology Quarterly*, delivered the special report "A Third Industrial Revolution" in 2012. Thus, issues about the third industrial revolution have become the focus of scholars.

McCraw [3] pointed out that the three industrial revolutions could be classed according to the technological revolution and Kondratieff wave characteristics. The first industrial revolution began in the late 18th century, marked by the emergence of the steam engine. The second industrial revolution came in the middle of the 19th century; the main symbol is the power and the oil technology revolution. Now the third revolution is under way; the main signs are widely used in computer and Internet technology. Table 1 indicates the three revolutions with the main technologies, infrastructures, industries, and prevailing techno-economic paradigms [4].

The first and the second industrial revolutions greatly improved the production capacity of enterprises to fit the growing needs of human consumption. People were richer and more urban. But they began to put forward more and more personalized requirements after their essential needs of life had been met. The gaps between supply and demand were highlighted once again. Many customers with professional skills are beginning to join in the production process to become producers and innovators. So a new production method is emerging; we call it peer production.

In this paper, we explore peer production and analyze its characteristics. The remainder of the paper is organized as follows. In Sect. 2, we review the literature about peer production and Open Source Software (OSS). In Sect. 3, we analyze two characteristics of this new production method. Finally, conclusions are stated in Sect. 4

2 Peer Production and Theory Literature

2.1 The Phenomenon of Open Source and Peer Production

OSS has gotten substantial attention in the last decade, as such programs as Linux, Apache, and Firefox have been adopted widely. Open source can be understood either narrowly or broadly. OS is a new development method like Agile Methodologies to resolve the software crisis narrowly. It is an amazing phenomenon and broadly a new method of production.

Recently open source hardware and open source auto have followed on the heels of OSS. Many new events, such as Wikipedia and Open-Source-Shakespeare, have begun to emerge. A new mode of economic production is vividly portrayed. The mode shows different characteristics from traditional production; it is called commons-based peer production by Harvard law professor Benkler [5].

Peer production is also known as mass collaboration. It is a new way to produce goods and services that relies on self-organizing communities of individuals. In such communities, thousands of individuals effectively cooperate with the help of the Internet, and the roles of producers and consumers will begin to blur and merge. Peer production is playing a role in the field of manufacturing industry by sharing open source hardware designs, while digital manufacturing technologies such as RepRap 3-D printers are widely applied. The number of such designs is growing exponentially on free design sites such as Youmagine [6].

2.2 Theory Literature

Today software engineers, economists, sociologists, anthropologists, organizations, and innovation theorists are all exploring OS phenomenon and peer production. What it presents is a fascinating sociotechnical problem which can only be understood and appreciated by bringing multiple disciplines to bear. The research can be classified into two major streams: motivation of contributors and mode of collaboration. We briefly review relevant work in each stream in what follows.

Inevitably, the most important puzzle is why thousands of developers contribute voluntarily to OSS projects without reward. Hertel et al. [7], Franke and Shah [8], von Hippel and von Krogh [9] found that developers were motivated by the need to improve the product for their own use. Source code becomes public good under an OS environment, so it may lead to a free-rider problem and the under-supply of a public good. But the practice of OSS development doesn't show these projects to have been impacted by this problem. It can be interpreted by selective incentive theory [10]. These selective benefits are often characterized by two ideal types of motivation: intrinsic and extrinsic. Extrinsic motives include benefits gained from extended functionality and reputation. Intrinsic motives include benefits gained from enjoyment and pro-social motives. For example, some developers participate in the project to increase their labor market value and get better jobs; some just do so for fun or to learn. There are also a portion of participants who like the sense of "helping others" or "giving something back" to like-minded people. von Hippel and von Krogh [9] synthetically presented a "private-collective" model of innovation incentives.

The second focus is how thousands of individuals coordinate on a project. OS presents a platform to share knowledge; every developer can view, modify, and redistribute source code freely. These individuals often interact in open source communities and contribute to the project by self-organization rather than by authority within a firm and the price mechanism [11]. Few core developers account for most of the contributions and control the development of the majority of the source code; most developers present a small part of the code. The vast majority of contributors are involved in only one or very few projects; few contributors join in many [12]. Healy and Schussman [13] found through empirical analysis that the developers–projects relationship follows power–law distributions. Crowston and Howison [14] analyzed the centralization and the hierarchy of project teams based on a social network analysis method.

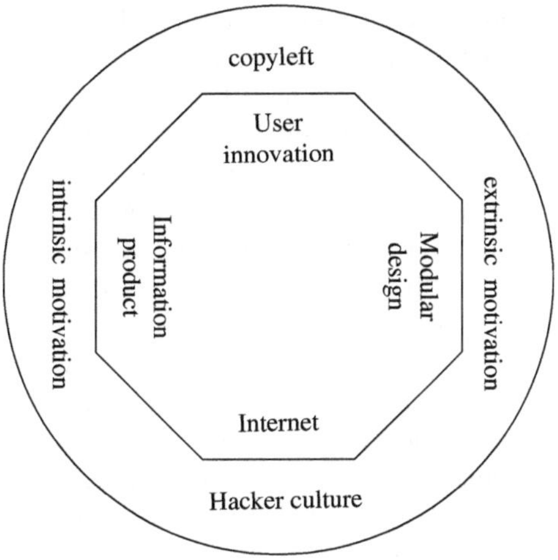

Fig. 1 Conditions of peer production

3 Characteristics of the Peer Production Method

3.1 Conditions

OSS is a complex and heterogeneous phenomenon. Scholars often describe its characteristics as crucial, namely to be (a) an information product, (b) reliant on the Internet, (c) a user–developer innovation process, (d) the product of a highly modular design. These characteristics lead to a low-cost situation in which large-scale users easily participate in software code development. This triggers OSS emergence. At the same time, there are some external forces such as incentive, institution, and culture, which appear to promote OSS. All of the above factors can be depicted as in Fig. 1. The inner octagon includes necessary conditions of OSS emerging. In fact, all kinds of products which meet these conditions, such as Wikipedia and Open-Source-Shakespeare, can be produced through the peer production method.

But whether the peer production method will be used or not is decided by those sufficient conditions in outer circularity. Here intrinsic and extrinsic motivations can solve the first order social dilemma during the offering of public goods. Hacker culture helps to strengthen intrinsic motivations of contributors. Copyleft licenses will enforce the voluntary action that conditional cooperators contribute to public goods and overcome the second order dilemma. The four factors form sufficient terms for the endurance of OSS.

3.2 Dynamic

Under dynamic view, peer production is driven by supplementary force and driving force. The supplementary force comes from the institutional environment which includes hacker culture and copyleft licenses. It implies in what environment the new production method can be antecedently used. For example, it is hacker culture within the software industry that makes the new method used initially to develop software, and copyleft licenses help OSS to succeed as an institutional innovation.

The driving force comes from factors that are closely related to production, which includes the properties of product (information product), modular design, the popularity of knowledge, and an effective instrument of communication (the Internet). "Information product" is non-rivalry; that is, consumption by the donor is not precluded as other persons consume these products. This property of non-rivalry can apply to all non-physical objects, such as ideas, designs, etc. Modularity is a technical and organizational way to manage complexity. Modularity makes complex work easier and enables a large number of contributors to work simultaneously on a project, keeping integration and coordination costs low. The popularity of knowledge will influence many aspects of society. First, knowledge is no longer peculiar to the minority; this makes individual ideas become more diversiform. Consumers' preferences have become more dissimilar and demand more personalized. Second, users have competence to participate in the contributions as they catch up on necessary knowledge. The Internet is an effective instrument of communication and allows developers to exchange information (software code) widely, easily, and cheaply. Thus the producers of information products can implement mass collaboration.

3.3 Characteristics

These four factors that influence the driving force jointly determine the two characteristics of peer production method, namely customers as innovators and mass collaboration, as shown in Fig. 2.

The futurist Toffler [15] in his book *The Third Wave* predicted that the roles of producers and consumers would begin to blur and merge, and called them prosumers. With the popularity of knowledge, customers have more information about

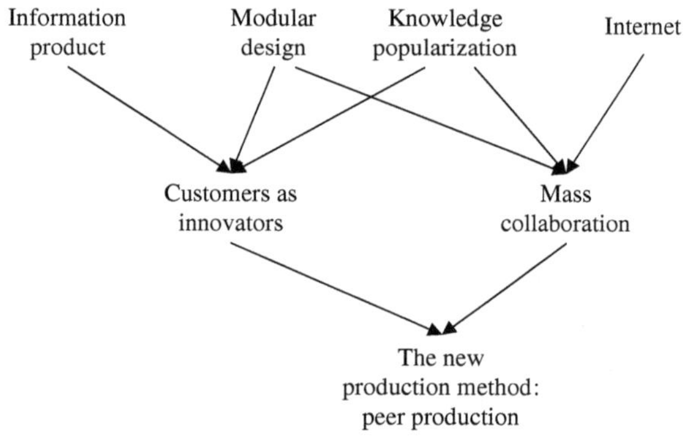

Fig. 2 Characteristics of peer production

products and become more captious. Manufacturers hardly fully understand and meet customers' needs. At the same time, users can be a rich source of innovation. They more easily detect the shortcomings of products and develop new ideas to help to make the products suit their needs better. In fact, there are many examples of customers as innovators. For example, many important innovations in snowboarding, windsurfing, and skateboarding equipment are developed by users.

Of course, not all customers are innovators. The customers who participate in development may even be customers in a different sense. Practically speaking, these customers can be classed into three kinds: lead users, active users, passive users. Lead users commonly have two characteristics: they (a) expect attractive innovation-related benefits from a solution to their needs, and (b) experience needs that will become general in a marketplace, but experience them months or years earlier than the majority of the target market [16]. So lead users generally are captious customers and have professional knowledge about products. Active users cannot develop new ideas about products, but they provide use-cases and feed the information back to developers, as well as testing new designs. Passive users are those orthodox customers; they consume the products that they enjoy, or on the contrary choose to vote with their feet.

The second characteristic of the peer production method is mass collaboration. Mass collaboration is a form of collective action that occurs when large numbers of people work independently on a single project. Its central characteristic is that groups of individuals successfully collaborate on projects following a diverse cluster of motivational drives and social signals, rather than either market prices or managerial commands. The result is the emergence of a vibrant, innovative, and productive collaboration. From Amazon.com to MySpace to craigslist, the most successful Web companies are building business models based on user-generated content. The tools of production are fully democratized, and the engine for growth is the spare cycles, talent, and capacity of regular folks, who are, in aggregate, creating a distributed labor force of unprecedented scale.

The peer production method has comparative advantage by reason of the two characteristics. Customers as innovators make products better meet customers' needs; mass collaboration helps to detect all bugs in products. Thus the peer production method can rapidly succeed. Furthermore, a certain culture and institutional innovation maintain it.

4 Conclusion

The practice of economic development indicates that every technical innovation and industrial revolution was driven by demand. Now that traditional mass production has been unable to meet the more diversified and personalized needs of consumers, it is triggering a new industrial revolution. OS phenomena that began in the late 20th century show that a new production method is on the rise. Thousands of individuals from all over the world produce goods and services in self-organizing communities; these individuals include not only producers but also consumers. In other words, the roles of producers and consumers will begin to blur and merge. Thus a new production method is emerging; it is peer production.

The goods and services which are produced under the peer production method are information product and have modular structure. This makes it very easy for thousands of individuals to participate in its designing and producing process, based on the Internet. At the same time, these participants have more knowledge and technical ability to make contributions to the projects that interest them, with knowledge becoming more and more popular. Thus more customers are becoming producers and innovators. They produce goods and services by self-organizing in the community. In this process, external environmental factors such as copyleft and hacker culture play an important part in mass collaboration. Peer production will be a kind of very effective production method in the era of the knowledge economy.

References

1. Rifkin J (2011) The third industrial revolution: how lateral power is transforming energy, the economy, and the world [M]. Macmillan, London
2. Markillie P (2012) A third industrial revolution [J]. The economist, 21 April 2012
3. McCraw TK (ed) (1997) Creating modern capitalism: how entrepreneurs, companies, and countries triumphed in three industrial revolutions [M]. Harvard University Press, Massachusetts
4. Perez C (2010) The financial crisis and the future of innovation: A view of technical change with the aid of history [R]. TUT Ragnar Nurkse School of Innovation and Governance
5. Benkler Y (2003) Freedom in the commons: Towards a political economy of information [J]. Duke Law J. 1245–1276
6. Wikipedia. Peer production [G/OL]. 2015(4.10) http://en.wikipedia.org/wiki/Peer_production
7. Hertel G, Niedner S, Hermann S (2003) Motivation of software developers in the F/OSS projects: an Internet-based survey of contributors to the Linux kernel [J]. Res Policy 327:1159–1177
8. Franke N, Shah S (2003) How communities support innovative activities: an exploration of assistance and sharing among innovative users of sporting equipment [J]. Res Policy 32 (1):157–178
9. von Hippel E, von Krogh G (2003) Open source software, the private—collective innovation model [J]. Issues Org Sci 14(2):216–217
10. Osterloh M, Rota S (2007) Open source software development—Just an other case of collective invention? [J]. Res Policy 36:157–171
11. Kuwabara K (2000) Linux: a bazaar at the edge of chaos [J]. First Monday, 5(3)
12. Krishnamurthy S (2002) Cave or community? an empirical examination of 100 mature F/OSS projects [J]. First Monday, (76)
13. Healy K, Schussman A (2003) The ecology of F/OSS software development [J]. First Monday, 6(7)
14. Crowston K, Howison J (2005) The social structure of free and open source software development [J]. First Monday, 10(2)
15. Toffler A (1981) The third wave [M]. Bantam books, New York
16. Von Hippel E (1986) Lead users: a source of novel product concepts [J]. Manag Sci 32 (7):791–805

Research on Development Countermeasures for New Energy Vehicles in China

Jinsong Zhang, Ze Zhu, Jie Shen and Minghao Yan

Abstract Considering a comprehensive review of the current situation of the development of new energy automotive industry in China, in order to solve the core problem of insufficient demand, this paper explores the factors influencing Chinese consumer perception and shifting behavior towards hybrid cars. On the basis of the current industry status and the empirical conclusions, suggestions for practical countermeasures are put forward in the article.

Keywords New energy vehicles · Industry development · Consumer behavior · Countermeasures

1 Introduction

According to the China Automotive News, ownership of cars in the country reached 85 million in 2010, while the fuel consumption of the auto industry has reached 4.39 tonnes, accounting for 1/3 of the total amount of national oil consumption [1]. The ministry of industry and information technology forecasts China's car ownership will be more than 200 million in 2020. Evidence has shown that the promotion of new energy vehicles can reduce carbon emissions by nearly 1/5, reducing by 50 % the air pollution in the city [2]. Therefore, the promotion of energy-saving and emission reduction relative to traditional cars, and so the industrialization of new energy vehicles, have become major issues for China's automobile industry. In contrast with the traditional vehicle industry, new energy car manufacturers haven't reached the level of economies of scale [3]. The problems of the development of the energy vehicles industry can be divided into R&D and market. The definition of product quality is based on the perception of the market; the deficiency level of

J. Zhang (✉) · Z. Zhu · J. Shen · M. Yan
South-Central University for Nationalities, Wuhan 430074, China
e-mail: zhangjinsong@scuec.edu.cn

M. Li et al. (eds.), *Proceedings of 2015 2nd International Conference on Industrial Economics System and Industrial Security Engineering*,
DOI 10.1007/978-981-287-655-3_61

R&D naturally causes the result that products are not accepted by consumers in the market. The lack of core technologies, of technical innovation, and of sufficient accumulation is at the heart of the problem.

2 Study Design Based on Consumer Perception

In previous studies, research on the similarities and differences between the purchase intentions of consumers based on gender, age, and educational background has been rare and hard to find. Consumers' value orientation, consumer population pressure, and other related factors are also neglected. Based on the review of related literature, this study proposes the following research variables of consumer perception of products (Table 1).

Table 1 Variable names and sources of research

Variable	Source
Power performance	Ewing et al. (2000)
Endurance performance	Chan (1991); Cheron and Zins (1997); Golob et al. (2000); Moore (2001); Axsen and Kurani (2009)
Handling stability	Axsen and Kurani (2009)
Fuel economy	Adler (2003); Ewing (2000); Axsen and Kurani (2009); Hyundo and Inha (2010)
Space inside the car	Xu et al. (2010)
Maintainability	Hyundo and Inha (2010)
Appreciation of purchase behavior	Li (2009)
Income level	Hogberg (2007)
Related infrastructure	Cheron and Zins (1997): Ewing (2000); Achtnicht (2008); Kalhammer (2009); Axsen and Kurani (2009)
Potential performance deficiencies	Kalhammer (2009)
Repair of risk	Hyundo and Inha (2010)
Social responsibility	Li (2009)
Improving the quality of life	Li (2009)
Whether value for money	Chan (1991); Kalhammer (2009)
Making a good impression	Li (2009)
Search cost	Our research paper Yu (2009)
Suggestions and behavior of people nearby	Mannering and Winston (1991, 1995)
Brand Subsidies and preferential	Zhou and Shi (2013); Ma et al. (2013); Tang and Wu (2012)

2.1 *Questionnaire Design*

The questionnaire contains buying influence factors and basic information of consumers. With reference to related literature, my research introduced the consumer value orientation. At the same time, we ultimately determined 26 questions by surveying a driving school. Using the Li Kete five subscales development questionnaire, the score of each option is sequential from 1 to 5.

2.2 *Data Collection*

We used two main types of surveys: one sent over the network and the other one commissioned. The actual commission issued 100 copies, with 91 valid questionnaires recovered (so the recovery rate was 91 %). The total recovery was 121 valid questionnaires issued through the network.

2.3 *Reliability Testing*

The 212 questionnaires were processed using SPSS19.0. Cronbach's Alpha value was 0.771; the questionnaire showed high reliability. The KMO was 0.791; Bartlett's test of sphericity yielded 1194.46, with 120 degrees of freedom, and the significance was 0, showing that the original variables are suitable for factor analysis.

3 Empirical Analysis and Conclusions

3.1 *Basic Characteristics of Sample*

In the survey of 212 consumers, the number of consumers less than 40 years old was 190. Some 70 % of the respondents are willing to buy new energy vehicles. Our sample of 129 males and 83 females account for 60.85 and 39.15 %, respectively. Of the men surveyed, 90 expressed a purchase intention; these represent 69.77 % of the total men interviewed. Furthermore, there are 58 of female respondents with purchase intention; these account for 69.88 % of the total women interviewed.

3.2 Factor Analysis

According to Bartlett's test of sphericity, the value obtained is 1194.46; the corresponding P value is 0, significantly less than the given level. And the KMO value is 0.791.According to the Kaiser given common KMO measures, this research is suitable for factor analysis. The results of data processing show that the four factors explain 61.408 % of the original variables, reflecting the original data contained to a very high degree.

In Table 2, in the case of factor 1, the coefficients of endurance performance, handling stability, maintainability, and fuel economy are high, being about 0.8 each; the average score of the four projects is about 4.5. (The first factor was named the "product quality factor".)

In the second component, the coefficients of appreciation, social responsibility, making a good impression, and improving the quality of life are about 0.8. The average score of appreciation is 4; the rest are about 3.6. (The second factor was named the "value orientation factor".)

In the third component, the coefficients of infrastructure, repair risk, and potential performance defects are approximately 0.78. The average comprehensive score is 3.8. (We named the third factor "use cost factor".)

In the fourth component, the coefficients of view of crowd nearby and purchase behavior recognition degree are significant. As group members, consumers have

Table 2 Rotated component matrix

Component	Factor			
	1	2	3	4
Endurance performance	0.839	0.059	0.159	−0.043
Handling stability	0.828	0.108	0.023	0.026
Maintainability	0.801	0.142	0.132	−0.007
Fuel economy	0.794	0.172	0.291	−0.110
Appreciation of purchase behavior	0.264	0.813	0.114	−0.039
Social responsibility	0.029	0.811	0.120	−0.088
Making a good impression	0.099	0.791	−0.015	0.190
Improving the quality of life	0.077	0.779	−0.053	0.245
Related infrastructure	0.188	−0.041	0.788	−0.082
Repair of risk	0.000	0.086	0.757	0.142
Potential performance deficiencies	0.222	−0.008	0.685	0.025
Whether value for money	0.103	0.060	0.461	0.150
Suggestions and behavior of people nearby	−0.032	−0.001	−0.046	0.848
Degree of recognition of people nearby	−0.282	0.219	0.213	0.558
Income level	0.405	−0.010	0.106	0.536
Search cost	−0.101	0.245	0.396	0.500

their behavior affected by the population effect to a great extent; strong influences may include relatives and friends, colleagues, and people in industry. (The fourth factor was therefore named the "group influence factor".)

3.3 Empirical Conclusions

(1) From the data collected from the survey, males and females each represent a 70 % expectation of buying new energy vehicles. There is no difference based on gender in acceptance and willingness to buy the new energy automotive products.
(2) The 16 factors can be extracted as four main categories: "product quality factor," "value orientation factor," "use cost factor," and "group effect factor." The cumulative contribution rate is 61.41 %.
(3) The average scores of social development trend and income level were 4.25 and 4.02, respectively. In the relevant measures that should be taken, the average score of increasing infrastructure construction such as of charging stations was the highest, scoring 4.61. The average score of relief purchase taxes and strengthening purchase subsidies was 4.2. These scores may largely explain the reasons why consumers adopt a shift in their consuming behavior.

4 Development Strategies

4.1 Increase R&D Investment and Public Procurement

The influence factors of product quality, endurance performance, handling stability, maintainability, and fuel economy are among the factors closely related to the level of R&D. The new energy automotive industry in China is in its infancy, so it's necessary to increase investment in research. In addition, inciting the Chinese government to replace its cars with new energy cars can be a way of promoting new energy cars to the Chinese population.

4.2 Standardize Auto Industry Information Criteria

With respect to the reference data related to energy-saving and new energy vehicles in Yearbook 2011, only one-third of 51 enterprises published data on energy consumption. Businesses complying with the requirement to release their energy consumption information are lining up with the government policy supporting the new energy car industry. And this regulation is critical, because without it the new energy car industry will not reach its maturity development stage.

4.3 Policy Support Based on the Entire Life Cycle

Usage cost, supporting facilities, maintenance risk, performance, price and other effects of defects are closely related to policy measures. Therefore, based on the influence factors of consumer behavior, in order to solve the existing problems when developing new energy vehicles, the government and enterprises should continue to increase R&D efforts and play the role of back-end market incentives. Meanwhile, they should also re-examine the prospective implementation effect and its completeness of policy.

4.4 Gradually Enhance the Criteria of Conventional Cars

The existing industrial assistance policy in the new energy automotive industry puts emphasis on the promotion of new energy automobile products, ignoring the emissions standards and the restrictions of conventional fossil fuel cars. The development process of the foreign auto industry conveys the point that strict limits on cars' exhaust gas emissions involve additional production costs in the initial policy. But it still affords the chance to participate in international market competition in the future. So, it's imperative to enhance the performance standard of traditional automobiles by a gradual method.

5 Conclusions

Based on consumer perspective, the key of this article is to recognize and process the factors which evidently impact consumers' adoption of new energy vehicles. Undoubtedly, the government and the Industry Association should undertake the responsibility to complete the industry standards, accomplish demonstrations, and cultivate the new energy automobile consumption market. In the future, one of the feasible research orientations will be to focus on the affecting factors of the consumer's behavior at different times: traditional car versus new energy car.

References

1. Liping W, Rui X (2012) Automotive industry development model in the low-carbon economy background [J]. Forum Sci Technol China 58(05):53–56
2. Zhou A, Jinglin L (2012) New exploration of new energy automobile on the influence of city energy-saving emission reduction [J]. Acad Exch 07:101–105
3. Energy-saving and new energy vehicles in the yearbook 2011

Research on Safe Evacuation Simulation of University Students' Dormitory Buildings Based on Pathfinder

Jing Zhao and Shouxin Song

Abstract In this paper, a study on simulation of a boys' dormitory building evacuation in case of emergency is carried out, using the evacuation simulation software Pathfinder 2011. With parameters set on the basis of the establishment of building models and their respective personnel, with the Steering model selected, and with different numbers of exits, the simulation results show that the time and evacuation numbers linearly correspond in a particular relationship; the rate of change of evacuation time along with the number of evacuations undergo dynamic growth in a certain number of intervals. Using the condition of saturation of evacuees, and carrying out the simulation and analysis of different numbers of exits, evacuees, and evacuation styles, where and when the highest risks are can be demonstrated, so as to optimize evacuations.

Keywords Safety science and engineering · Evacuation · Dormitory building · Simulation

1 Introduction

In recent years, with the popularization of higher education, the increasing number of university students and dormitories has led to overcrowding, and security problems have been gradually emerging. Once there is an emergency in a dormitory building, such as fire, terrorist attack, or natural disaster such as earthquake,

J. Zhao (✉) · S. Song
School of Economics and Management, Beijing Jiaotong University,
Beijing 100044, China
e-mail: 406826137@qq.com; zhaojing@bjtu.edu.cn

S. Song
e-mail: songsx@yeah.net

M. Li et al. (eds.), *Proceedings of 2015 2nd International Conference on Industrial Economics System and Industrial Security Engineering*,
DOI 10.1007/978-981-287-655-3_62

secondary damage may easily be caused; stampede, for example, represents a great threat to safety at an accident scene, and becomes a major hidden threat of security on campus in particular.

With the development of computer technology, evacuation simulation technology has been applied in the security field, which has promoted the development of the analysis of the safety of building structures and provided an effective tool for personnel safety evacuation. This paper studies the results of using the evacuation simulation software Pathfinder 2011 to simulate a boys' dormitory building emergency evacuation. By studying evacuation processes, it provides a reference for the improvement of the internal structure of buildings and helps reduce accidents.

2 Software Overview

Pathfinder was developed by American Thunderhead Engineering Company as a simulator, which is based on the movement of personnel entering and leaving. It provides analog design and execution of the graphical user interface and the analysis results of 3D visualization tools [1, 2].

The Steering personnel movement model is a human movement behavior model; it uses path planning, guiding mechanisms, and collision handling to control movement. The core of this kind of path generation algorithm is based on principles of human behavior; if the distance between personnel and the path of the closest point gets bigger than a certain threshold, the algorithm will generate a new path to adjust the manner of personnel movement [3, 4].

3 Software Simulation Research

3.1 University Students' Dormitory Building Mode

The simulation objects are based on a university boys' dormitory building, having an overall length of 62 m, width 14.58 m, single layer 3 m high, 6 floors, total building height 18.92 m. Two main entrances are located in the south building, with symmetrical distribution between both sides. The first layer is equipped with 21 dorm rooms (5.54 m long, 3.54 m wide, room door 0.9 m wide), and a duty room (the size same as a dorm room); 2−6 layers equipped with 24 dorm rooms dorms and an activity room. The building channel width is 1.8 m; layers are connected to the east and west by two staircases of which the width is 1.6 m, the step height is 0.15 m, and the depth is 0.3 m (as shown in Figs. 1 and 2).

Fig. 1 Building showing overall 3D view

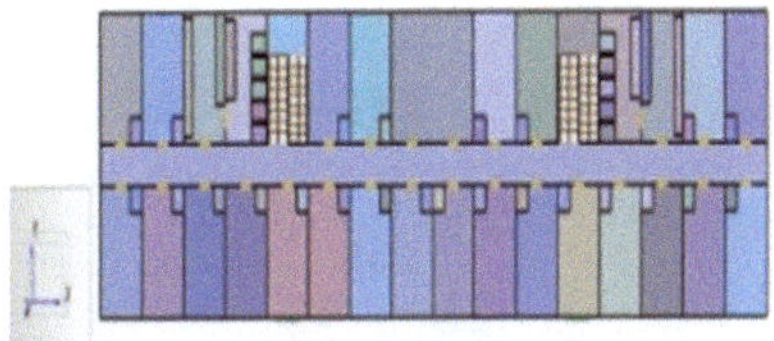

Fig. 2 First floor of dormitory building, top view

3.2 Personnel Parameter Setting

For the boys' dormitory, the personnel movement speed is in the range 0.8 ~ 1.4 m/s, with an average of 1.1 m/s. The simulation data are from a survey of 50 students randomly selected by the school; the study assumes that personnel get out the evacuation message at the same time, response lag time 0.4 to 1.6 s, personnel shoulders 38 cm to 46 cm wide.

3.3 Simulation of Different Numbers of Evacuations

The simulation uses the Steering model, with a base number of evacuees being set, each layer of personnel being randomly distributed (Table 1, with the different rows reflecting, respectively, 1, 2, 3, and 4 people in each dorm room).

Through analysis of different numbers of evacuees, the results in Fig. 3 show the distribution of the times required for the complete evacuation process. Comparisons of time changes and the growth rate of time for different exit numbers are shown in Fig. 4.

It showed that the times needed for building evacuation and evacuees have a linear relation, where the evacuation time along with the changing rate of evacuees have a dynamic growth in certain numbers of intervals. Depending on the type of building, 300 and 500 evacuees become greatly increased control points.

Table 1 Number of occupants on each layer

Total number	First floor	Second floor	Third floor	Fourth floor	Fifth floor	Sixth floor
141	21	24	24	24	24	24
282	42	48	48	48	48	48
423	63	72	72	72	72	72
564	84	96	96	96	96	96

Fig. 3 Evacuation times with different numbers of evacuees

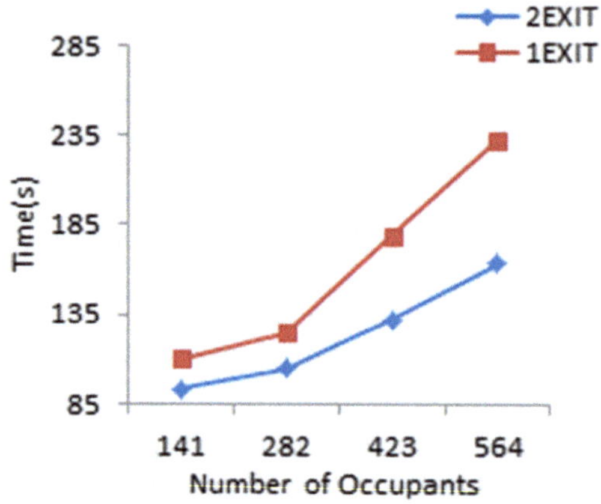

Fig. 4 Evacuation time growth rate with different numbers of evacuees

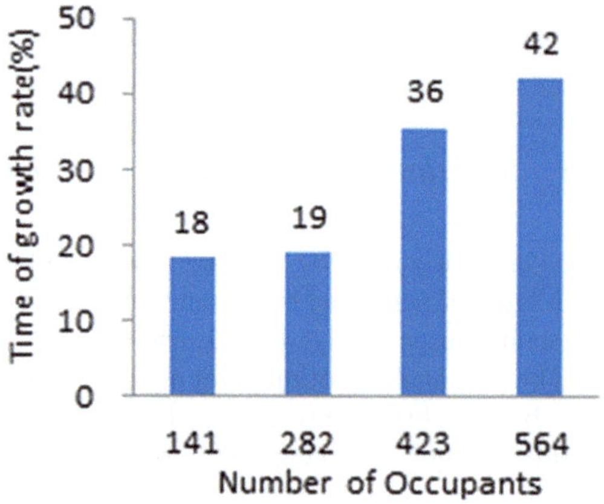

3.4 Evacuation Simulation with Saturation Condition

(1) Simulation of different waiting times
With the simulation adopting the Steering mode, average distribution of the initial position in a room, each room holding 4 people, 2 exits open. With increasing waiting time without delay, 10, 15, or 20 s for each floor, to simulate the four

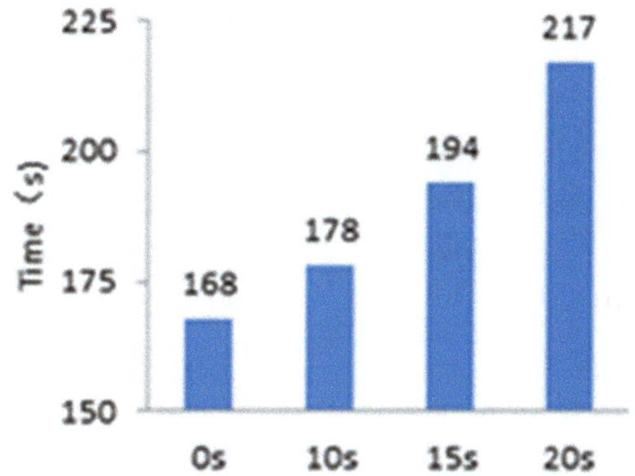

Fig. 5 Time for evacuees to complete evacuation

different waiting times. Complete evacuation times are shown in Fig. 5; the waiting time of each layer of evacuees is shown in Table 2.

For each layer on the east side of the stairs, the number of people changes over time, as shown in Fig. 6.

With each layer experiencing waiting time in equal increments, the simulation result shows that with the increasing waiting time in each layer, the whole evacuation time increases linearly, and the growth rate from 5 to 10 %. But with the increase in evacuation time, the number of flights of stairs with congested segments decreases. The biggest change is for 5 to 6 layers of stairs, and the smallest is the east hall; this explains how when 2 exits are open at the same time, congestion may not occur. The reason for the above phenomenon is that the layers of congestion segments spread exit pressure.

(2) The researchers using orderly evacuation software
The initial positions of personnel for this simulation are the same as in the simulation with different waiting times, and all evacuees start to evacuate at the same time. When the doors and stairs create congestion where the arriving evacuees line up, they can pass through in orderly fashion after waiting for congestion relief. Complete evacuation takes 203 s; the result is similar to that of incremental step-by-step waiting times of 15 and 20 s. Comparing the simulation with the model of waiting time without delay, 10 s, the results are shown in Fig. 7.

Analysis results show that the fewest evacuees carry out an orderly evacuation with minimal stairs congestion segments. With increasing floor numbers, the number of people who can be safely evacuated decreases; the differences narrow

Table 2 The waiting time of each layer of evacuees

Waiting time	First floor	Second floor	Third floor	Fourth floor	Fifth floor	Sixth floor
0	0	0	0	0	0	0
10	10	20	30	40	50	60
15	15	30	45	60	75	90
20	20	40	60	80	100	120

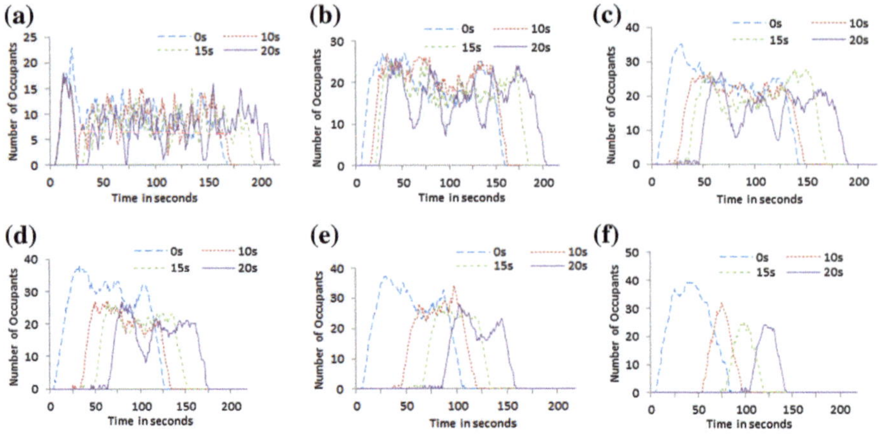

Fig. 6 Number of evacuees and evacuation time of east hall (**a**) and stairs (**b**. 1–2 floor, **c**. 2–3 floor, **d**. 3–4 floor, **e**. 4–5 floor, **f**. 5–6 floor)

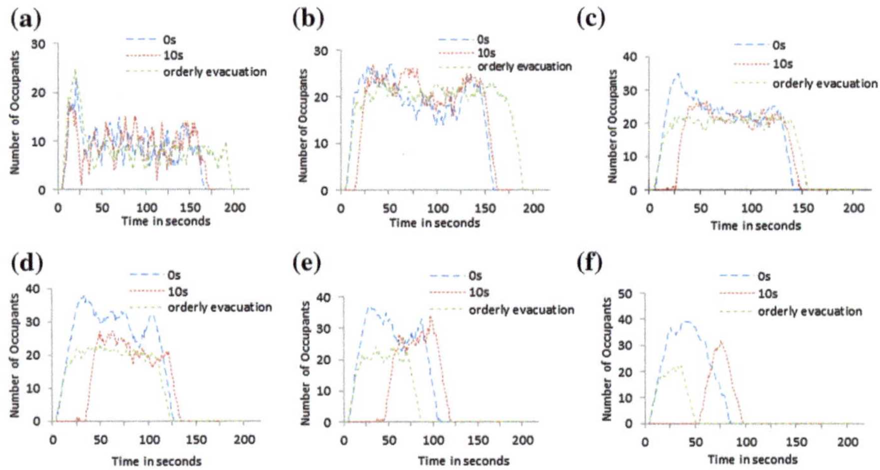

Fig. 7 Different numbers of evacuees and evacuation time of east hall (**a**) and stairs, (**b**. 1−2 floor, **c**. 2−3 floor, **d**. 3–4 floor, **e**. 4–5 floor, **f**. 5–6 floor)

between layers. With 3−4 flights of stairs for the node, the lower changes weakly; however, the upper changes significantly.

Students in daily safety education should carry out increased evacuation drills and other simulation exercises; there can also be layered evacuation by way of radio when danger arises. The simulation shows that it is more reasonable to build in a 10 s increment for each layer. Reducing the number of people in congested locations can reduce the risk of congestion occurring, thereby reducing secondary injuries due to congestion.

References

1. Shu-ran L, Kai Y (2014) Fire and evacuation simulation [M], vol 218, Chemical Industry Press, Beijing
2. Xiao-ming Z, Zhong-ri H, Feng L et al (2007) The discussion an appliance research of evacuation models [J]. Fire Sci Technol 26(4):386–390
3. Shu-ping Z, Yao-ze Z, Zhao-ming L et al (2005) Research of human behavior reaction in the building fires [J]. Fire Sci Technol 24(5):563–566
4. Wei-guo S, Yan-fei Y, Tao C (2003) Influences of exit conditions on pedestrian evacuation [J]. Fire Safety Sci 12(2):100–104

References

Erratum to: Economic Effects of the Modern Information Service Industry in China: Evidence from Beijing

Fenfei Chen, Chenchen Wang and Yuan Ren

Erratum to: Chapter "Economic Effects of the Modern Information Service Industry in China: Evidence from Beijing" in: M. Li et al. (eds.), *Proceedings of 2015 2nd International Conference on Industrial Economics System and Industrial Security Engineering*, DOI 10.1007/978-981-287-655-3_28

The book was inadvertently published with an incorrect surname of one of the authors as "Anderdassen" whereas it should be "Andergassen" in the reference section. The erratum book and the reference has been updated.

The updated original online version for this chapter can be found at
DOI 10.1007/978-981-287-655-3_28

F. Chen (✉) · C. Wang · Y. Ren
School of Economics and Management, Beijing Jiaotong University,
Beijing 100044, China
e-mail: ffchen@bjtu.edu.cn

C. Wang
e-mail: 12120510@bjtu.edu.cn

Y. Ren
e-mail: 13120498@bjtu.edu.cn

F. Chen
Beijing Center for Industrial Security and Development Research,
Beijing Jiaotong University, Beijing 100044, China

M. Li et al. (eds.), *Proceedings of 2015 2nd International Conference on Industrial Economics System and Industrial Security Engineering*,
DOI 10.1007/978-981-287-655-3_63

MIX
Papier aus verantwortungsvollen Quellen
Paper from responsible sources
FSC® C105338

If you have any concerns about our products,
you can contact us on
ProductSafety@springernature.com

In case Publisher is established outside the EU,
the EU authorized representative is:
Springer Nature Customer Service Center GmbH
Europaplatz 3, 69115 Heidelberg, Germany

Printed by Libri Plureos GmbH
in Hamburg, Germany